JN119596

01601

Table of contents

Nanami

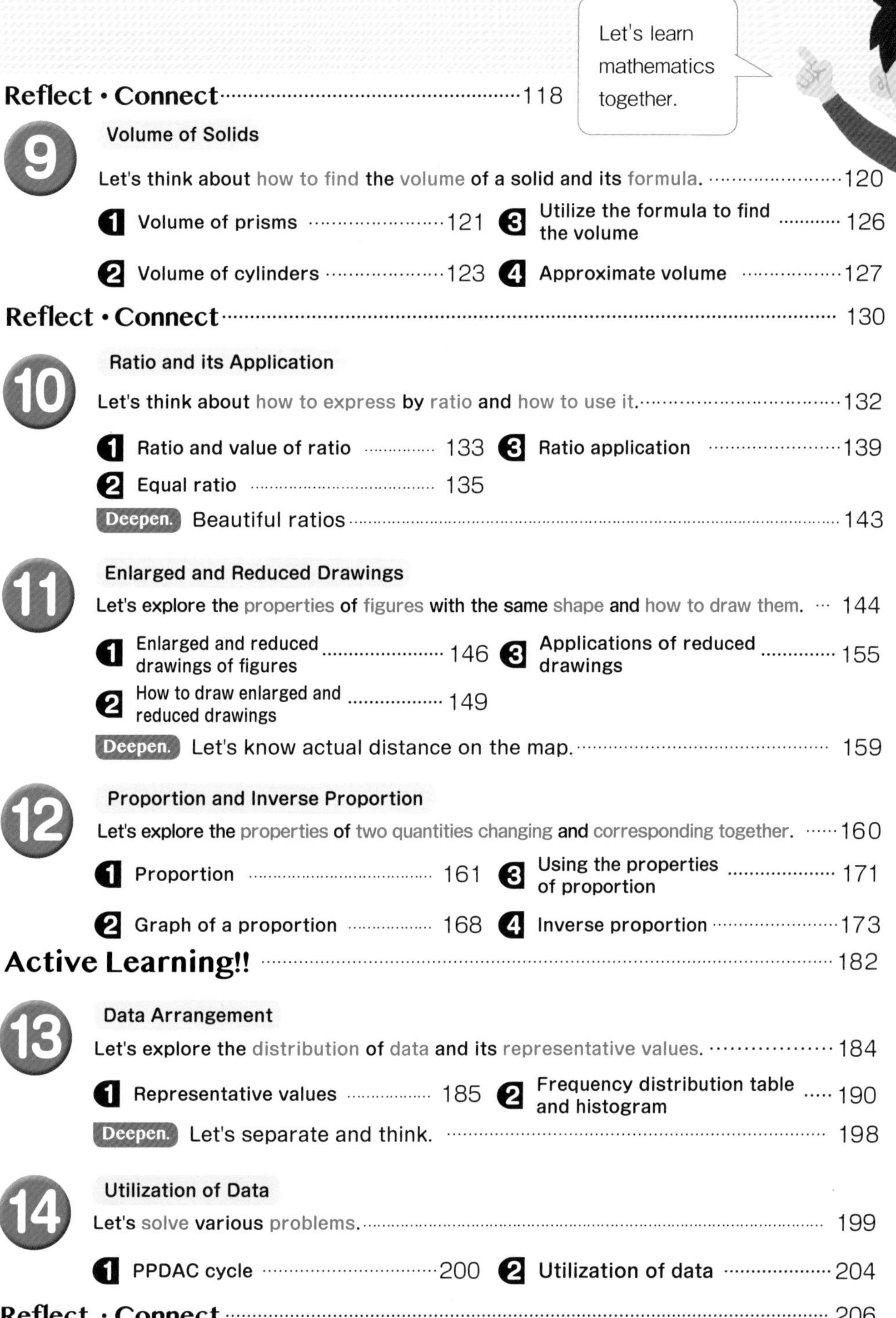

Let's learn mathematics together.

Daiki

Yui

3 learning competencies to develop

1 Thinking Competency

Competency to think the same or similar way
Competency to find what you have learned before and think in the same or similar way

Competency to find rules
Competency to analyze numbers and various expressions that change together, and investigate any rules

Competency to explain the reason
Competency to explain the reason why, based on learned rules and important ideas

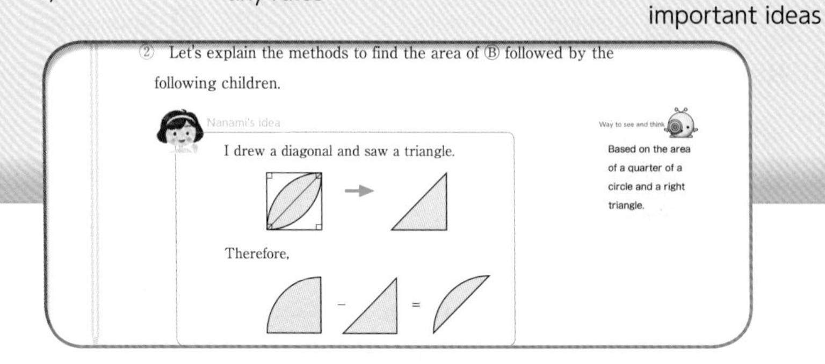

2 Judgement Competency

Competency to find mistakes
Competency to find the over generalization of conventions and rules

Competency to categorize by properties and patterns
Competency to categorize by observing the number's and shape's properties and patterns

Competency to compare ideas and ways of thinking
Competency to find the same or different ways of thinking by comparing friends' ideas and your own ideas

3 Representation Competency

Competency to represent sentences with diagrams or expressions
Competency to read problem sentences, draw diagrams, and represent using expressions

Competency to represent data in graphs, tables or representative values
Competency to express the explored data in tables or graphs, in simpler and easier ways, according to the purpose

Competency to communicate with friends and yourself
Competency to communicate your ideas to friends in simpler and easier manners and to write notes in simpler and easier ways for you

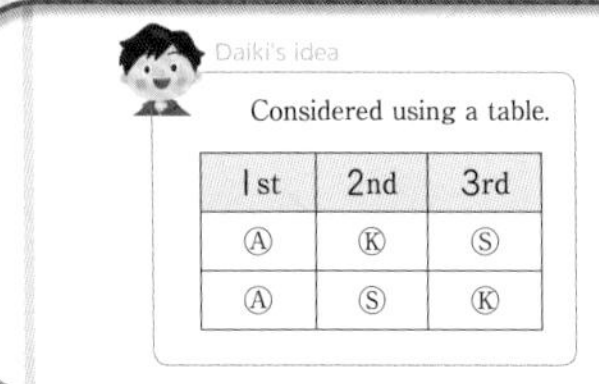

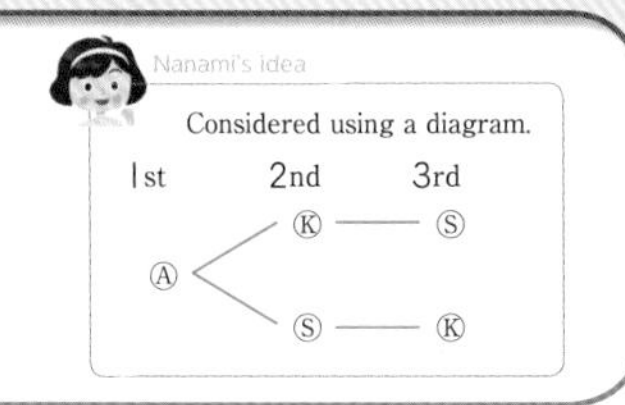

Let's find monsters.

Monsters which represent ways of thinking in mathematics

Unit

Setting the unit.

Once you have decided one unit, you can represent how many.

Separate

If you try to separate...

Decomposing numbers by place value and dividing figures make it easier to think about problems.

Other way

If you represent in other way...

If you represent in other expression, diagram, table, etc., it is easier to understand.

Looks same

Can you do the same or similar way?

If you find something the same or similar, then you can understand.

Why

You wonder why?

Why this happens? If you communicate the reasons in order, it will be easier to understand for others.

Align

If you try to arrange...

You can compare if you align the number place and align the unit.

Summarize

If you try to summarize...

It is easier to understand if you put the numbers in groups of 10 or summarize in a table or graph.

Change

If you try to change the number or figure...

If you try to change the problem a little, you can better understand the problem or find a new problem.

Rule

Is there a rule?

If you examine a few examples, then you can find out whether there is a rule.

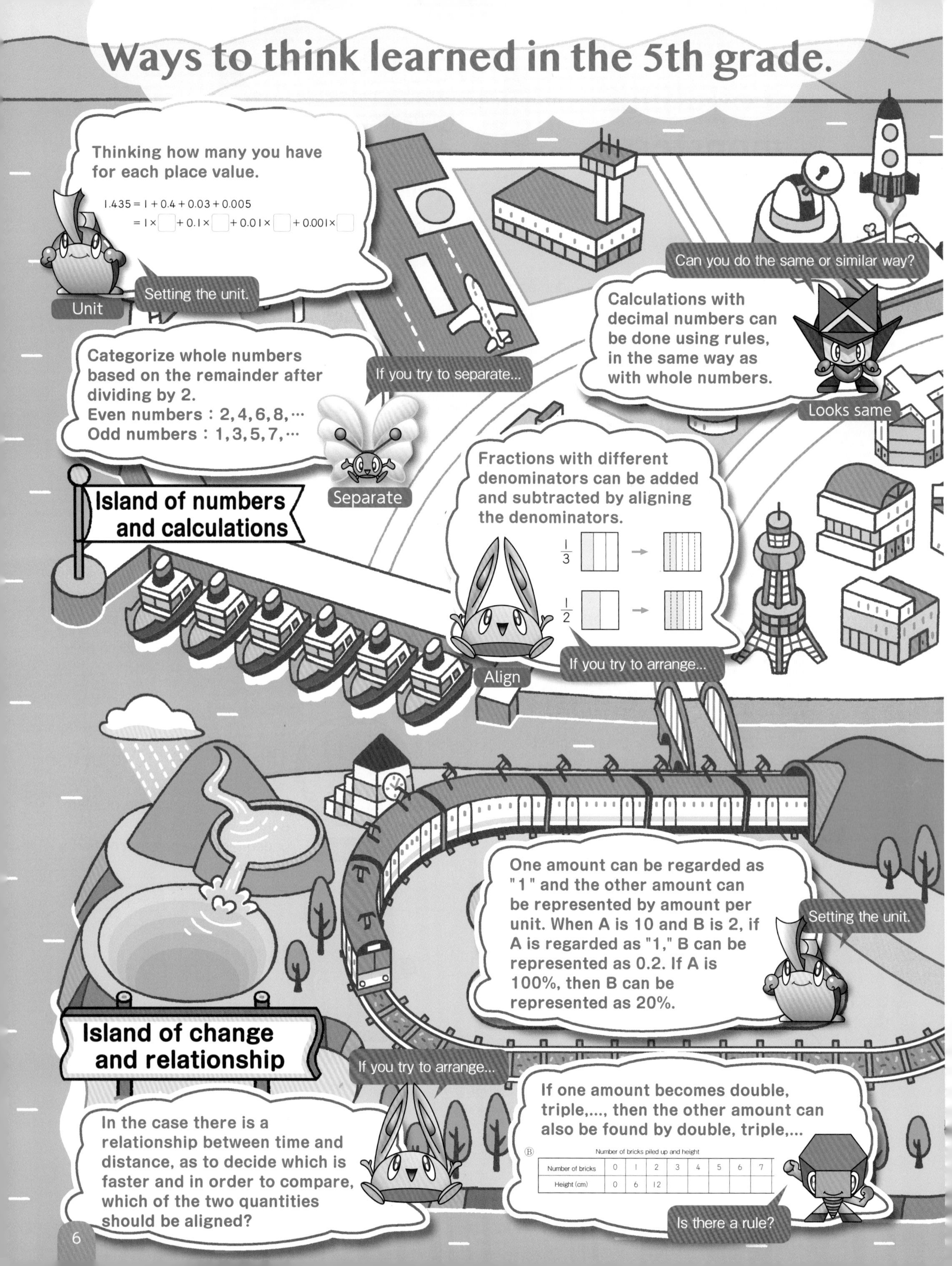

Ⓑ Number of bricks piled up and height

Number of bricks	0	1	2	3	4	5	6	7
Height (cm)	0	6	12					

There are many other "ways to see and think."
Can you do the same or similar way?
Understanding the rule between the circumference and diameter. Circumference=diameter×3.14
Circle Ⓐ
Circle Ⓑ
Circle Ⓒ
Circumference (cm)
Diameter (cm)
4
8
12
Based on the sides and angles, you can find two figures with the same size and shape.
2.4cm
95°
130°
4cm
3cm
70°
65°
5cm
A
B
C
D
E
F
G
H
You wonder why?
Why
Finding the rules while confirming with various methods.
Is there a rule?
A
B
C
Rule
Transform the figure whose area you want to find into a figure you studied before.
D A J
E H
B F K G C
If you try to change the number or figure...
Change
Island of figures
The same as in length and area, make one unit to represent volume as a number.
1 cm
1 cm
1 cm
Setting the unit.
Island of data
According with what you want to know and represent, decide the type of graph.
Men's favorite Olympic events
0 10 20 30 40 50 60 70 80 90 100(%)
Soccer
Baseball
Volleyball
Tennis
Other
Sports where Japanese athletes have won gold medals at the Summer Olympic Games
0(100)
10
20
30
40
50
60
70
80
90
Judo
Wrestling
Gymnastics
Swimming
Athletics
Volleyball
Other
Other way
If you represent in other way...

Let's progress through 3 ways of learning.

Self directed learning: Learning on your own initiative.

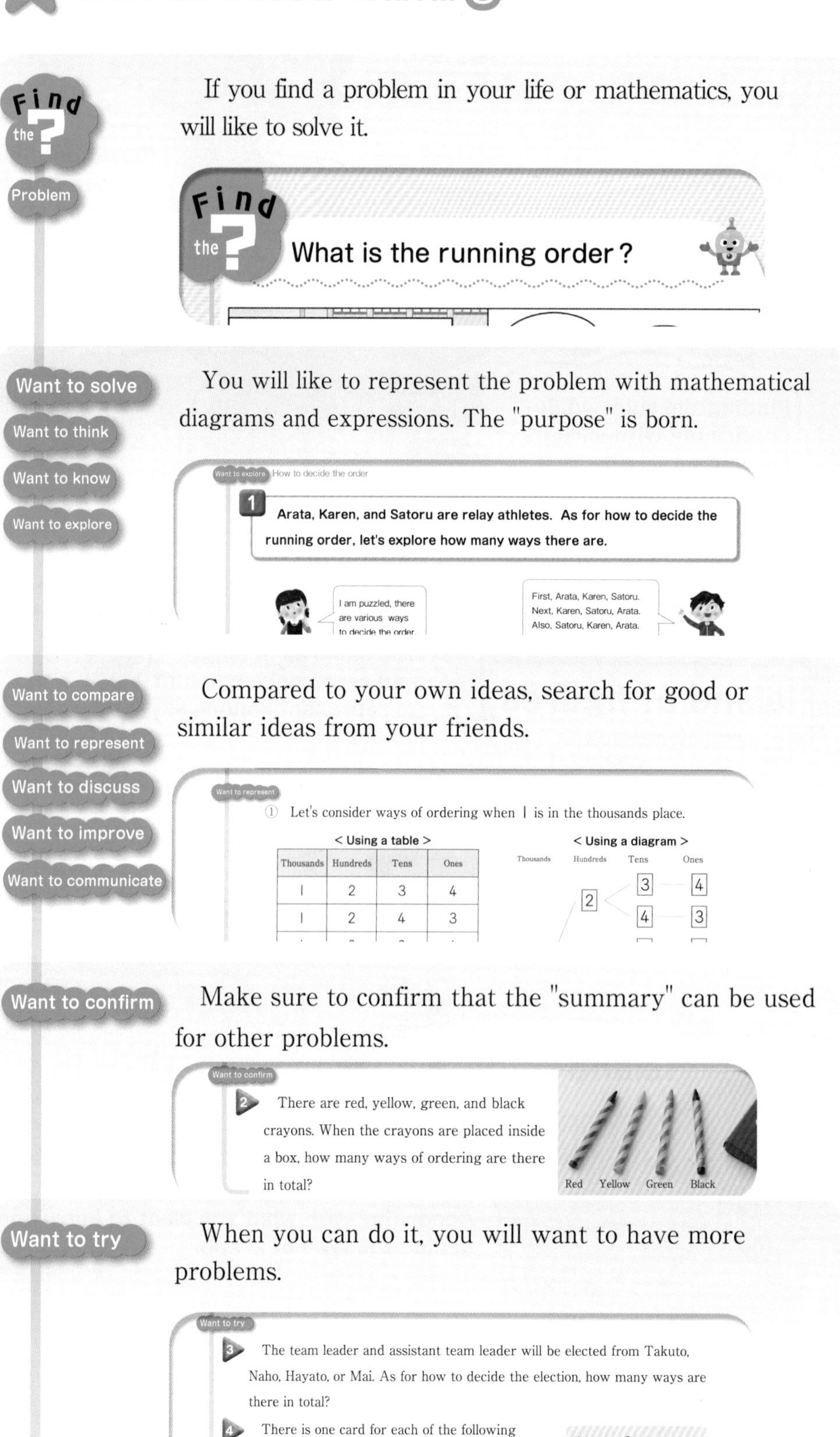

If you find a problem in your life or mathematics, you will like to solve it.

You will like to represent the problem with mathematical diagrams and expressions. The "purpose" is born.

Compared to your own ideas, search for good or similar ideas from your friends.

Make sure to confirm that the "summary" can be used for other problems.

When you can do it, you will want to have more problems.

As learning progresses, you will want to know what your friends are thinking. Also, you will like to share your own ideas with your friends. Let's discuss next to each other, in groups, or with the whole class.

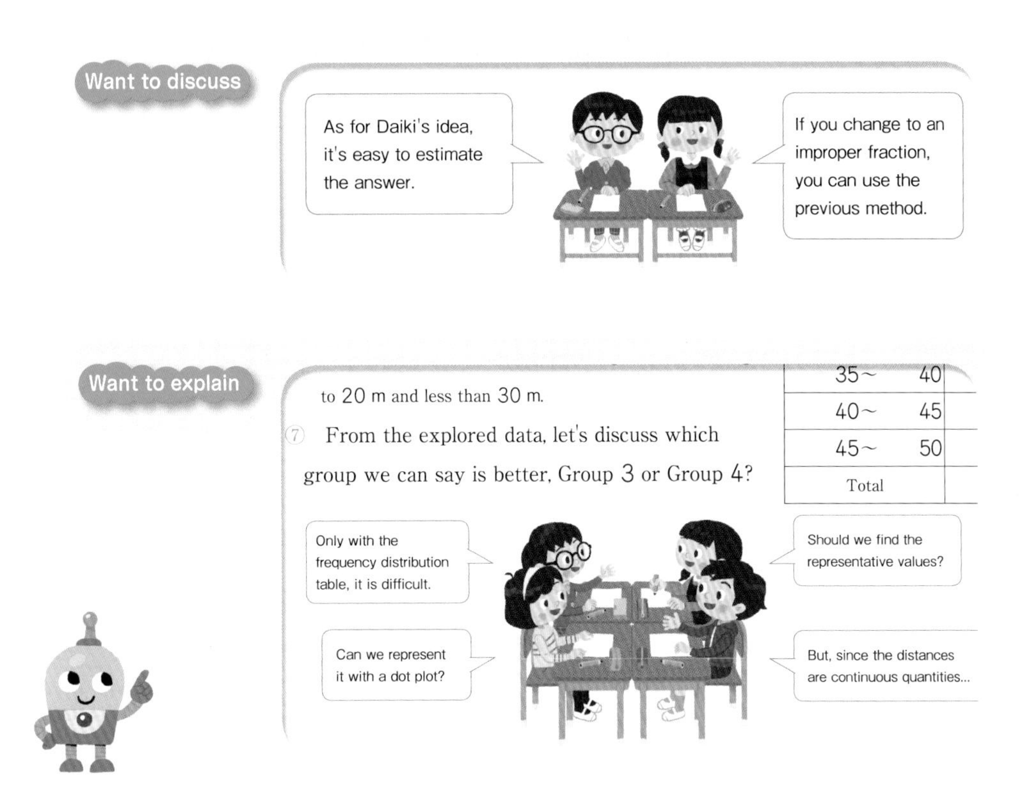

3 Deep learning: Usefulness and efficiency of what you learned.

Let's cherish the feeling "I want to know more." and "Can I do this in another case?" Let's deepen learning in life and mathematics.

What is the running order?

As for the running order of 3 children, how many ways are there?

Ways of Ordering and Combinating

1 Let's arrange without missing or overlapping.

1 Ways of combinating

Want to explore How to decide the order

Arata, Karen, and Satoru are relay athletes. As for how to decide the running order, let's explore how many ways there are.

I am puzzled, there are various ways to decide the order.

First, Arata, Karen, Satoru.
Next, Karen, Satoru, Arata.
Also, Satoru, Karen, Arata.
Mmm, are there still more?

Purpose How can we count without missing or overlapping?

Want to think

① Consider the case where Arata runs first. How will the running order of Karen and Satoru be decided? Let's think about ways of ordering.

Arata…Ⓐ
Karen…Ⓚ
Satoru…Ⓢ

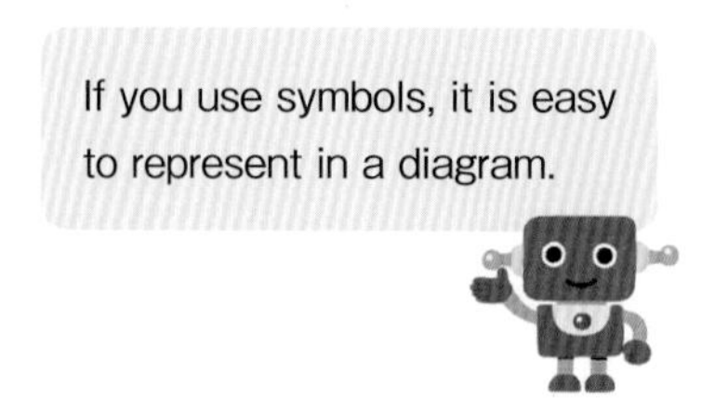

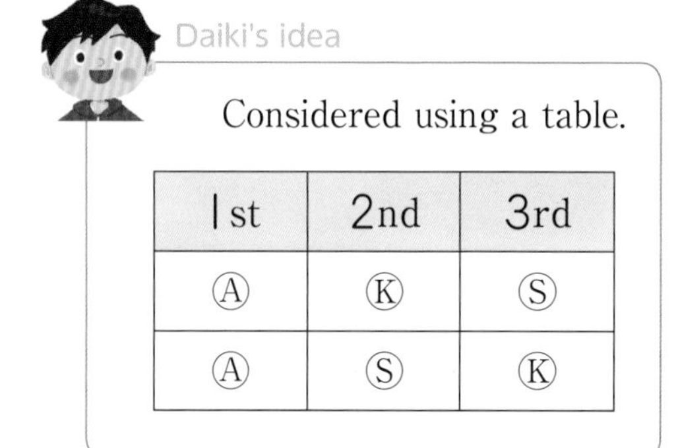

1st	2nd	3rd
Ⓐ	Ⓚ	Ⓢ
Ⓐ	Ⓢ	Ⓚ

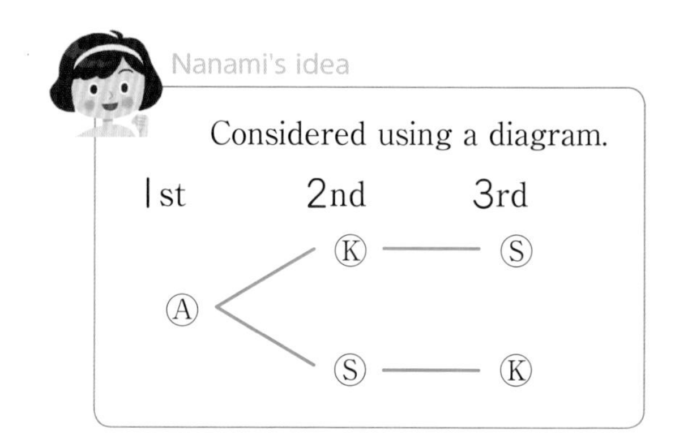

② Let's think about the ways of ordering when the first runner is Karen or Satoru respectively.

③ In total, how many ways of ordering are there?

Summary

Using tables and diagrams, you can count how many ways there are without missing or overlapping.

Want to confirm

At the amusement park, you ride one time on the Go Karts, Ferris Wheel, and Merry Go Round. As for the riding order, how many ways are there in total?

Want to solve

2

There is one card for each of the following numbers: [1], [2], [3], [4].
With these cards, create 4-digit whole numbers. How many whole numbers can you make in total?

Want to represent

① Let's consider ways of ordering when 1 is in the thousands place.

< Using a table >

Thousands	Hundreds	Tens	Ones
1	2	3	4
1	2	4	3
1	3	2	4
1	3	4	2
1	□	□	□
1	□	□	□

□ ways

< Using a diagram >

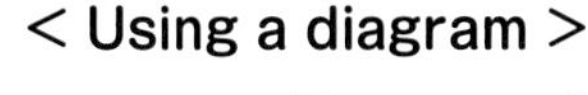

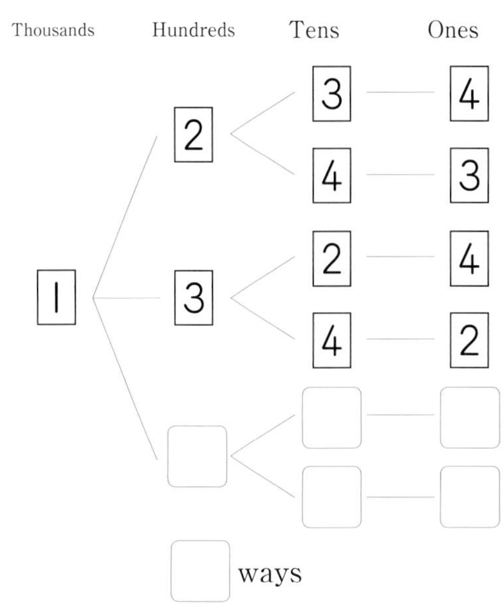

□ ways

② How many ways are there when 2, 3, and 4 are placed in the thousands place? As in ①, let's think by writing a diagram or table.

③ From the results on ① and ②, how many whole numbers can you make in total?

Want to confirm

There are red, yellow, green, and black crayons. When the crayons are placed inside a box, how many ways of ordering are there in total?

Want to solve

3 **There is one card for each of the following numbers: 1, 3, 5, 7. From these 4 cards, use 3 cards to create 3-digit whole numbers. How many whole numbers can you make in total?**

Hundreds	Tens	Ones
□	□	□

① Let's consider ways of ordering when 1 is in the hundreds place.

<Using a table>

Hundreds	Tens	Ones
1	3	5
1	3	□
1	5	□
1	□	□
1	□	□
1	□	□

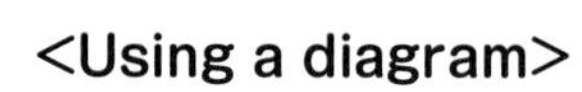
<Using a diagram>

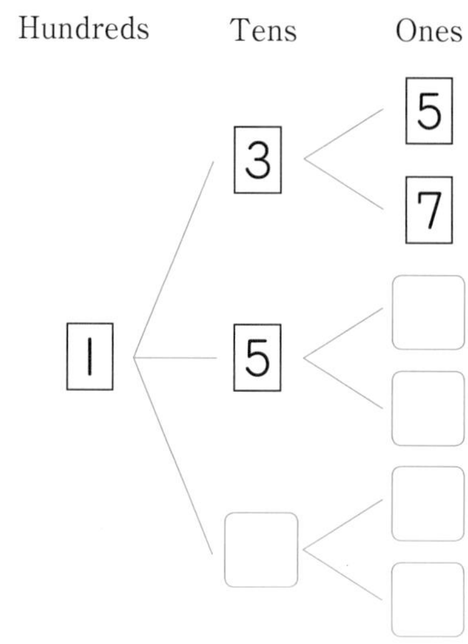

② How many 3-digit whole numbers can you make in total? Let's explain the thinking methods.

Want to try

3 The team leader and assistant team leader will be elected from Takuto, Naho, Hayato, or Mai. As for how to decide the election, how many ways are there in total?

4 There is one card for each of the following numbers: 0, 2, 4, 6. From these 4 cards, use 3 cards to create 3-digit whole numbers. How many whole numbers can you make in total?

If you place 0 in the hundreds place, you cannot make a 3-digit whole number.

You are shooting soccer penalty kicks. When 3 consecutive penalty kicks are shot, what different cases are there?

① Place ○ in case of goal and × in case of miss.

Let's explore using a table and diagram the case in which the 1st kick is a goal.

1st kick is ①,
2nd kick is ②, and
3rd kick is ③.

< Using a table >

①	②	③
○	○	
○	○	
○	×	
○	×	

< Using a diagram >

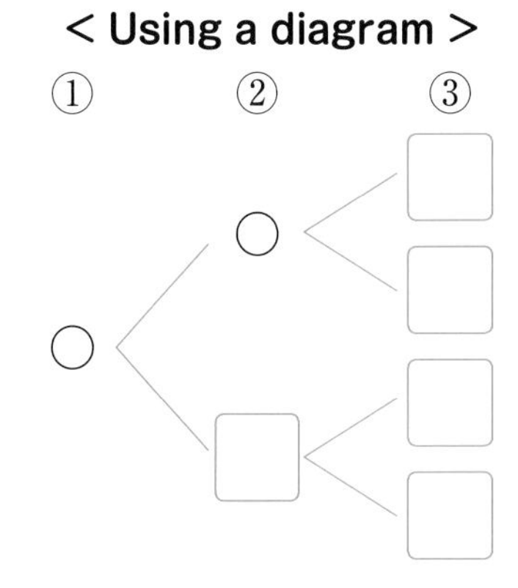

② Let's also explore using a table and diagram the case in which the 1st kick is a miss.

③ As for the results of the penalty kicks, how many ways are there in total?

Want to confirm

A 500 yen coin is thrown three consecutive times. As for the front and back of the coin coming out, how many ways are there in total?

Front

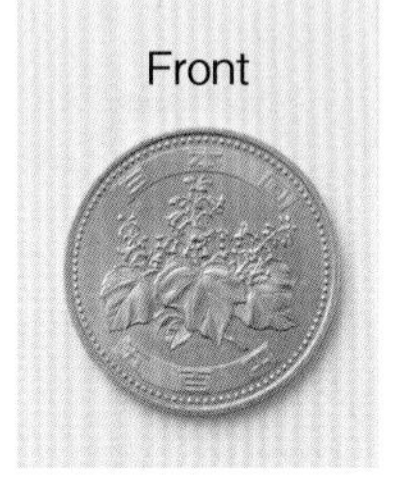

Back

That's it

Password

Passwords that utilize numerals are used in various places in daily life.

① From your surroundings, let's try to explore which things use passwords.

② If you create a password using three numerals from 0~9, how many ways are there in total?

000, 001, 002, ..., 998, 999.

Want to explore Number of games

Activity

1 **Four teams of 6th graders play basketball games. If each team competes with the other teams only one time, how many games will be played in total?**

Daiki

The 1st team will play a game with the 2nd, 3rd, and 4th team.

Since 1st team vs 2nd team and 2nd team vs 1st team is the same game...

Nanami

Purpose How can we count without missing or overlapping?

Want to explain

① Let's explain the ideas of the following children.

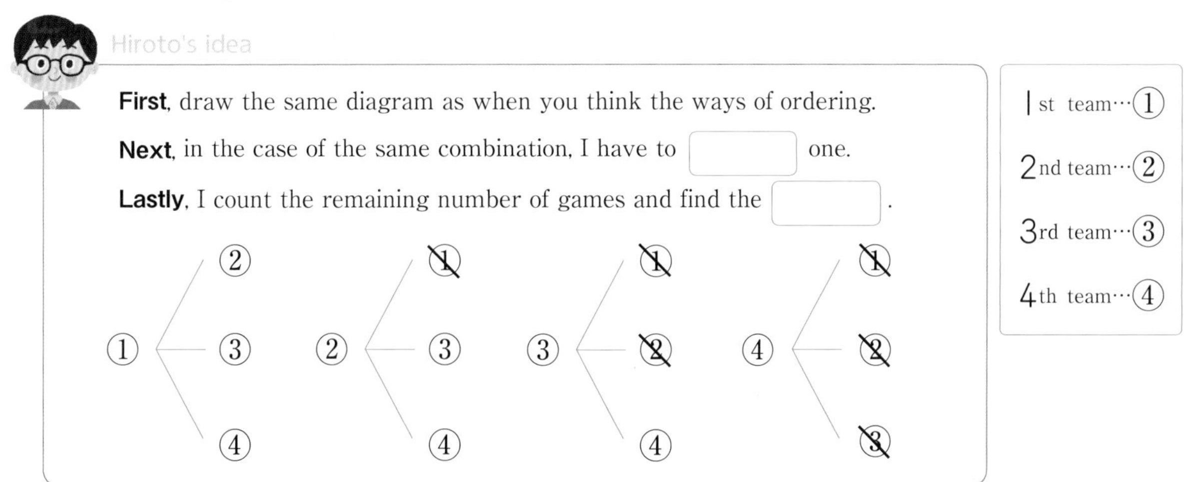

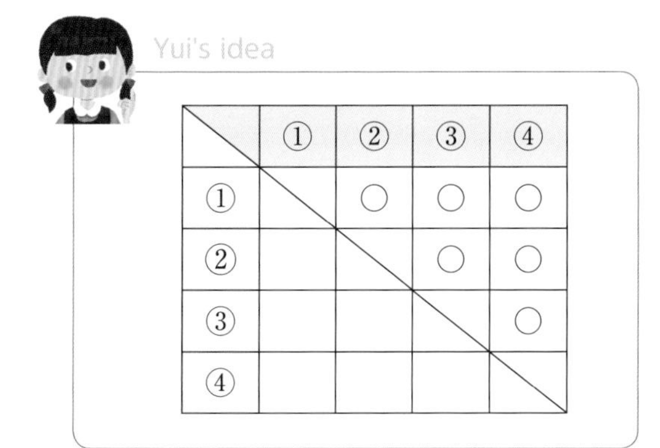

	①	②	③	④
①		○	○	○
②			○	○
③				○
④				

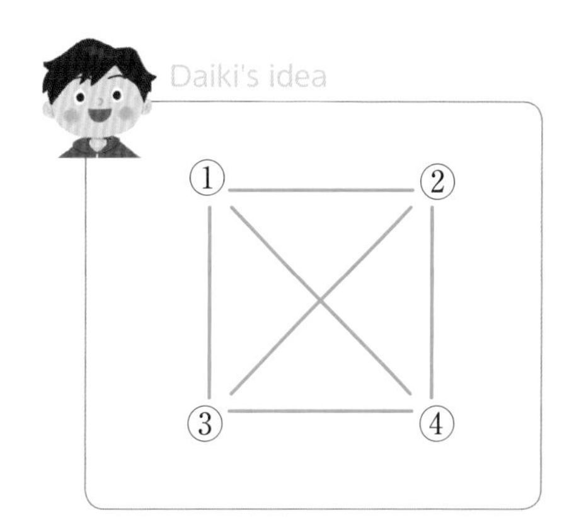

② As for games, how many ways of combinating are there in total?

Summary

The same as when exploring ways of ordering, use a table or diagram to erase one of the repeated combinations and then count the total cases.

Want to confirm

Five teams play baseball games. If each team competes with the other teams only one time, how many games will be played in total?

Words

【First.... Next.... Lastly....】

These are convenient words to use when explaining in order.

2

From the following 5 types of pastries, 2 types were bought. How many combinations are there in total?

① As for how to choose 2 types of pastries, let's draw a table or diagram to explore the existing cases.

② How many combinations are there in total?

Want to try

Three members of the breeding committee are chosen among 4 children: Fumiko, Genta, Hitomi, and Issei. How many ways to choose the members are there in total?

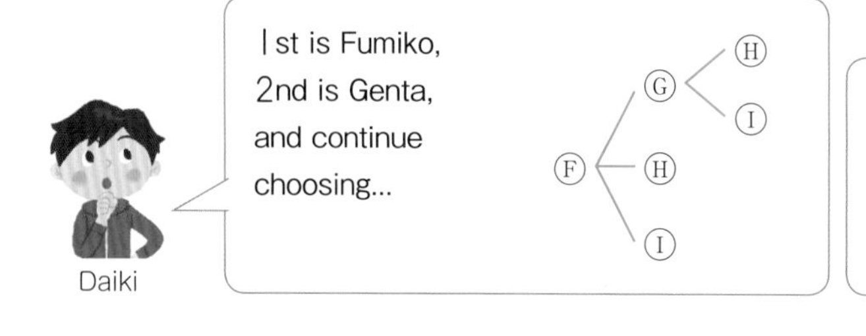

Want to confirm

There is one card for each of the following numbers: 1, 2, 3, 4, 5. From these 5 cards, choose 4 cards to find the sum. As for the sums, how many can you find in total?

What you can do now

☐ **Can count the ways of ordering without missing or overlapping.**

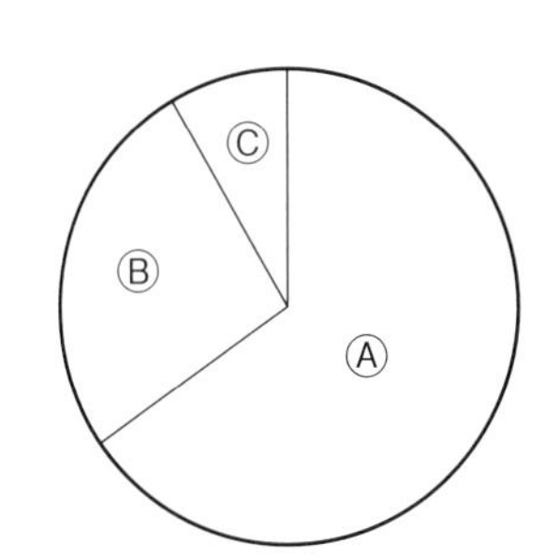

1 A circle graph, like the one on the right, was drawn. Using the colors red, yellow, and blue, the sections Ⓐ, Ⓑ, and Ⓒ can be colored. Let's write down all the coloring ways.

☐ **Understanding ways of combination.**

2 There is one coin for each of the following: 1 yen, 10 yen, 50 yen, and 100 yen. From these 4 coins, choose 3 coins and find the total amount of money. Let's write down all the different amounts of money.

☐ **Can create a combination of numbers.**

3 There are three number cards: [3], [4], [5]. Let's answer the following questions.

① Write down all the 3-digit numbers that can be created with these cards. How many can you make in total?

② As for choosing 2 cards out of 3, how many combinations are there? Also, let's write all of the combinations.

③ When you make a 2-digit number choosing 2 cards out of 3, what is the third largest number you can make?

Supplementary problems → p.222

Number of games in a knock-out competition?

Eight teams will play a soccer knock-out competition. What is the total number of games until the winning team is decided?

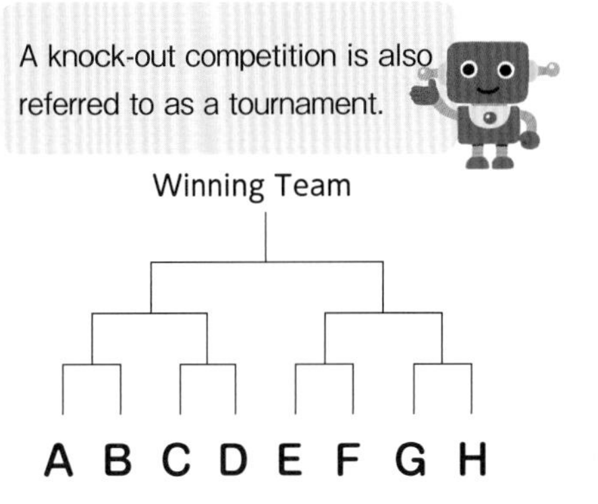

Usefulness and efficiency of learning

1 There is a road as shown in the following diagram. How many ways to go from position A to position B are there in total?

☐ Can choose the ways of ordering without missing or overlapping.

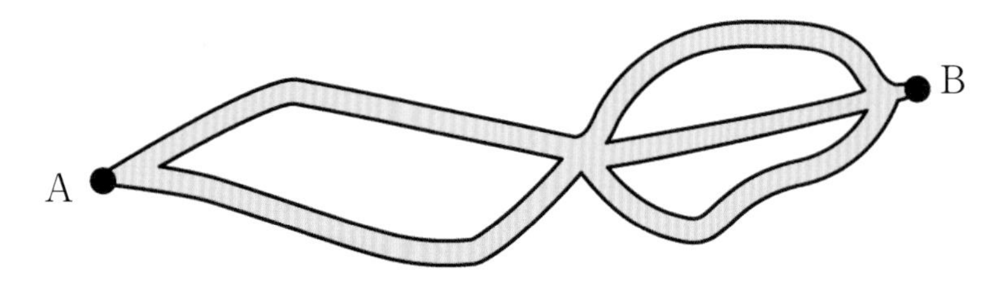

2 There is one card for each of the following numbers: [0], [1], [2], [3]. From these 4 cards, create 4-digit whole numbers. Let's answer the following questions.

☐ Can create a combination of numbers.

① How many whole numbers can you create? Also, write down all of them.

② From the created whole numbers, how many are even numbers? Also, write down all the even numbers in ascending order.

3 Juri, Rena, Sakito, and Manaka are sitting on a bench. As for the sitting ways in which Juri and Manaka sit next to each other, how many are there in total?

☐ Understanding ways of combination.

If we consider Juri and Manaka as one pair, then we can think of it as a group of 3 elements.

Yui

Become a writing master.

Notebook for thinking
Let's write how you thought.

April 15

> Write today's date.

Let's explore how many ways for the running order of Arata, Karen, and Satoru.

> Write the problem of the day that you must know.

〈My idea〉

Represent in a table using symbols.

> Let's write a prospect for the solution.

1st	2nd	3rd
Ⓐ	Ⓚ	Ⓢ
Ⓚ	Ⓢ	Ⓐ
Ⓢ	Ⓚ	Ⓐ

It looks like there are others. Let's try to think as Ⓐ - Ⓢ - Ⓚ .

> Let's keep previous thoughts even when the way of thinking changes or a better idea is recognized.

1st	2nd	3rd
Ⓐ	Ⓚ	Ⓢ
Ⓐ	Ⓢ	Ⓚ
Ⓚ	Ⓐ	Ⓢ
Ⓚ	Ⓢ	Ⓐ
Ⓢ	Ⓐ	Ⓚ
Ⓢ	Ⓚ	Ⓐ

6 ways

I decided the 1st person and then thought the order of the rest.

> Let's keep a summary of thoughts.

Find the ?

Which math sentence is it?

Problem When representing scenes of everyday life by math sentences, are there other ways to represent □ or ○?

2 Mathematical Letter and Sentence

Let's use letters to represent quantities and relationships in math sentences.

1 Various quantities and math letters

Want to solve

Let's think a math expression for the following prices.

Ⓐ The price when buying 6 buns .

Ⓑ The price when buying a ribbon that costs 80 yen per meter.

Purpose When there are unknown quantities, how should we represent math expressions?

① Let's write a math sentence that represents the price, considering □ yen as the cost for 1 bun.

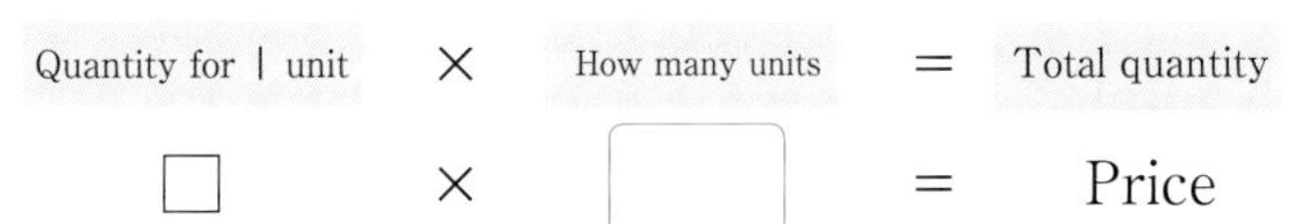

Quantity for 1 unit	×	How many units	=	Total quantity
□	×		=	Price

Therefore, the price is (□ × 6) yen.

② Let's write a math expression that represents the price, considering □ m as the length of the bought ribbon.

The price is (× □) yen.

If you apply the meaning of multiplication to a math expression, you can understand.

In mathematics, numbers and quantities can be represented using mathematical letters such as x or a other than □ and ○.

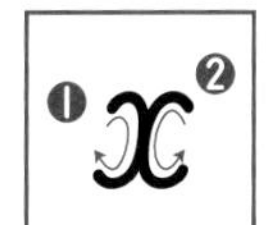

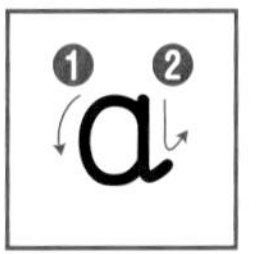

Using mathematical letters, the prices in Ⓐ and Ⓑ can be represented as follows.

Ⓐ The price when buying 6 buns that cost x yen each, $(x \times 6)$ yen.

Ⓑ The price when buying a meters of a ribbon that costs 80 yen per meter, $(80 \times a)$ yen.

Summary

When there are unknown quantities, each quantity can be represented by x or a in the math expression.

2

There are 2 boxes and 4 apples. Each box has the same number of apples. Let's explore about the total number of apples.

① If each box has 10 apples, how many apples are there in total?

② Let's write a math expression that represents the total number of apples, considering x as the number of apples in each box.

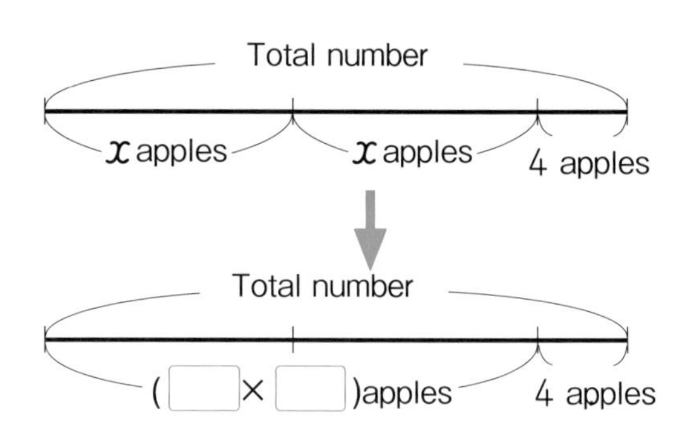

[ ]

③ When each box has 15 apples, how many apples are there in total?

Want to confirm

1 There are 2 boxes and 6 caramels. Let's write a math expression to represent the total number of caramels considering x as the number of caramels inside each box.

2 There are 3 bottles and 2 dL of juice. Let's write a math expression to represent the total amount of juice considering x as the amount of juice inside each bottle.

Also, if each bottle has 5 dL of juice, how many deciliters are there in total?

2 Math sentence for changing quantities

Want to represent Math sentences that use two mathematical letters

1

Let's represent, with a math sentence, the relationship between the length of one side and the surrounding length of an equilateral triangle.

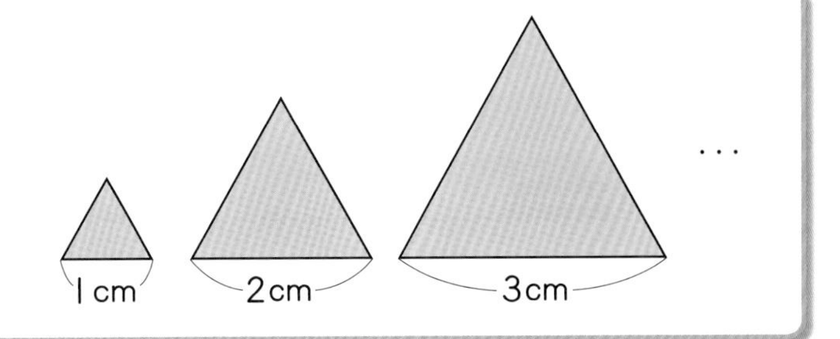

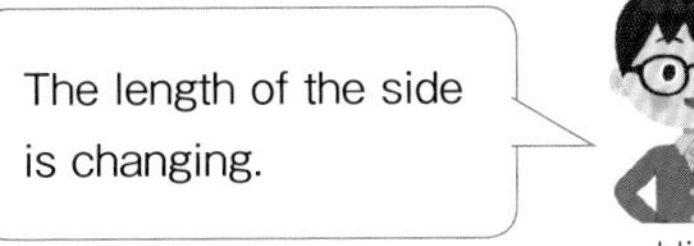

Purpose As for the relationship of two quantities changing together, how should we represent math sentences?

① The length of one side and the surrounding length were summarized in the following table. Let's write the numbers inside the blank spaces.

Length of one side (cm)	1	2	3	4	
Surrounding length (cm)	3				

② Let's represent, with a math sentence, the relationship between □ and ○, when □ cm is the length of one side and ○ cm is the surrounding length.

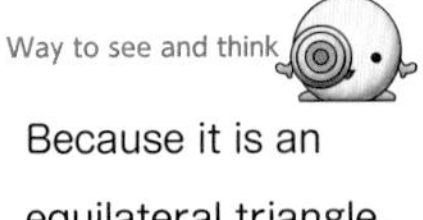

Because it is an equilateral triangle, the surrounding length is length of one side × 3.

	Length of one side	×	Number of sides	=	Surrounding length
Case for 1 cm ···	1	×	3	=	3
Case for 2 cm ···	2	×	3	=	6
Case for 3 cm ···	3	×	3	=	9
⋮					
Case for □ cm ···	□	×	3	=	○

In this math sentence, the two quantities changing together are "ength of one side" and "surrounding length."

③ Let's write a math sentence considering x cm as the length of one side and y cm as the surrounding length.

Represent □ as x and ○ as y.

Length of one side	×	Number of sides	=	Surrounding length
x	×	□	=	y

The relationship of two quantities that change together can be represented in math sentences by using x, y, etc.

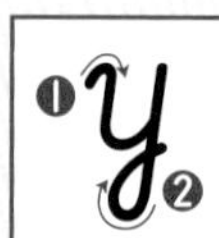

1 Let's explore about the area of the open section of a window that has a height of 90 cm.

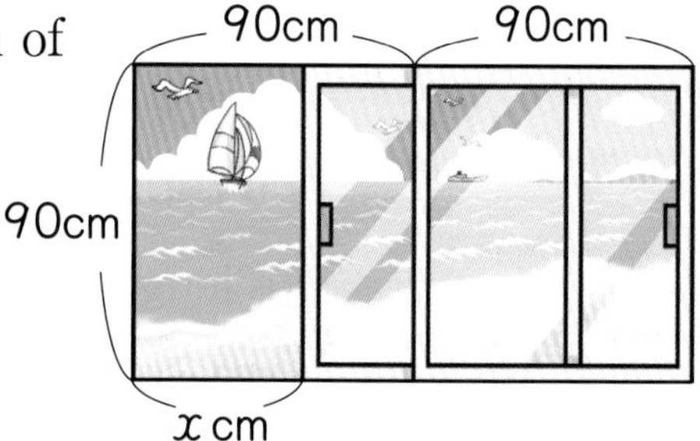

① As for the open window, the following table summarizes the width of the open section as x cm and the area of the open section as y cm². Let's write the numbers inside the blank spaces.

Width of open section x (cm)	5	10	12.5	90
Area of open section y (cm²)	450			

Way to see and think

Since the area of the open section is a rectangle with a length of 90 cm, the formula for area can be used.

② Let's write a math sentence to find the area of the open section y cm² when the width of the open section is x cm.

y = □ × □

2 The length of a circumference can be found as diameter × 3.14.
Let's write a math sentence to find the length of the circumference y cm when the radius is x cm. Also, find y when x is 2.

3 Number that satisfies the math sentence

Want to solve Math sentence represented by addition

There were a certain number of origami sheets. When 7 sheets were added, the number of sheets became 35. Let's answer the following questions.

① Let's write a math sentence considering that the original number of sheets is x and the total number is 35.

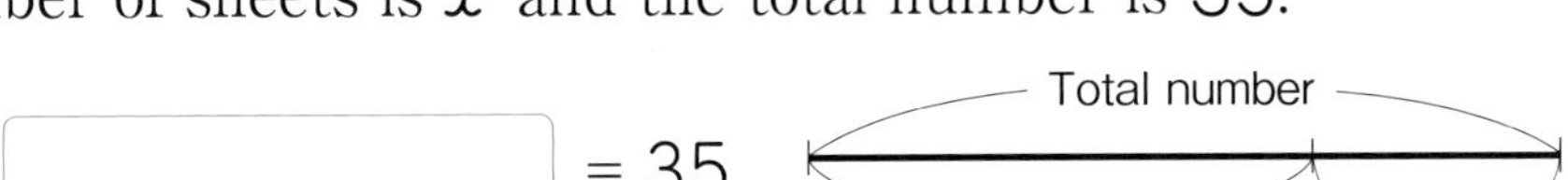

The reason to deduce an addition sentence can be understood by the number line or addition for total number.

② In order to find the original number of origami sheets, let's think a number x that satisfies $x + 7 = 35$.

Purpose How can we find the number that applies for x?

③ Daiki considered the following. Let's explain Daiki's idea.

Daiki's idea

$$x + 7 = 35$$
$$x = 35 - 7$$
$$x = 28$$

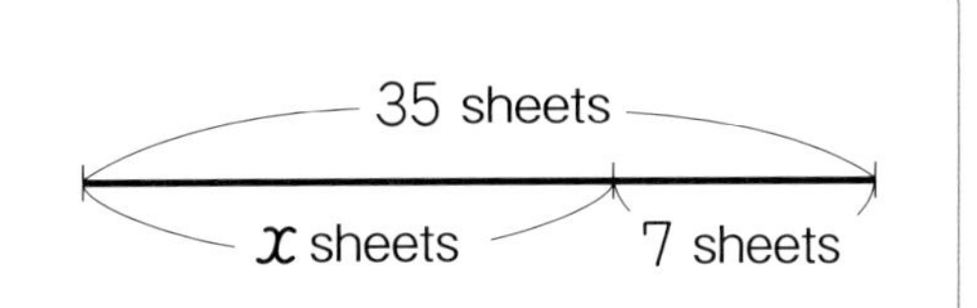

It's easier to see if the equality sign is vertically aligned.

Summary

In the case of an addition sentence such as $x + 7 = 35$, the number that applies for x can be found using subtraction as the inverse operation of addition.

Want to confirm

Let's find the number that applies for x.

① $x + 4 = 22$ ② $38 + x = 54$

③ $x - 6 = 15$ ④ $x - 2.7 = 1.8$

2 **As shown on the right, let's find the height of a parallelogram with an area of 18 cm² and a base of 5 cm.**

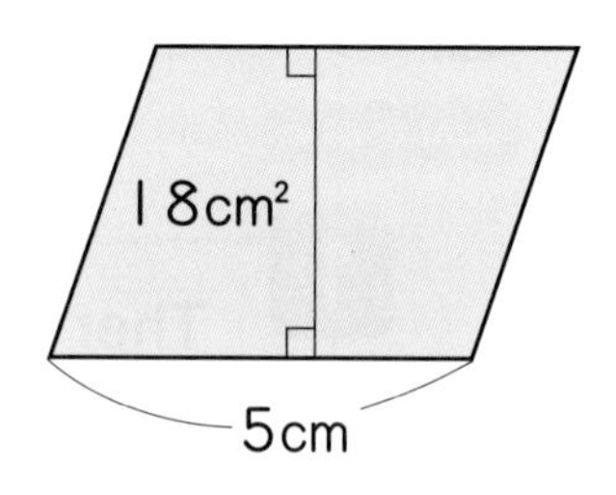

The math sentence to find the area of a parallelogram is a multiplication.

Since the height is unknown, let's consider it as x cm.

Way to see and think

The area of a parallelogram can be found as base × height.

① Let's write a math sentence to find the area, considering the height as x cm.

$\square = 18$

Purpose How can we find the number that applies for x?

② Based on the math sentence from ①, let's find the height of the parallelogram.

$\square \times x = 18$

$x = 18 \div \square$

$x = \square$

x represents not only whole numbers but also decimal numbers.

Summary

In the case of a multiplication sentence such as $5 \times x = 18$, the number that applies for x can be found using division as the inverse operation of multiplication.

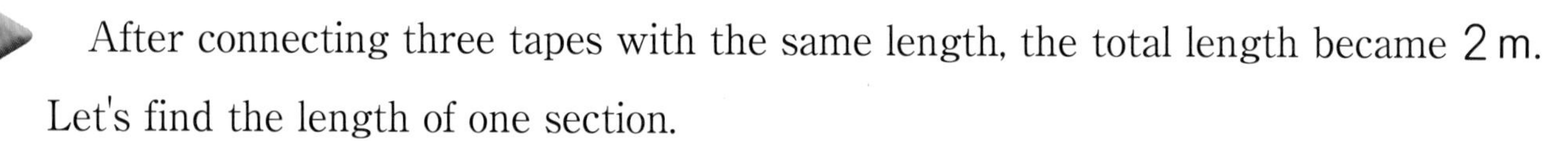

Want to confirm

2 After connecting three tapes with the same length, the total length became 2 m. Let's find the length of one section.

① Let's write a math sentence to find the total length, considering x m as the length of one section.

$\square = 2$

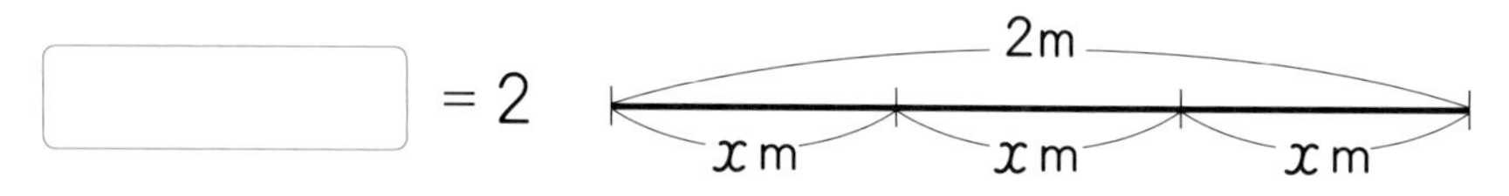

x represents not only whole numbers or decimal numbers but also fractions.

② How many meters is the length of one section of tape?

3 Let's find the number that applies for x.

① $8 \times x = 20$ ② $7 \times x = 5$ ③ $x \div 4 = 8$ ④ $x \div 6 = 3$

Want to think Math sentence that incorporates various operations

3

There are 2 boxes and 3 chocolates. Each box has the same number of chocolates. Let's answer the following questions.

① Let's write a math expression to find the total number, considering that the number of chocolates inside one box is x.

$\square \times \square + \square$

② If all the chocolates are taken out of the boxes, including the individual ones, the total number is 29 chocolates. Let's represent this situation in a math sentence.

$\square = 29$

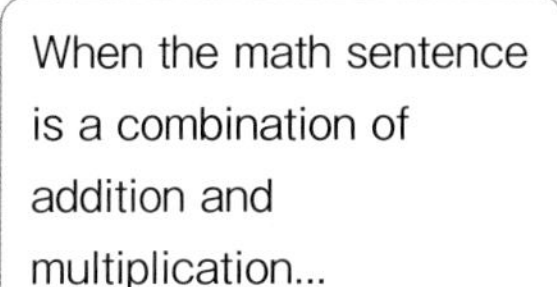

Nanami

③ How many chocolates were inside one box?

Let's place 10, 11, 12,... as x and explore the total number of chocolates using the following table.

x	10	11	12					
$x \times 2$	20							
$x \times 2 + 3$	23							

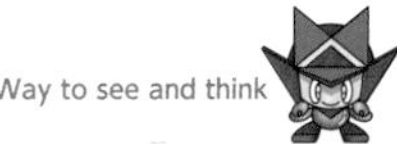

The same as for □ or ○, there is a way to find the number that applies for x by placing various numbers as x.

Want to confirm

4 There are 8 bundles and 3 sheets of colored paper. Let's answer the following questions.

① Let's write a math sentence to represent the total number of sheets of paper, considering that the number of sheets of paper in one bundle is x.

② In total, there were 115 sheets of paper. How many sheets of paper were gathered in one bundle? Let's place 10, 11, 12, ... as x and explore the total number of sheets of colored paper.

When exploring various numbers as x, let's find an estimate answer using numbers as 10 or 20 that are easy to calculate.

Want to try

5 Let's find the number that applies for x by placing 8, 9, 10, ... as x in the following math sentences.

① $x \times 3 + 4 = 37$ ② $x \times 8 - 5 = 67$

Develop in Junior High School

That's it

What's the number that applies for x ?

The following can be done to find the number x that satisfies $x \times 8 + 3 = 115$.

$$\begin{aligned} x \times 8 + 3 &= 115 \\ x \times 8 &= 115 - 3 \\ x \times 8 &= 112 \\ x &= 112 \div 8 \\ x &= 14 \end{aligned}$$

Way to see and think

If you think of $x \times 8$ as one unknown quantity, it becomes an addition sentence. Therefore, you can use subtraction as the inverse operation of addition.

I can see that 3 is subtracted from both $x \times 8 + 3$ and 115.

It looks like there are some calculation rules.

Considering exercise ② of **3** from the previous page, let's try to find the number x that satisfies the math sentence using the above finding method.

4 Reading math sentences

Want to explain

1 **Shun went to the vegetable shop. The costs for the vegetables were the following: one carrot for x yen, one tomato for 50 yen, and one radish for 120 yen. What do the following math expressions ①～④ represent?**

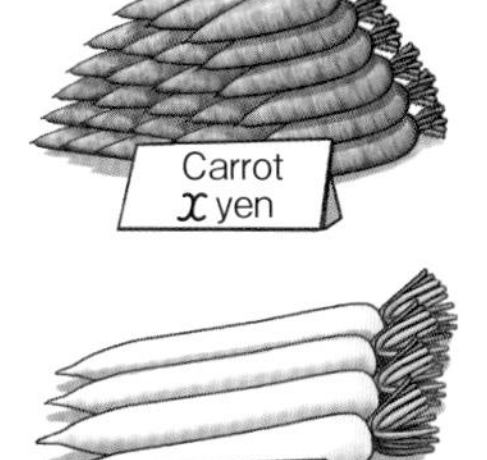

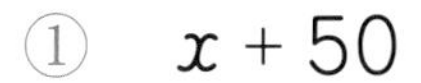

① $x+50$　　② $x\times 7$

③ $x\times 5+120$　　④ $x\times 4+50\times 4$

Want to confirm

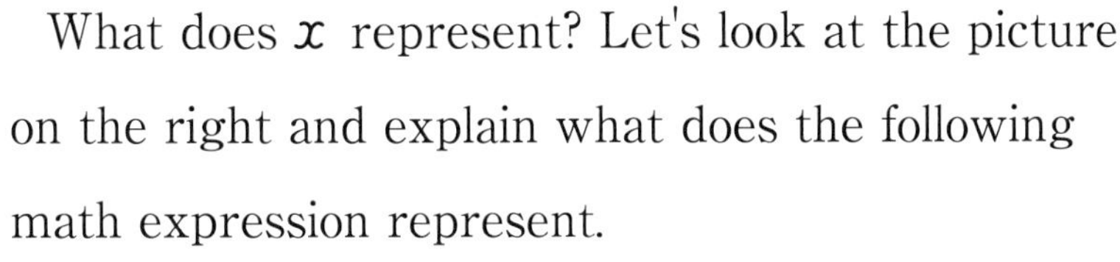

1 What does x represent? Let's look at the picture on the right and explain what does the following math expression represent.

① $70\times x$　　② $70\times x+200\times 4$

Want to try

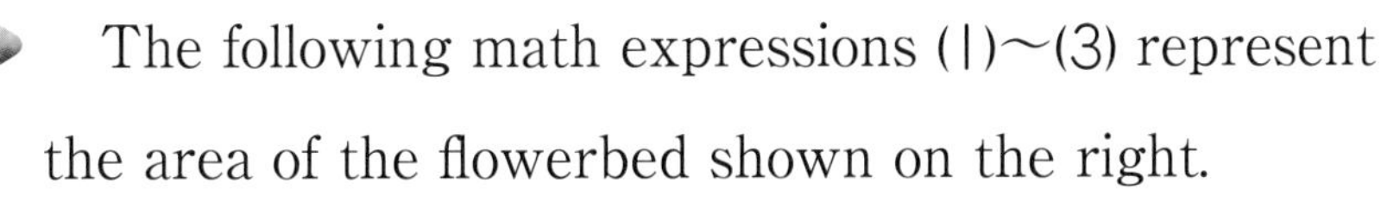

2 The following math expressions (1)～(3) represent the area of the flowerbed shown on the right.

(1) $10\times 5+(10-x)\times 7$

(2) $10\times 12-x\times 7$

(3) $x\times 5+(10-x)\times 12$

Let's answer the following questions.

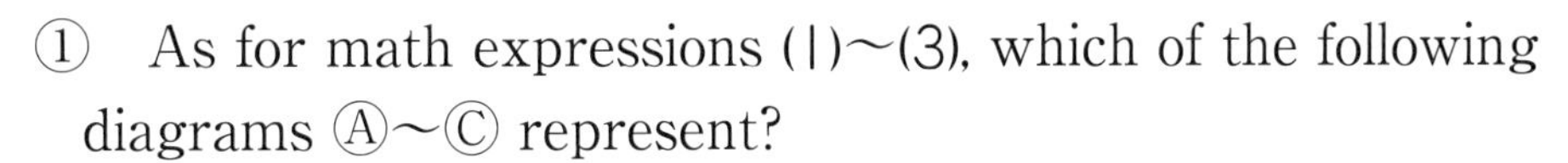

① As for math expressions (1)～(3), which of the following diagrams Ⓐ～Ⓒ represent?

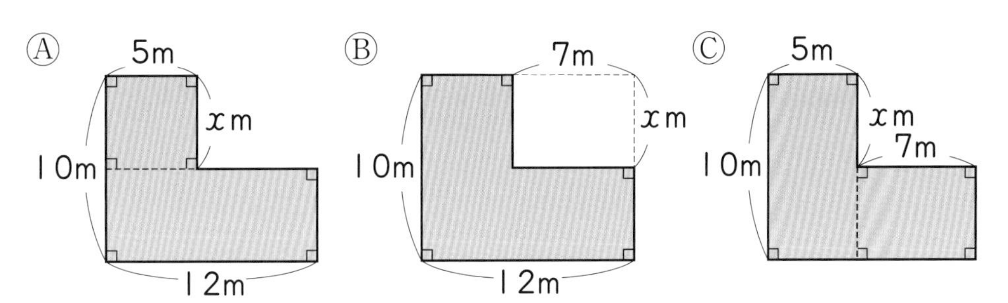

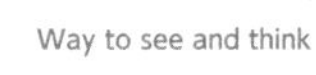

In mathematics, it is important to represent a diagram by math expressions, but it is also important to represent a math expression with diagrams.

② Let's explain the reasons for the answer in ①.

What you can do now

☐ **Can represent sentences in math sentences by using mathematical letters.**

1 Let's represent in math sentences by using x.

① There are x sheets of origami. If 30 sheets were added, the number became 44 sheets.

② There were 15 chocolates. Since you ate x chocolates, 6 chocolates were left.

③ The area of a rectangle with a length of 8 cm and a width of x cm is 48.8 cm².

④ If x kg of flour is divided into 9 bags, each bag weighs 0.6 kg.

☐ **Can find the number that satisfies the math sentence.**

2 Let's find the number that applies for x.

① $x + 8 = 12$ ② $7 + x = 13$

③ $x - 9 = 11$ ④ $x - 19 = 13$

⑤ $4 \times x = 28$ ⑥ $x \div 7 = 8$

⑦ $x - 3.5 = 7$ ⑧ $x \times 3 = 4.2$

☐ **Can read and understand the meaning of a math expression.**

3 There is a bottle with x L of juice inside it.

Which of the following scenes Ⓐ～Ⓓ represent the math expressions ①～④?

① $x + 6$ ② $x - 6$ ③ $x \times 6$ ④ $x \div 6$

Ⓐ The amount of juice per person when x L is divided by 6 people.

Ⓑ The amount of juice when x L is combined with 6 L.

Ⓒ The total amount of juice when there are 6 bottles with x L of juice per bottle.

Ⓓ The remaining amount of juice when 6 L were drunk from x L.

Supplementary problems p.223

Usefulness and efficiency of learning

1 Let's write math sentences by using x and find the number that applies for x.

① A bundle of envelopes cost x yen and 6 bundles of envelopes cost 720 yen.

② The cost of 1 notebook is x yen and the cost of 5 notebooks is 650 yen.

③ There are 20 marbles. If x marbles were added, the number became 52 marbles.

④ There is a ribbon with a length of x cm. When 50 cm were used, the remaining length was 60 cm.

- [ ] Can represent math sentences by using mathematical letters.
- [ ] Can find the number that satisfies the math sentence.

2 Let's explore the number that applies for x by placing 6, 7, 8, ... as x in the following math sentences.

① $x \times 4 + 7 = 39$ ② $x \times 5 - 9 = 36$

- [ ] Can find the number that satisfies the math sentence.

3 Which of the following scenes Ⓐ~Ⓓ represent the math expressions ①~④?

① $x + 30$ ② $x \times 30$ ③ $x \div 30$ ④ $x - 30$

- [ ] Can read and understand the meaning of a math expression.

Ⓐ If x candies are distributed equally among 30 people, how many candies will each person receive?

Ⓑ If a x m ribbon is connected with a 30 m ribbon, how many meters is the total length of the ribbon?

Ⓒ A rectangle with an area of 30 cm² has been subtracted from a square with an area of x cm². How many square centimeters is the remaining area?

Ⓓ How many square centimeters is the area of a parallelogram that has a base of x cm and a height of 30 cm?

Find the ? Can fractions also be multiplied?

Problem Until now, we have learned addition, subtraction, multiplication, and division of whole numbers. We are also able to calculate the addition and subtraction of fractions, but can we also perform multiplication and division?

Multiplication and Division of Fractions and Whole Numbers

Let's think about the meaning of multiplication and division and how to calculate.

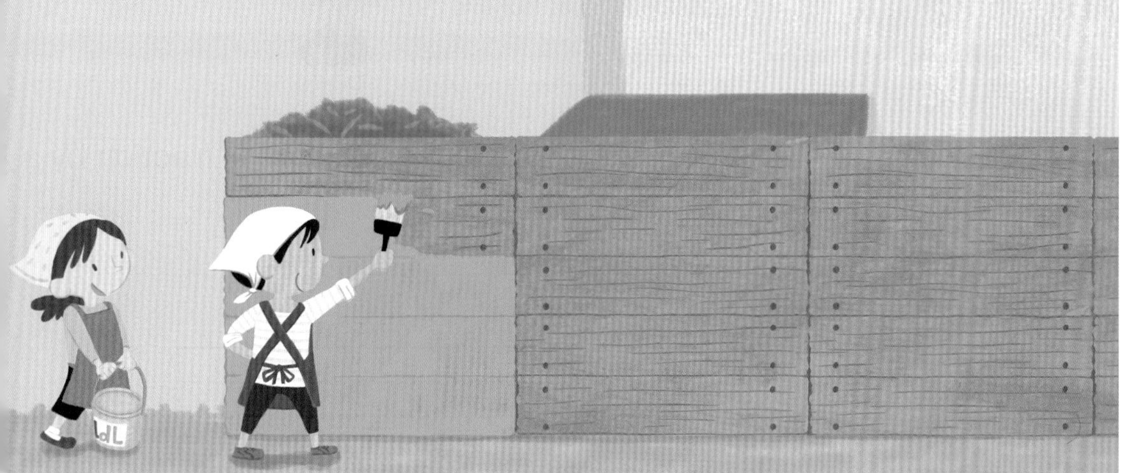

<Procedure for problem solving>

Read the problem.
↓
Represent with a table or diagram.
↓
Write a math expression.
↓
Think how to calculate.
↓
Try to confirm.
↓
Think a better method.
↓
Try to confirm.

1 The case: fraction × whole number

Want to solve

Activity

1 **The fence is being painted with green paint. If $\frac{4}{5}$ m² can be painted per deciliter of paint, how many square meters can be painted with 3 dL?**

Measure per unit quantity — Total measurement

Painted area	0	$\frac{4}{5}$	(x m²)
Amount of paint	0	1	3 (dL)

How many units

Measure per unit quantity	Total measurement
$\frac{4}{5}$ m²	x m²
1 dL	3dL

How many units

Since you know "Measure per unit quantity" and "How many units," you can use a multiplication expression.

① Let's write a math expression.

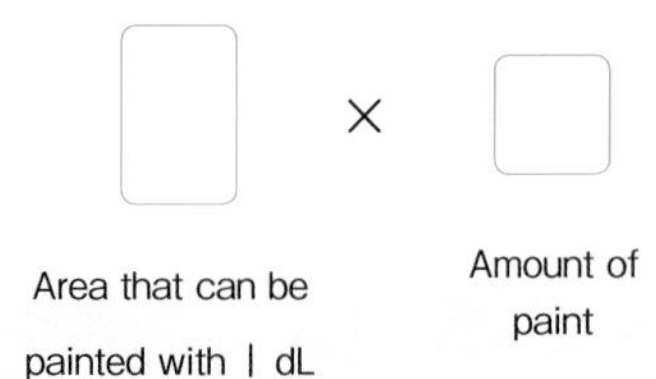

Want to think

② Let's think about how to calculate.

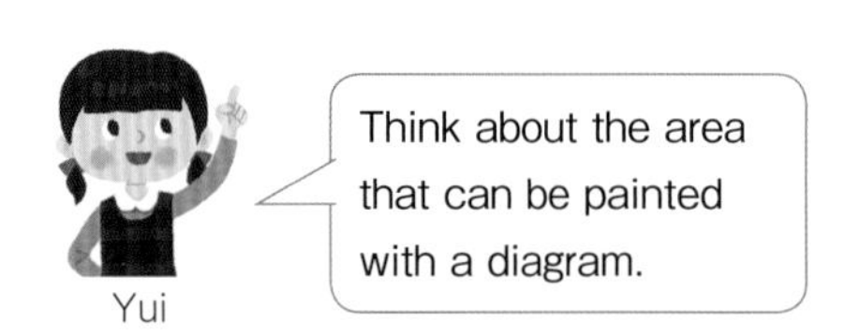

Since $\frac{4}{5}$ has four sets of $\frac{1}{5}$ ...

Daiki

Purpose What should we do to multiply an improper fraction or proper fraction by a whole number?

Yui's idea

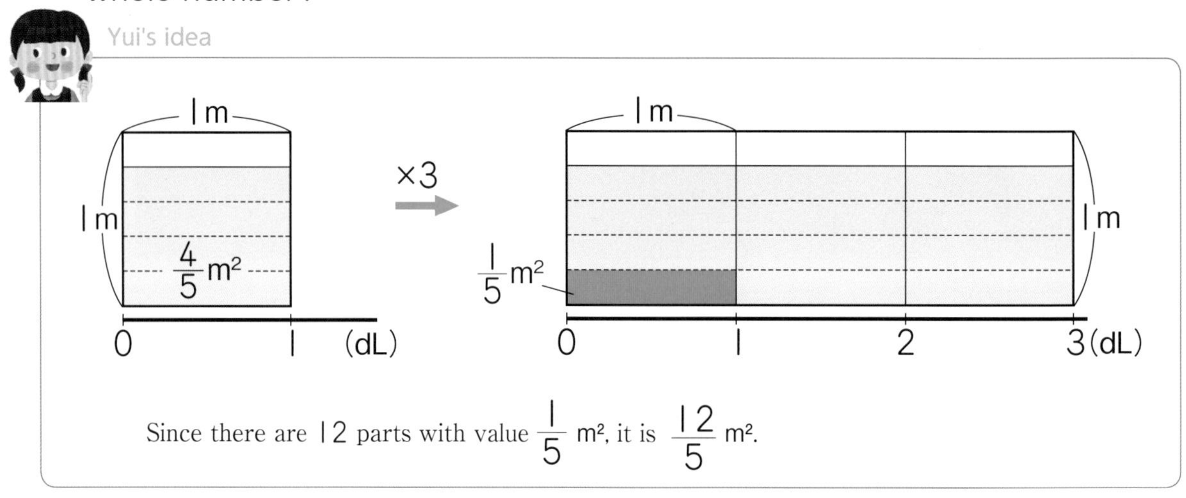

Since there are 12 parts with value $\frac{1}{5}$ m², it is $\frac{12}{5}$ m².

Daiki's idea

If I consider $\frac{1}{5}$ as a unit, $\frac{4}{5}$ has four sets of $\frac{1}{5}$.

As for $\frac{4}{5} \times 3$, since it's three sets of $\frac{4}{5}$, it becomes (4×3) sets of $\frac{1}{5}$.

$$\frac{4}{5} \times 3 = \frac{4 \times 3}{5} = \frac{12}{5}$$

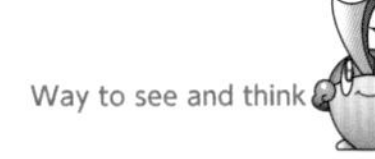

Both considered by "how many units" of $\frac{1}{5}$.

Want to compare

③ Let's try to compare the calculation methods of Yui and Daiki.

Summary

When you multiply an improper fraction or proper fraction by a whole number, leave the denominator as it is and multiply the numerator by the whole number.

$$\frac{b}{a} \times c = \frac{b \times c}{a}$$

When comparing the ideas of the children, consider the following:
- What did each child consider?
- What is different between them?
- What is the same between them?

Want to confirm

Let's solve the following calculations.

① $\frac{2}{5} \times 2$　② $\frac{4}{9} \times 2$　③ $\frac{7}{11} \times 4$　④ $\frac{6}{7} \times 5$

⑤ $\frac{11}{8} \times 3$　⑥ $\frac{7}{6} \times 5$　⑦ $\frac{9}{4} \times 7$　⑧ $\frac{11}{5} \times 2$

Want to improve

The operation $\frac{2}{9} \times 3$ was solved as shown below. Let's explain how each child calculated.

Hiroto's idea

$$\frac{2}{9} \times 3 = \frac{2 \times 3}{9} = \frac{\overset{2}{\cancel{6}}}{\underset{3}{\cancel{9}}} = \frac{\square}{\square}$$

Nanami's idea

$$\frac{2}{9} \times 3 = \frac{2 \times \overset{1}{\cancel{3}}}{\underset{3}{\cancel{9}}} = \frac{\square}{\square}$$

Summary

Operations become easy if fractions are reduced in the middle of the calculation.

Want to confirm

Let's solve the following calculations.

① $\frac{3}{8} \times 2$　② $\frac{5}{18} \times 3$　③ $\frac{9}{16} \times 8$　④ $\frac{7}{12} \times 6$

⑤ $\frac{7}{6} \times 4$　⑥ $\frac{13}{10} \times 25$　⑦ $\frac{3}{2} \times 4$　⑧ $\frac{4}{3} \times 6$

Activity

2

4 pieces of tape will be made, each with a length of $1\frac{2}{5}$ m. How many meters of tape will be needed?

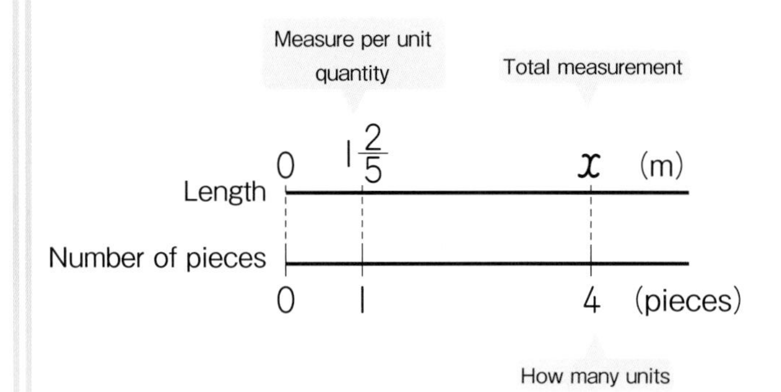

Measure per unit quantity — Total measurement

$1\frac{2}{5}$ m	x m
1 piece	4 pieces

How many units

Way to see and think

If you think about decomposing $1\frac{2}{5}$ into a whole number and a fraction, the approximate answer is easy to understand.

① Let's write a math expression.

② About how many meters are needed?

Want to think

③ Let's think about how to calculate.

Purpose What should we do to multiply a mixed fraction by a whole number?

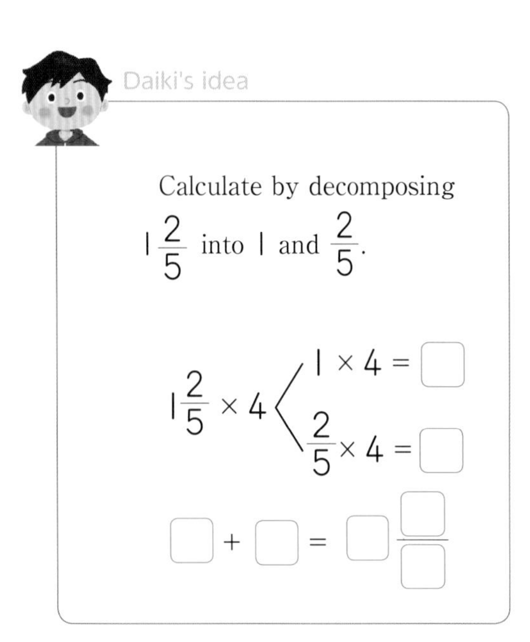

Nanami's idea

Calculate by changing $1\frac{2}{5}$ into an improper fraction.

$$1\frac{2}{5} \times 4 = \frac{7}{5} \times 4$$
$$= \square$$
$$= \square\frac{\square}{\square}$$

Way to see and think

Decompose mixed fractions into whole numbers and fractions or make them improper fractions.

Want to compare

④ Let's try to compare the methods from Daiki and Nanami.

Also, let's present the good points of their ideas.

Summary

When you multiply a mixed fraction by a whole number, if the mixed fraction is changed into an improper fraction then it can be solved as learned before.

Want to confirm

4 Let's solve the following calculations.

① $1\frac{3}{7} \times 2$ ② $2\frac{2}{3} \times 2$ ③ $1\frac{5}{8} \times 2$ ④ $2\frac{5}{6} \times 12$

Want to try

5 $2\frac{3}{10}$ dL of soda is used to make one fruit punch. How many deciliters of soda will be needed to make four fruit punches?

6 How many square meters is the area of a rectangular flowerbed with a length of $4\frac{2}{3}$ m and a width of 6 m?

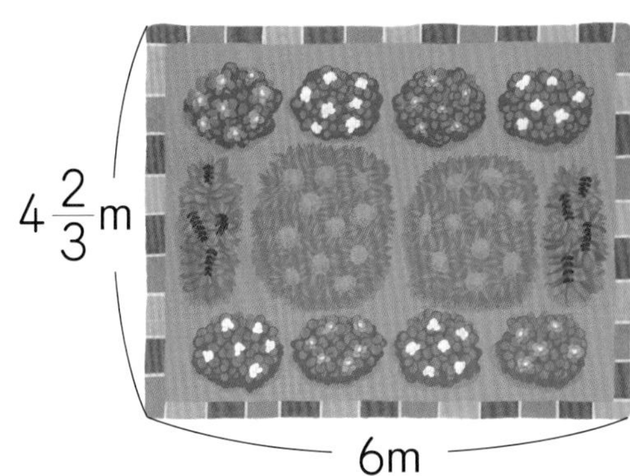

2 The case: fraction ÷ whole number

Want to solve

Activity

1 **2 dL of blue paint is used to paint a $\frac{4}{5}$ m² wall. How many square meters can be painted per deciliter of paint?**

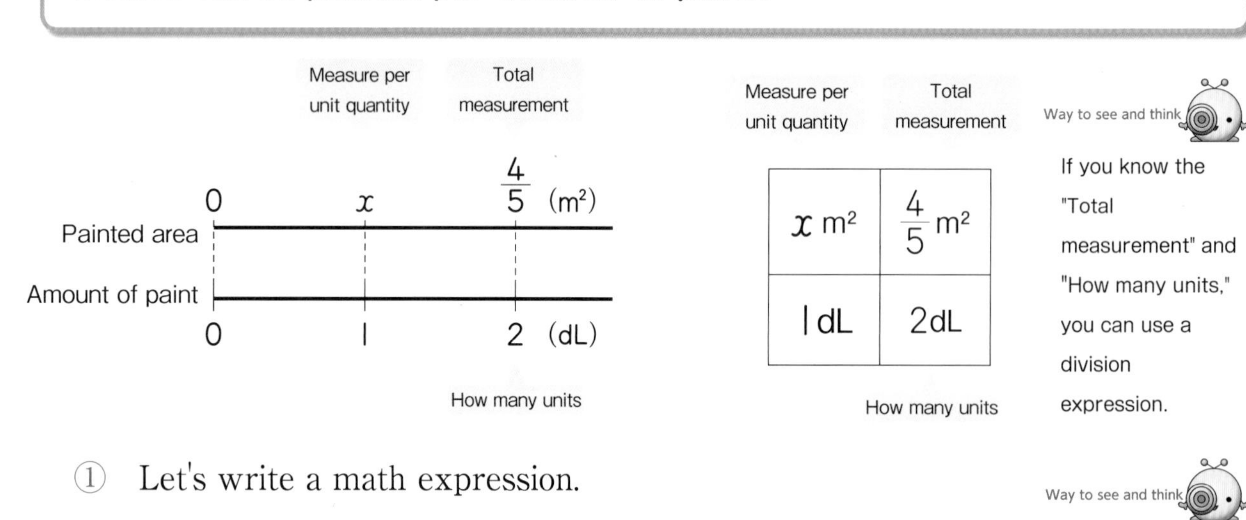

Way to see and think

If you know the "Total measurement" and "How many units," you can use a division expression.

① Let's write a math expression.

☐ ÷ ☐

Painted area　Amount of paint

Way to see and think

If you can paint xm² per deciliter of paint, it can be represented as $x \times 2 = \frac{4}{5}$.

Want to think

② Let's think about how to calculate.

Purpose　What should we do to divide a fraction by a whole number?

③ Let's explain the ideas of the following children.

Yui's idea

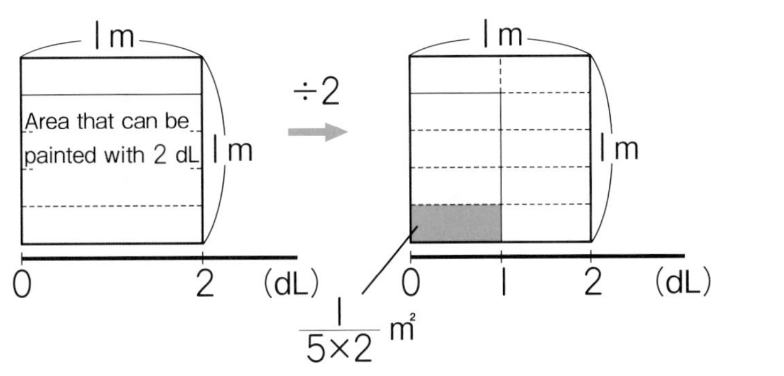

Since there are 4 sets of ▭ ,

$$\frac{4}{5} \div 2 = \frac{1}{5 \times 2} \times 4$$
$$= \frac{4}{5 \times 2}$$
$$= \frac{\square}{\square}$$

Using a diagram, think about how many units of $\frac{1}{5\times2}$ there are.

Nanami's idea

Considering the same as the multiplication of fractions, as for the case of a division, divide the numerator.

$$\frac{4}{5} \div 2 = \frac{4 \div 2}{5} = \frac{\square}{\square}$$

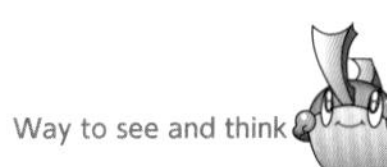

Considering how many units of $\frac{1}{5}$ $\frac{4}{5}$ has , take half of it.

Daiki's idea

If I use the rules of division and solve the calculation of whole number ÷ whole number,

$$\frac{4}{5} \div 2 = \left(\frac{4}{5} \times 5\right) \div (2 \times 5)$$
$$= 4 \div (2 \times 5)$$
$$= 4 \div (5 \times 2)$$

If represented using fractions, $\frac{4}{5} \div 2 = \frac{4}{5 \times 2} = \frac{\square}{\square}$

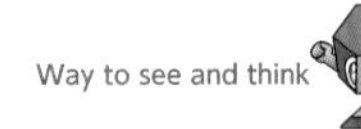

Use the rules of division, and think of it as a calculation of whole numbers.

Want to solve

Activity

2

3 dL of red paint is used to paint a $\frac{4}{5}$ m² wall. How many square meters can be painted per deciliter of paint?

① Let's write a math expression.

Want to think

② Let's think about how to calculate.

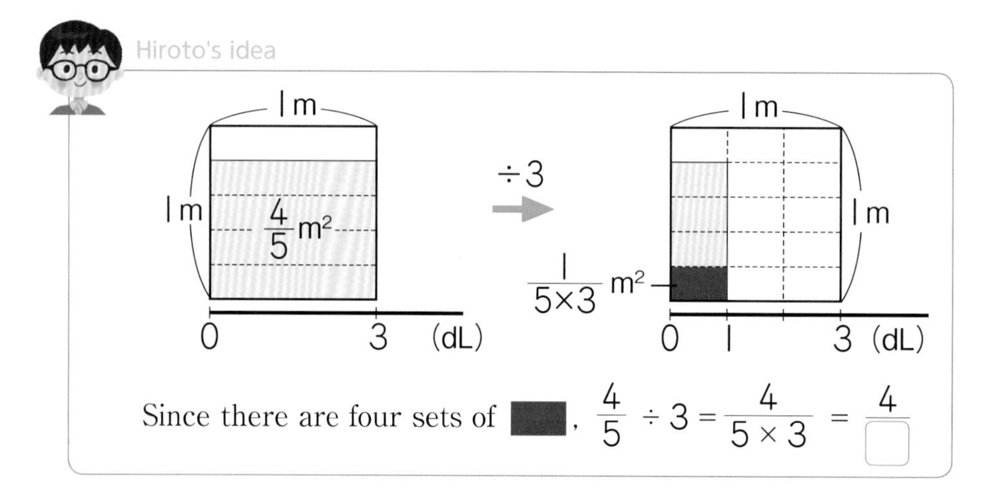

Since there are four sets of ■, $\frac{4}{5} \div 3 = \frac{4}{5\times3} = \frac{4}{\square}$

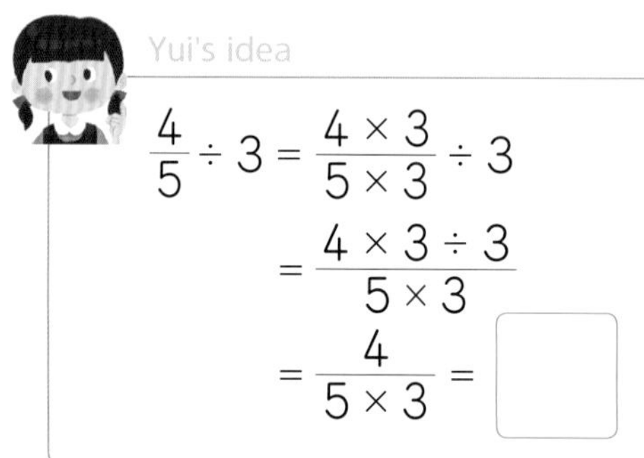

$$\frac{4}{5} \div 3 = \frac{4\times3}{5\times3} \div 3$$
$$= \frac{4\times3\div3}{5\times3}$$
$$= \frac{4}{5\times3} = \square$$

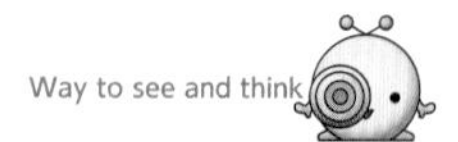

As for Yui's idea, since 4 cannot be divided by 3, it first uses the properties of fractions and then is able to divide.

Summary

When you divide an improper fraction or proper fraction by a whole number, the denominator is multiplied by the whole number and the numerator is left as it is.

$$\frac{b}{a} \div c = \frac{b}{a\times c}$$

③ Which child in the previous page has the same idea as Hiroto's and Yui's in exercise ②?

Want to confirm

Let's solve the following calculations.

① $\frac{1}{2} \div 4$ ② $\frac{3}{4} \div 2$ ③ $\frac{5}{6} \div 4$ ④ $\frac{7}{8} \div 5$

⑤ $\frac{3}{2} \div 5$ ⑥ $\frac{5}{3} \div 7$ ⑦ $\frac{8}{5} \div 3$ ⑧ $\frac{7}{6} \div 2$

Want to improve

As for the calculation of $\frac{10}{7} \div 4$, let's compare methods Ⓐ and Ⓑ.

Ⓐ $\frac{10}{7} \div 4 = \frac{10}{7 \times 4}$

$= \frac{\overset{5}{\cancel{10}}}{\underset{14}{\cancel{28}}}$

$= \square$

Ⓑ $\frac{10}{7} \div 4 = \frac{\overset{5}{\cancel{10}}}{7 \times \underset{2}{\cancel{4}}}$

$= \square$

It is easier to calculate if fractions can be reduced in the middle of the calculation.

Want to confirm

Let's solve the following calculations.

① $\frac{6}{7} \div 3$ ② $\frac{3}{4} \div 12$ ③ $\frac{12}{5} \div 4$ ④ $\frac{8}{3} \div 6$

Want to try

There is a field of $\frac{12}{13}$ ha. If this field is equally divided among 6 people, how many hectares will each person receive?

Want to solve Operation of mixed fraction ÷ whole number

4 **There is an iron bar that weighs $2\frac{1}{4}$ kg and has a length of 3 m. How many kilograms is the weight of this iron bar per meter?**

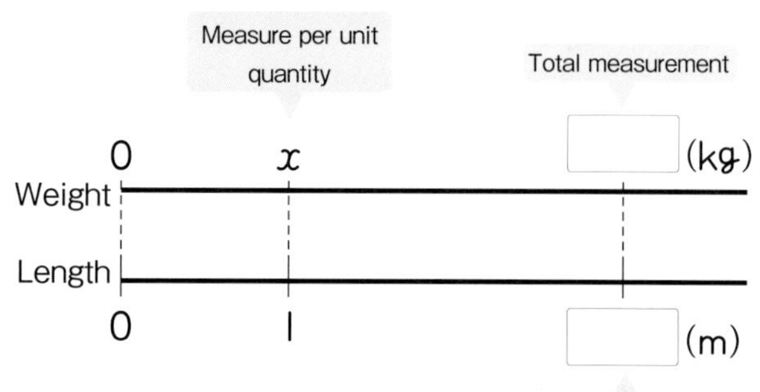

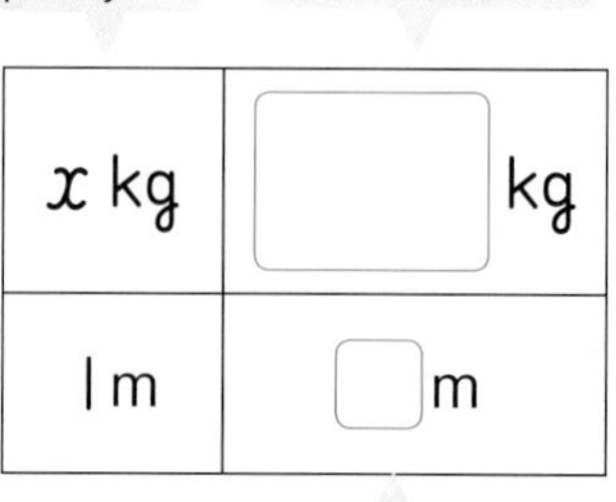

x kg	☐ kg
1 m	☐ m

Way to see and think

As for mixed fraction × whole number, consider the following methods:

- Decompose the mixed fraction into a whole number and fraction.
- Change the mixed fraction into an improper fraction.

① Let's write a math expression.

② As for the weight per meter, is it heavier than 1 kg? Or lighter?

Purpose How can we divide a mixed fraction by a whole number?

Want to compare

③ Let's compare the calculation methods of the following children.

Yui's idea

$$2\frac{1}{4} \div 3 = \frac{\square}{4} \div 3 = \frac{\square}{4 \times 3} = \frac{\square}{4}$$

Nanami's idea

$$2\frac{1}{4} \div 3 = \begin{cases} 2 \div 3 = \frac{2}{3} \\ \frac{1}{4} \div 3 = \frac{1}{4 \times 3} = \frac{1}{12} \end{cases}$$

$$\frac{2}{3} + \frac{1}{12} = \square + \square = \square$$

Summary

When you divide a mixed fraction by a whole number, if the mixed fraction is changed into an improper fraction then it can be solved as learned before.

Let's reduce fractions if possible.

Hiroto

Want to confirm

4 Let's solve the following calculations.

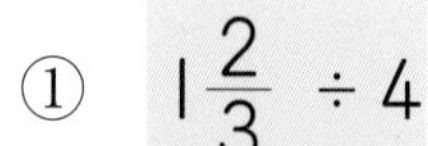

① $1\frac{2}{3} \div 4$ ② $2\frac{5}{8} \div 6$ ③ $2\frac{2}{7} \div 8$ ④ $3\frac{1}{2} \div 7$

What you can do now

☐ **Understanding how to calculate fraction × whole number and fraction ÷ whole number.**

1 Let's summarize how to calculate fraction × whole number and fraction ÷ whole number.

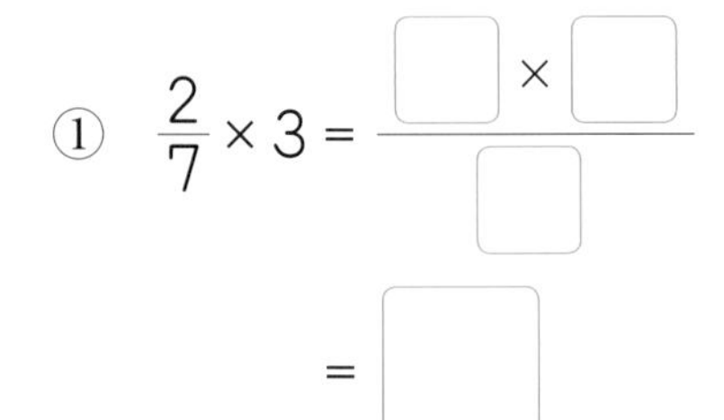

① $\frac{2}{7} \times 3 = \frac{\square \times \square}{\square}$

$= \square$

② $\frac{5}{7} \div 3 = \frac{\square}{\square \times \square}$

$= \square$

☐ **Can calculate fraction × whole number and fraction ÷ whole number.**

2 Let's solve the following calculations.

① $\frac{2}{9} \times 4$　② $\frac{7}{11} \times 5$　③ $\frac{4}{21} \times 3$　④ $\frac{7}{6} \times 8$

⑤ $\frac{3}{7} \times 28$　⑥ $1\frac{5}{12} \times 7$　⑦ $3\frac{3}{10} \times 30$　⑧ $2\frac{3}{4} \times 12$

⑨ $\frac{5}{8} \div 3$　⑩ $\frac{2}{5} \div 7$　⑪ $\frac{3}{2} \div 2$　⑫ $\frac{3}{10} \div 6$

⑬ $\frac{4}{5} \div 8$　⑭ $\frac{10}{7} \div 10$　⑮ $1\frac{3}{8} \div 3$　⑯ $2\frac{5}{8} \div 3$

☐ **Can create a math expression and find the answer.**

3 Every day, I drink $\frac{5}{6}$ L of milk.

How many liters of milk will I drink in 3 days?

4 $\frac{7}{6}$ L of milk will be equally divided into 3 bottles.

How many liters of milk will be poured into each bottle?

5 $2\frac{4}{7}$ dL of paint were used to paint a $3m^2$ wall.

How many deciliters of paint were used to cover a $1m^2$ wall?

Supplementary problems p.225

Usefulness and efficiency of learning

1 Let's find the mistake in the following calculations and write the correct answer.

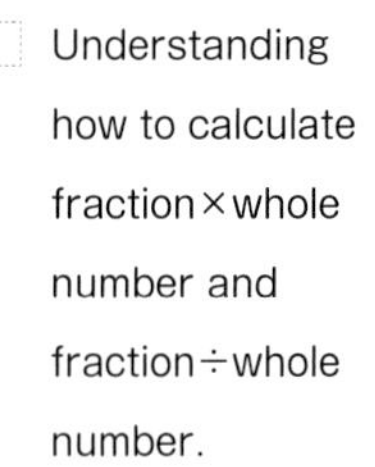

① $\frac{2}{5}\times 10=\frac{\overset{1}{\cancel{2}}}{5\times\underset{5}{\cancel{10}}}=\frac{1}{25}$

② $\frac{8}{7}\div 4=\frac{7\times\overset{1}{\cancel{4}}}{\underset{2}{\cancel{8}}}=\frac{7}{2}$

2 Let's answer the following questions.

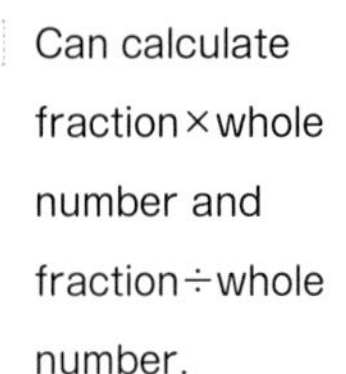

① There is one card for each of the following numbers [1] ~ [5]. Let's place one card inside □, so that the following math sentence holds true.

$\square\frac{\square}{\square}\times\square=6\frac{\square}{5}$

② There is one card for each of the following numbers [1] ~ [4]. Let's place one card inside □, so that the following math sentence holds true.

$\square\frac{\square}{\square}\div\square=\frac{5}{12}$

3 There is a wire that weighs $1\frac{5}{6}$ g per meter. How many grams is the weight for 4 m of this wire?

□ Can create a math expression and find the answer.

4 There is a rice field in which $4\frac{2}{3}$ kg of rice can be cultivated for every 4m². Let's answer the following questions.

① How many kilograms of rice can be cultivated per square meter?

② If this rice field is 300m², how many kilograms of rice can be cultivated?

If you can solve fraction × whole number and fraction ÷ whole number, then we can also solve fraction × fraction and fraction ÷ fraction.

Is there a calculation of fraction × fraction?

How can we solve calculations like $\frac{4}{5}\times\frac{1}{3}$?

Fraction × Fraction

Let's think about the meaning of the multiplication of fractions and how to calculate.

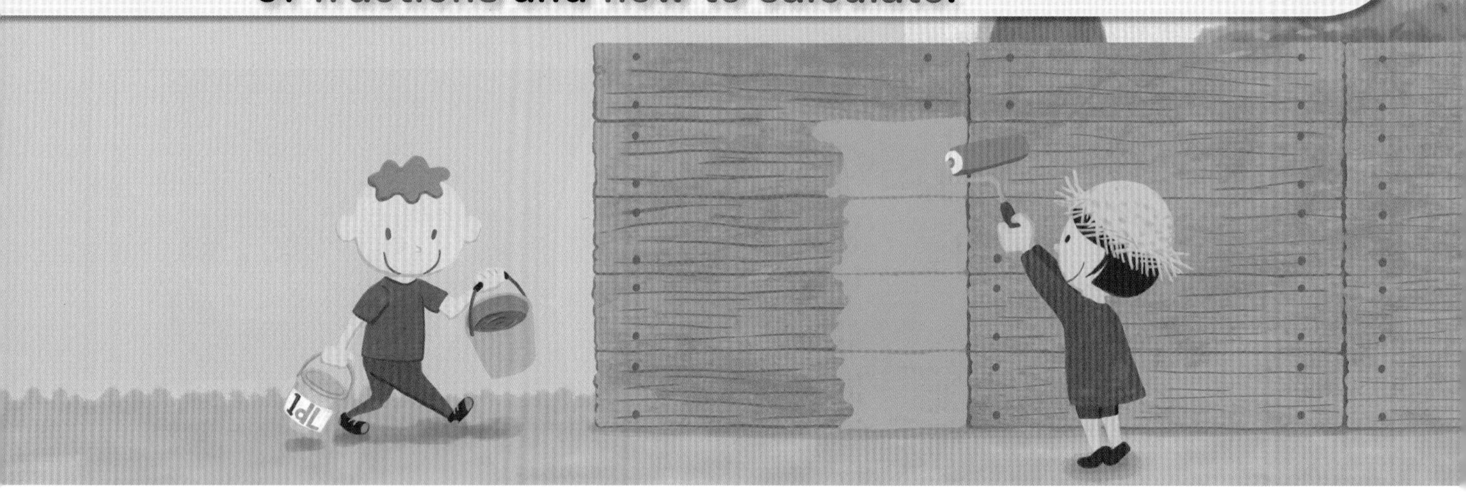

1 The case: fraction × fraction

Want to solve Operation of proper fractions

1 **A wall is painted with green paint. $\frac{4}{5}$ m² can be painted per deciliter of this paint. Let's think about the amount of paint and the painted area.**

① How many square meters can be painted with $\frac{1}{3}$ dL of paint? Let's write a math expression.

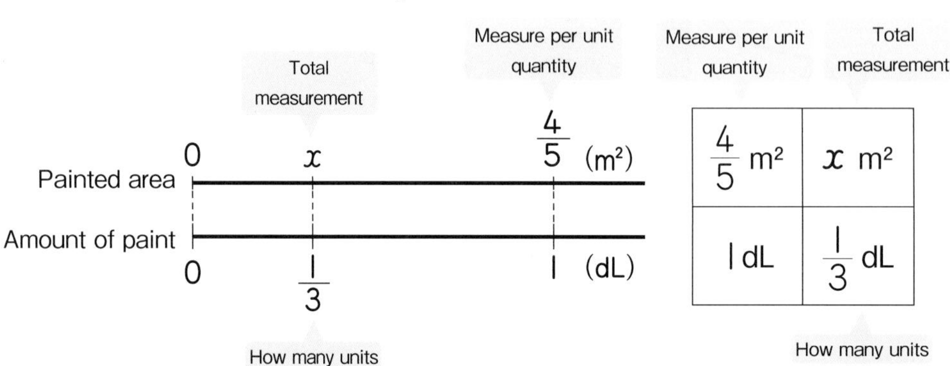

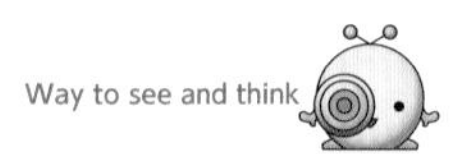

Since you know the "Measure per unit quantity" and "How many units," you can use a multiplication expression.

Purpose What should we do to multiply a proper fraction by a proper fraction?

Want to think

② Let's look at the following diagrams and think about how to calculate.

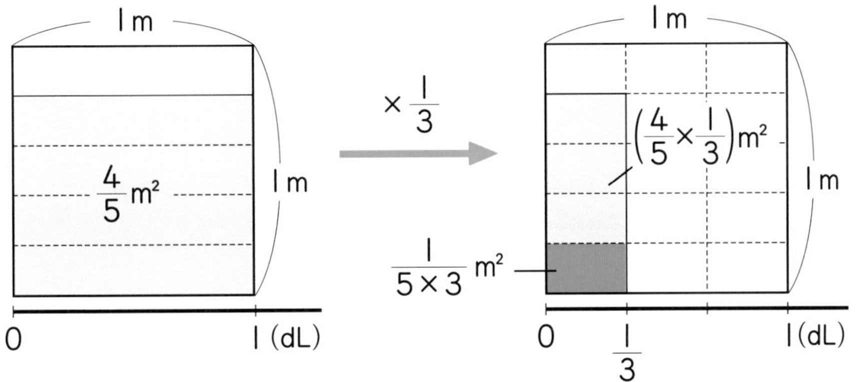

The multiplication by $\frac{1}{3}$ is the same as the division by 3.

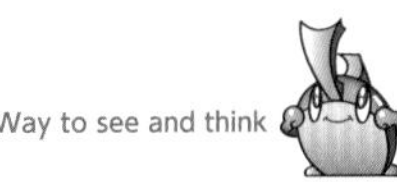

You can calculate by "How many units" of $\frac{1}{15}$ m².

Since there are 4 sets of $\frac{1}{5\times3}$ m², $\frac{4}{5}\times\frac{1}{3}=\frac{\square}{\square\times\square}$

$=\square$

Want to solve

Activity

2 How many square meters can be painted with $\frac{2}{3}$ dL of the paint from 1?

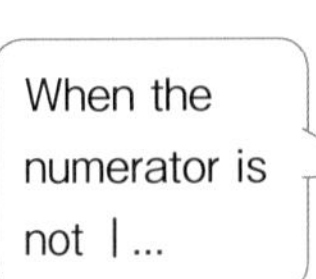

① Let's write a math expression.

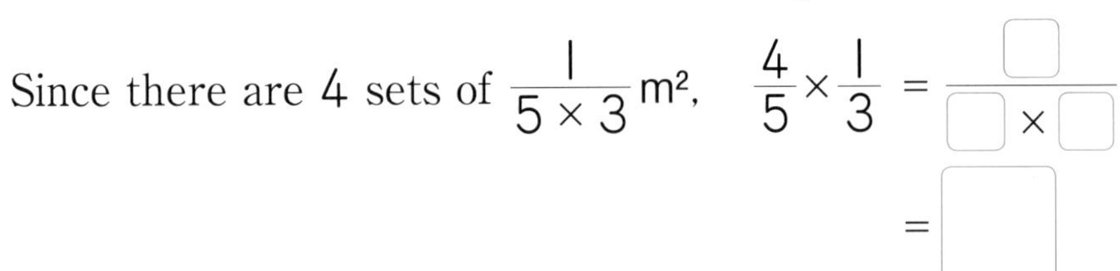

$\frac{4}{5}$ m²	x m²
1 dL	$\frac{2}{3}$ dL

Want to think

② Let's think about how to calculate.

Can I use the operation of fraction × whole number and fraction ÷ whole number?

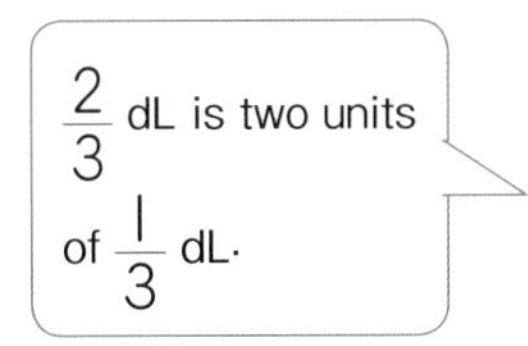

Want to compare

③ Let's explain the ideas of the following children and find the same resulting math sentence.

Way to see and think

Thinking based on the area diagram.

Hiroto's idea

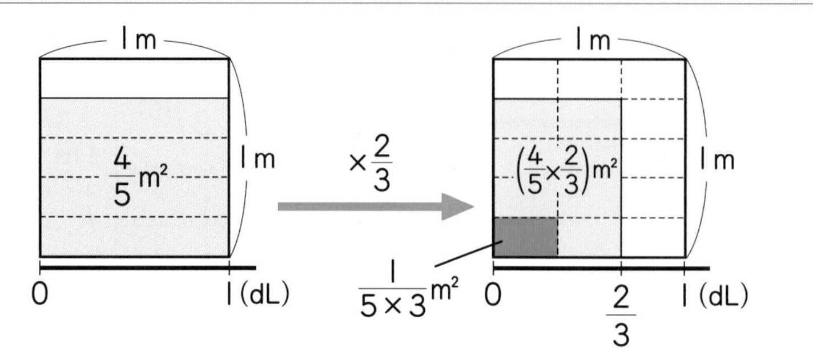

As for the painted area, since there are (4×2) sets of $\frac{1}{5\times3}$ m², it is $\frac{4\times2}{5\times3}$ m².

$$\frac{4}{5}\times\frac{2}{3}=\frac{\square\times\square}{\square\times\square}=\square$$

Daiki's idea

As for the painted area with $\frac{1}{3}$ dL, it was previously done as $\frac{4}{5}\times\frac{1}{3}=\frac{4}{5\times3}$.

$\frac{2}{3}$ dL is two units of that.

$$\frac{4}{5\times3}\times2=\frac{\square\times\square}{5\times3}$$

$$=\square$$

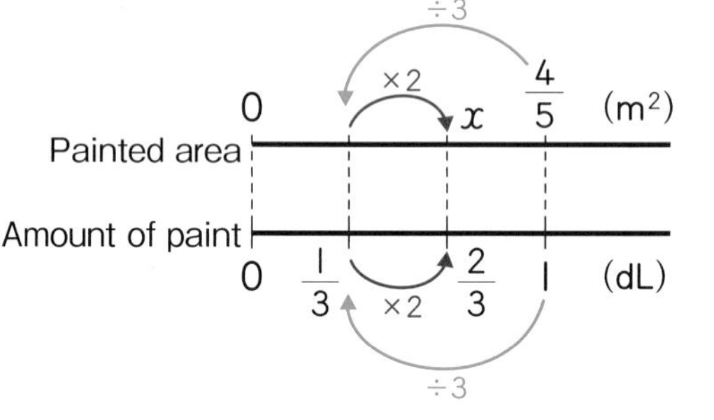

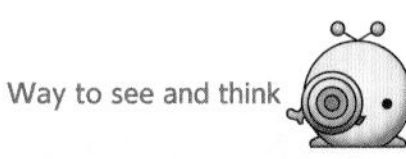

Way to see and think

Use the multiplication rules to change $\frac{2}{3}$ into a whole number.

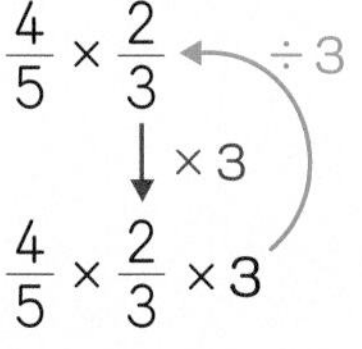

Nanami's idea

Using the operation of fraction × whole number and fraction ÷ whole number.

$$\frac{4}{5}\times\frac{2}{3}=\frac{4}{5}\times\left(\frac{2}{3}\times3\right)\div3$$

$$=\frac{4}{5}\times2\div3$$

$$=\frac{4\times2}{5}\div3=\frac{4\times2}{5\times3}=\square$$

Summary

When you multiply a proper fraction by a proper fraction, calculate by multiplying the two denominators and the two numerators.

$$\frac{b}{a} \times \frac{d}{c} = \frac{b \times d}{a \times c}$$

Want to solve Operation of improper fractions

3 **How many square meters can be painted with $\frac{4}{3}$ dL of a paint that covers $\frac{4}{5}$ m² per deciliter?**

① Let's write a math expression.

Purpose Even when multipliers are improper fractions, can we do the same calculation learned until now?

Want to think

② Let's look at the following diagrams and think about the answer.

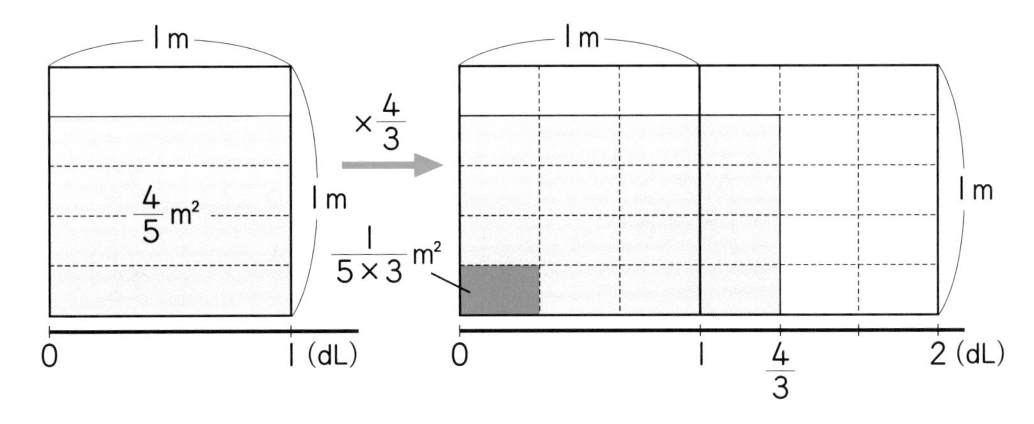

③ Let's find the answer and confirm that is the same as the one shown in ②.

Summary

Even when multipliers are improper fractions, also calculate by multiplying the two denominators and the two numerators.

Want to confirm

Let's solve the following calculations.

① $\frac{3}{4} \times \frac{1}{2}$　② $\frac{1}{3} \times \frac{5}{6}$　③ $\frac{3}{5} \times \frac{7}{8}$　④ $\frac{5}{9} \times \frac{2}{7}$

⑤ $\frac{1}{4} \times \frac{5}{3}$　⑥ $\frac{5}{7} \times \frac{3}{2}$　⑦ $\frac{8}{7} \times \frac{9}{5}$　⑧ $\frac{10}{9} \times \frac{4}{3}$

Want to try

1 kg of potatoes contains about $\frac{3}{4}$ L of water. How many liters of water does $\frac{5}{7}$ kg of potatoes contain?

Potato field (Furano City, Hokkaido)

Want to solve　Various operations

Let's think about how to solve the following calculations.

① $\frac{4}{15} \times \frac{5}{6} = \frac{\overset{2}{\cancel{4}} \times \overset{\square}{\cancel{5}}}{\underset{3}{\cancel{15}} \times \underset{\square}{\cancel{6}}}$

$= \square$

② $3\frac{1}{7} \times 2\frac{1}{10} = \frac{22}{7} \times \frac{21}{10}$

$= \frac{\overset{11}{\cancel{22}} \times \overset{3}{\cancel{21}}}{\underset{1}{\cancel{7}} \times \underset{5}{\cancel{10}}}$

$= \square$

Daiki

Calculations become easy if fractions are reduced in the middle of the calculation.

As for the multiplication of fractions, change the mixed fractions into improper fractions and then calculate.

Want to confirm

 Let's solve the following calculations.

① $\frac{5}{8}\times\frac{3}{10}$　② $\frac{1}{6}\times\frac{6}{7}$　③ $\frac{8}{9}\times\frac{3}{4}$　④ $\frac{6}{5}\times\frac{5}{12}$

⑤ $3\frac{1}{2}\times 1\frac{5}{9}$　⑥ $2\frac{5}{8}\times 2\frac{2}{9}$　⑦ $9\frac{1}{3}\times\frac{3}{8}$　⑧ $\frac{6}{7}\times 4\frac{2}{3}$

 There is 1 L of sand that weighs $1\frac{3}{5}$ kg. How many kilograms is the weight when there are $3\frac{3}{4}$ L of sand?

Want to try

 Let's think about how to solve the following calculations.

① $2\times\frac{3}{5}=\frac{2}{\square}\times\frac{3}{5}$

$=\square$

② $\frac{4}{7}\times 2=\frac{4}{7}\times\frac{2}{\square}$

$=\square$

As for the multiplication of a whole number and a fraction, if the whole number is changed into the fraction form, it becomes a operation of fraction × fraction.

Want to confirm

 Let's solve the following calculations.

① $4\times\frac{1}{5}$　② $4\times\frac{3}{5}$　③ $4\times\frac{6}{5}$

The answer gets bigger and bigger as the multiplier increases.

Yui

The answer in exercise ③ is larger than 4.

Hiroto

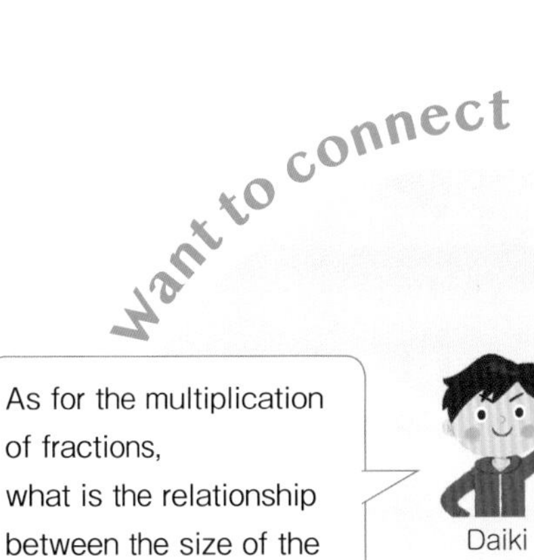

Want to think The size of the product

There is a wire that weighs 10 g per meter. Let's answer the following questions.

① How many grams is the weight of a wire with a length of $1\frac{1}{4}$ m or $\frac{2}{5}$ m?

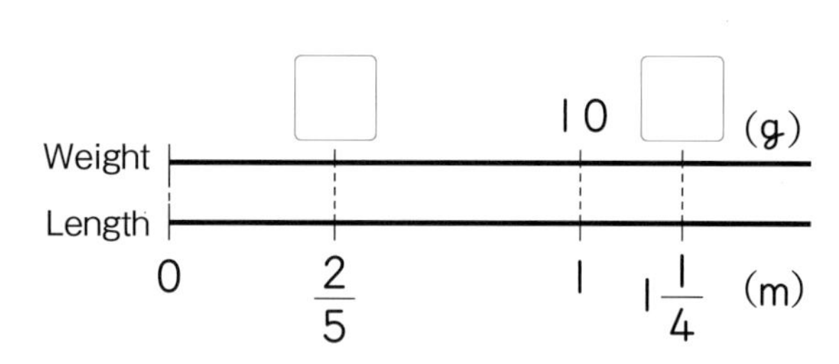

$10 \times 1\frac{1}{4} = \square$

$10 \times 1 = 10$

$10 \times \frac{2}{5} = \square$

Purpose What kind of relationship exists between the size of the multiplier and the product?

② As for the math expression $10 \times 1\frac{1}{4}$ and $10 \times \frac{2}{5}$, which has a product that became smaller than 10?

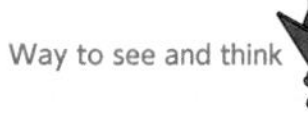

Way to see and think

The same can be said for the multiplication of decimal numbers.

When the multiplier is a fraction larger than 1, the product becomes larger than the multiplicand.

When the multiplier is a fraction smaller than 1, the product becomes smaller than the multiplicand.

When the multiplier is 1, the product becomes the same as the multiplicand.

Want to confirm

When the length of the wire in 5 is $\frac{7}{5}$ m, is it heavier than 10 g? Is it lighter?

Want to try

From the following Ⓐ~Ⓓ, let's choose the calculations that have a product smaller than $\frac{3}{4}$.

Ⓐ $\frac{3}{4} \times \frac{2}{3}$ Ⓑ $\frac{3}{4} \times \frac{3}{2}$ Ⓒ $\frac{3}{4} \times 1\frac{1}{5}$ Ⓓ $\frac{3}{4} \times \frac{3}{4}$

2 Various operations

Want to think

1 **The calculation $\frac{3}{4} \times \frac{1}{5} \times \frac{5}{6}$ was solved. Let's compare the following Ⓐ and Ⓑ calculation methods.**

Ⓐ $\frac{3}{4} \times \frac{1}{5} \times \frac{5}{6} = \frac{3 \times 1}{4 \times 5} \times \frac{5}{6}$

$= \frac{3}{20} \times \frac{5}{6}$

$= \frac{\overset{1}{\cancel{3}} \times \overset{1}{\cancel{5}}}{\underset{4}{\cancel{20}} \times \underset{2}{\cancel{6}}}$

$= \frac{1}{8}$

Ⓑ $\frac{3}{4} \times \frac{1}{5} \times \frac{5}{6} = \frac{\overset{1}{\cancel{3}} \times 1 \times \overset{1}{\cancel{5}}}{4 \times \underset{1}{\cancel{5}} \times \underset{2}{\cancel{6}}}$

$= \frac{1}{8}$

As for the multiplication of 3 or more fractions, calculate by multiplying the denominators together and the numerators together.

Want to confirm

Let's solve the following calculations.

① $\frac{3}{4} \times \frac{2}{3} \times \frac{3}{7}$ ② $\frac{5}{6} \times \frac{1}{5} \times \frac{2}{3}$ ③ $\frac{3}{8} \times 5 \times \frac{4}{5}$

Want to think

2 **Let's find the area of the rectangle shown on the right.**

$\frac{5}{7}$ m

$\frac{2}{3}$ m

Purpose Can we use the formula even when the length of the side is represented by a fraction?

① Daiki found the area as shown below.

Let's fill in the ☐ with numbers.

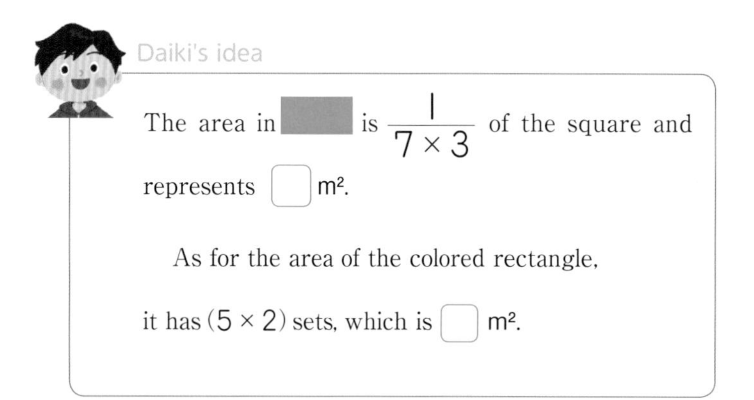

② Let's try to find the area of the rectangle using the formula.

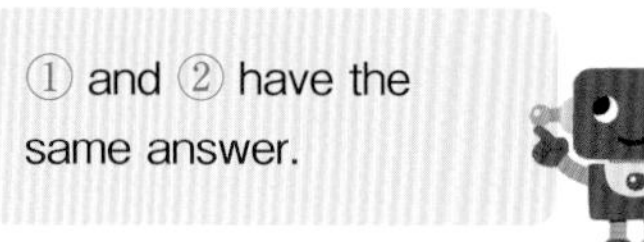

Want to try

2 Let's find the volume of the cuboid shown in the diagram on the right.

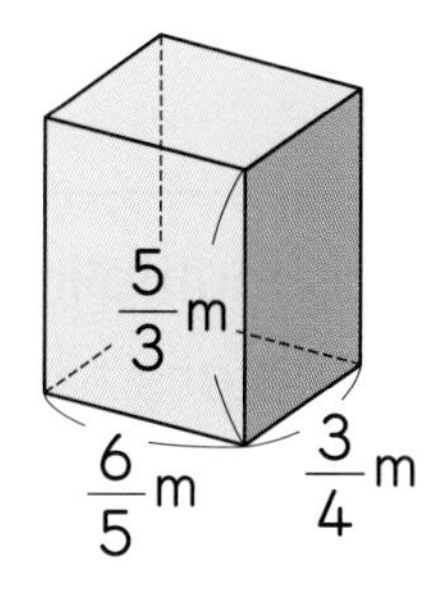

Way to see and think

The volume of a cuboid can be found by length × width × height.

Summary

The area and volume can be found by the formula that applies even when the length of the side is represented by a fraction.

Want to confirm

3 Let's find the area of the parallelogram in ① and the volume of the cube in ②.

①

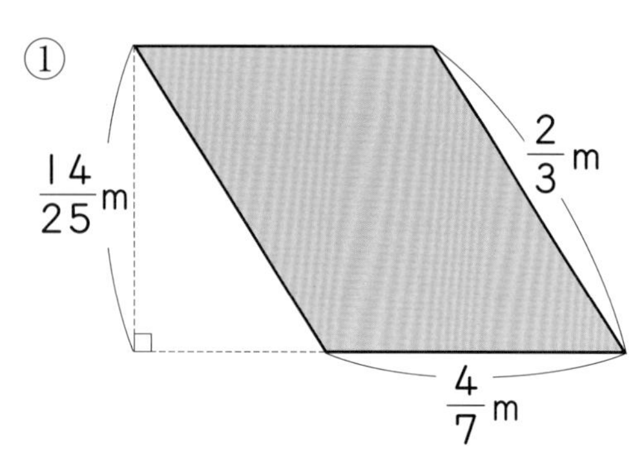

②

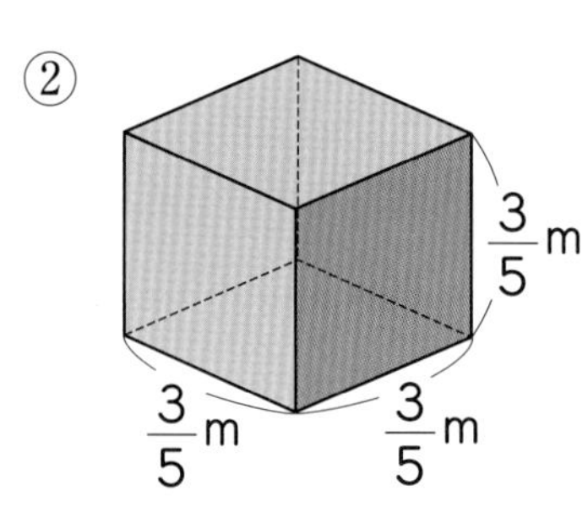

3 Rules of operations

Want to explore

1 **The following rules of operations are valid for whole numbers and decimal numbers. Let's explore whether these rules are also valid for fractions.**

Ⓐ $a\times b=b\times a$ Ⓑ $(a\times b)\times c=a\times(b\times c)$

Ⓒ $(a+b)\times c=a\times c+b\times c$ Ⓓ $(a-b)\times c=a\times c-b\times c$

① Let's explain that the rule of operation Ⓐ is valid using the area of the rectangle shown in the figure on the right.

$\frac{2}{5}\times\frac{3}{4}=\square$ $\frac{3}{4}\times\frac{2}{5}=\square$

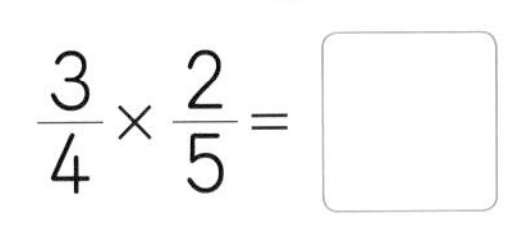

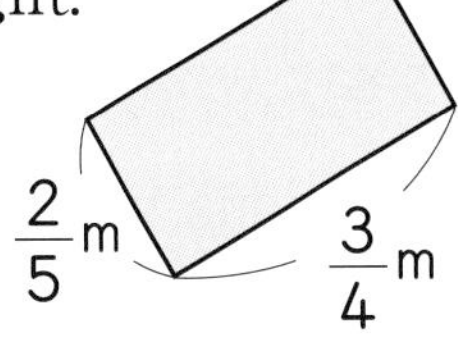

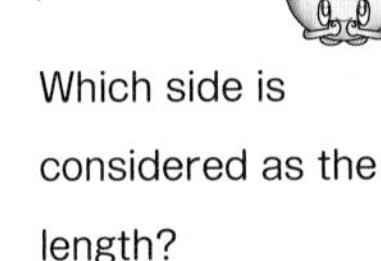

Way to see and think

Which side is considered as the length?

② Let's explain that the rule of operation Ⓑ is valid using the volume of the cuboid shown in the figure on the right.

$\frac{2}{3}$ m

$\frac{1}{2}$ m $\frac{6}{7}$ m

$\left(\frac{1}{2}\times\frac{6}{7}\right)\times\frac{2}{3}=\square$ $\frac{1}{2}\times\left(\frac{6}{7}\times\frac{2}{3}\right)=\square$

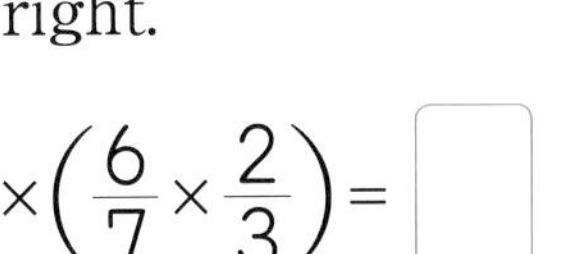

③ Using $a=\frac{2}{3}$, $b=\frac{1}{2}$, and $c=\frac{6}{7}$, let's confirm that rules Ⓒ and Ⓓ can be used.

The rules of operations, which are valid for whole numbers and decimal numbers, are also valid for fractions.

Want to deepen

Yui solved the calculation $\frac{3}{5}\times\frac{2}{3}+\frac{2}{5}\times\frac{2}{3}$ as follows. Let's explain Yui's idea.

Yui's idea

$\frac{3}{5}\times\frac{2}{3}+\frac{2}{5}\times\frac{2}{3}=\left(\square+\square\right)\times\frac{2}{3}=\square\times\frac{2}{3}=\frac{2}{3}$

4 Reciprocal

Want to try

1 **There are 2 cards for each number from [1] to [9]. Let's answer the following questions.**

① Let's use the cards to complete the following calculation.

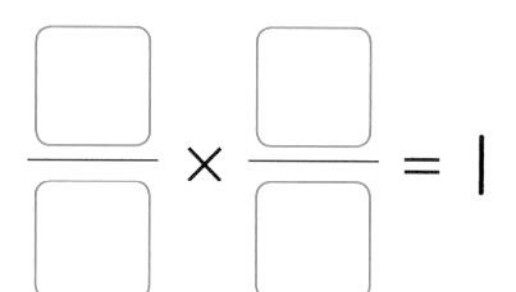

$\frac{\square}{\square} \times \frac{\square}{\square} = 1$

As for $\frac{\boxed{3}\times\square}{\boxed{5}\times\square}$, consider to reduce the fraction such that both the numerator and denominator are 1.

② Let's say what you noticed by looking at the multiplicand and multiplier of multiplications where the product is 1.

When the product of two numbers is 1, one number is called the **reciprocal** of the other number.

The reciprocal of $\frac{2}{3}$ is $\frac{3}{2}$. Also, the reciprocal of $\frac{3}{2}$ is $\frac{2}{3}$.

The reciprocal of a fraction is a fraction where the denominator and the numerator are interchanged. $\frac{b}{a} \rightleftarrows \frac{a}{b}$

Want to confirm

1 Let's find the reciprocals of 0.4 and 6.

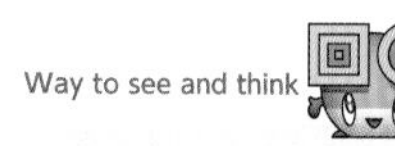

Way to see and think

When thinking about the reciprocals of whole numbers and decimal numbers, you can represent them in the form of fractions.

2 Let's find the reciprocals of the following numbers.

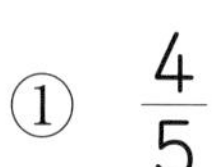

① $\frac{4}{5}$ ② $\frac{10}{3}$ ③ $\frac{1}{8}$

④ $1\frac{5}{6}$ ⑤ 0.6 ⑥ 11

What you can do now

Can solve calculations that have fractions as multipliers.

1 Let's solve the following calculations.

① $\frac{1}{5}\times\frac{3}{4}$　② $\frac{2}{5}\times\frac{6}{7}$　③ $\frac{5}{6}\times\frac{2}{3}$

④ $\frac{9}{14}\times\frac{7}{18}$　⑤ $2\frac{5}{6}\times\frac{2}{17}$　⑥ $1\frac{2}{3}\times1\frac{1}{5}$

⑦ $\frac{15}{8}\times\frac{6}{5}$　⑧ $7\times\frac{4}{5}$　⑨ $6\times\frac{9}{8}$

Understanding the relationship between multiplier and product.

2 Which calculation has a product smaller than 5?

Ⓐ $5\times1\frac{1}{12}$　Ⓑ $5\times\frac{5}{6}$　Ⓒ $5\times\frac{4}{3}$　Ⓓ $5\times\frac{9}{10}$

Understanding the formulas studied until now and the rules of operations that are valid for fractions.

3 Let's find the area of the following figures.

① Trapezoid

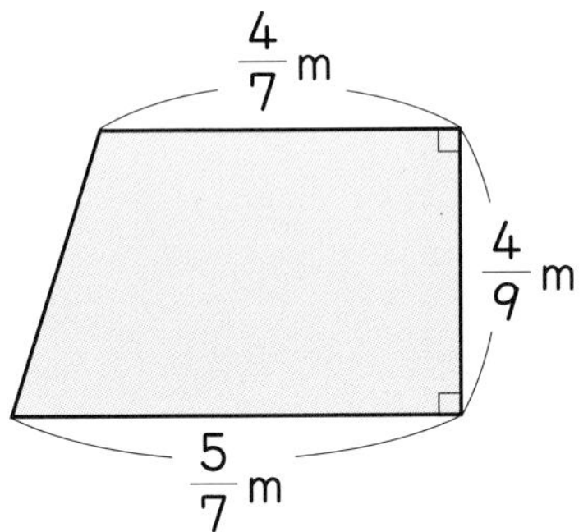

② Rectangle

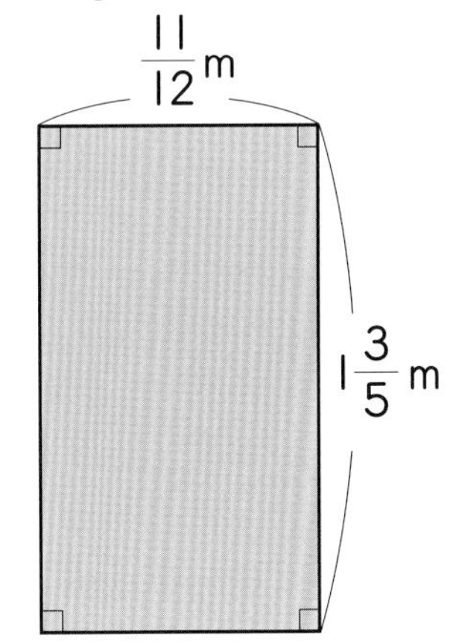

Understanding the meaning of reciprocals and how to find them.

4 Let's find the reciprocals of the following numbers.

① $\frac{1}{3}$　② $\frac{7}{2}$　③ $\frac{5}{6}$　④ $1\frac{1}{2}$　⑤ 9　⑥ 0.7

Supplementary problems p.226

Usefulness and efficiency of learning

1 There is a rice field that produces $\frac{4}{7}$ kg of rice per square meter.

How many kilograms of rice will $\frac{5}{8}$ m² of this field produce?

- Can solve calculations that have fractions as multipliers.
- Understanding the relationship between multiplier and product.

2 Let's find the area of the following figures.

①

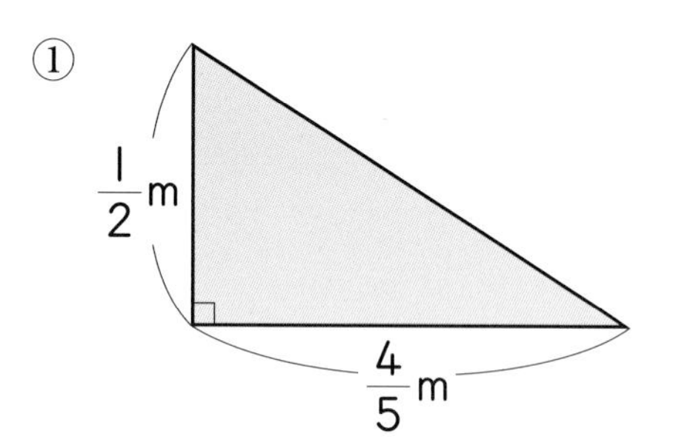

②

3m
$1\frac{1}{2}$m
$\frac{2}{3}$m
$1\frac{1}{6}$m
$1\frac{1}{6}$m

- Understanding the formulas studied until now and the rules of operations that are valid for fractions.

3 Let's find the volume of the following cuboid.

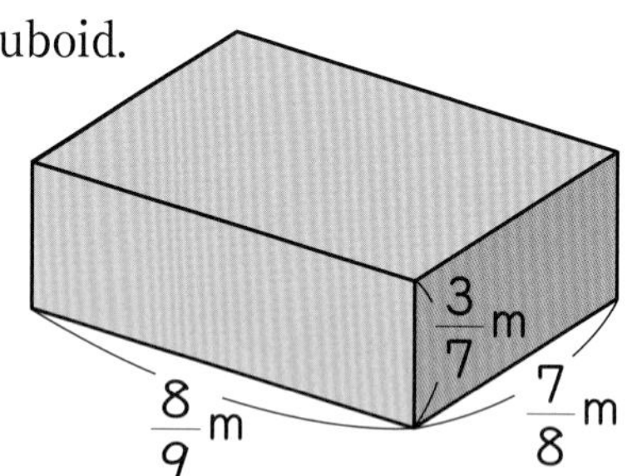

- Understanding the formulas studied until now and the rules of operations that are valid for fractions.

4 There are 2 cards for each number from [2] to [7].

[2] [2] [3] [3] [4] [4] [5] [5] [6] [6] [7] [7]

Let's create math expression by placing cards inside each ☐ shown on the right .

$\frac{\square}{\square} \times \frac{\square}{\square}$

① Let's create a math expression with answer 1.

② Let's create a math expression with answer 2.

- Understanding the meaning of reciprocals.

Want to connect

Since I can calculate fraction × fraction, I want to try to calculate fraction ÷ fraction.

Is there a calculation of fraction÷fraction?

We've been thinking about the problem of painting a wall.

1

When $\frac{4}{5}$ m² can be painted with 2 dL of paint,

how many square meters can be painted per deciliter of paint?

It can be found by $\frac{4}{5} \div 2$.

2

When $\frac{2}{5}$ m² can be painted with $\frac{1}{4}$ dL of paint, how much can I paint with 1 dL of paint?

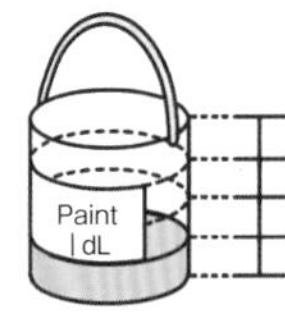

3

The area painted with 1 dL of paint can be found, as with whole numbers, by the division $\frac{2}{5} \div \frac{1}{4}$?

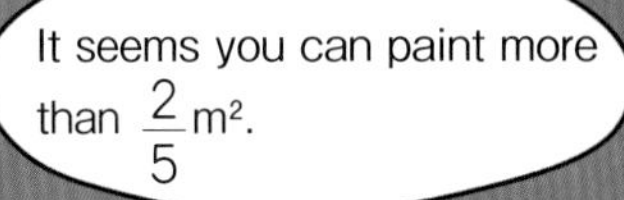

It seems you can paint more than $\frac{2}{5}$ m².

4

How can we solve calculations like $\frac{2}{5} \div \frac{1}{4}$?

5 Fraction÷Fraction

Let's think about the meaning of the division of fractions and how to calculate.

1 The case: fraction÷fraction

Want to solve　Operation of proper fractions

1 **$\frac{1}{4}$ dL of yellow paint was used to paint a $\frac{2}{5}$ m^2 wall. How many square meters can be painted per deciliter of this paint?**

① Let's write a math expression.

Painted area　　Amount of paint

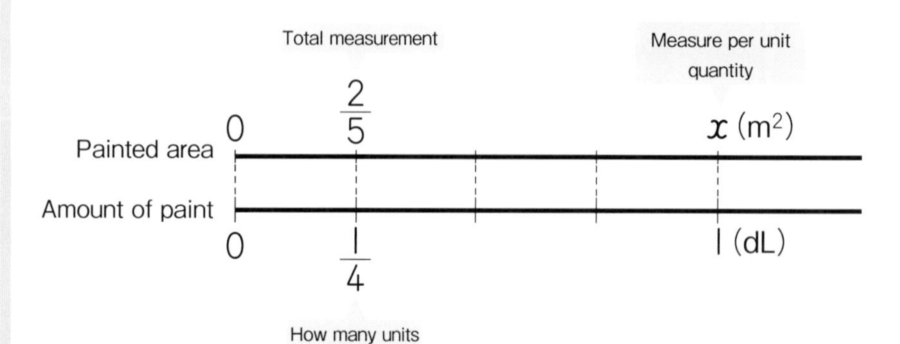

Measure per unit quantity　　Total measurement

x m^2	$\frac{2}{5}$ m^2
1 dL	$\frac{1}{4}$ dL

How many units

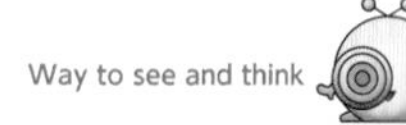

Way to see and think

Since you know "Total measurement" and "How many units," you can use a division expression.

Purpose. What should we do to divide a proper fraction by a proper fraction?

As for $\frac{2}{5} \div 4$, it can be found by $\frac{2}{5 \times 4}$.

If you think in the same way, you will have a fraction in the denominator.

Way to see and think

If you can paint x m^2 per deciliter, it can be represented by $x \times \frac{1}{4} = \frac{2}{5}$.

Want to think

② Let's look at the following diagram and think about how to calculate.

Since 1 dL has four sets of $\frac{1}{4}$ dL, it can be found by $\frac{2}{5} \times 4$.

Yui

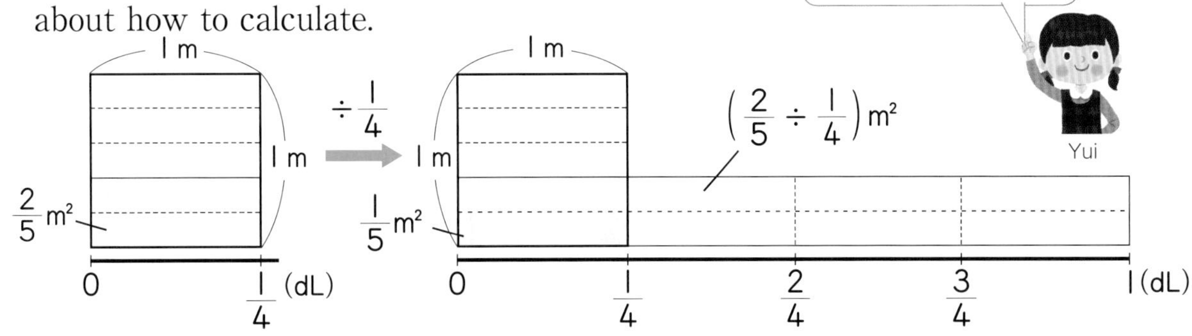

Since there are (2×4) sets of $\frac{1}{5}$ m², $\frac{2}{5} \div \frac{1}{4} = \frac{\square \times \square}{\square}$

$= \square$

Want to solve

Activity

2 **$\frac{3}{4}$ dL of blue paint was used to paint a $\frac{2}{5}$ m² wall.**
How many square meters can be painted per deciliter of this paint?

① Let's write a math expression.

Painted area: 0, $\frac{2}{5}$, x (m²)

Amount of paint: 0, $\frac{3}{4}$, 1 (dL)

x m²	$\frac{2}{5}$ m²
1 dL	$\frac{3}{4}$ dL

Want to think

② Let's think about how to calculate.

If I think how many square meters can be painted with $\frac{1}{4}$ dL of paint...

Can I use the rules of operations?

Want to compare

③ Let's explain the ideas of the following children and find the same resulting math expression.

Hiroto's idea

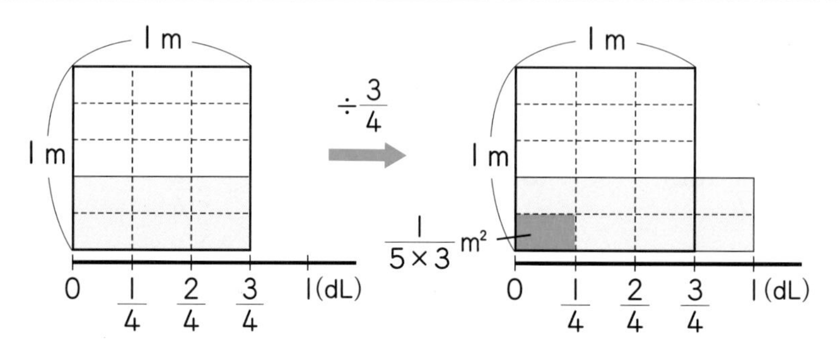

As for the painted area, since there are (2×4) sets of $\frac{1}{5 \times 3}$ m², it is $\frac{2 \times 4}{5 \times 3}$ m².

$$\frac{2}{5} \div \frac{3}{4} = \frac{1}{5 \times 3} \times (2 \times 4)$$

$$= \frac{2 \times 4}{5 \times 3}$$

$$= \square$$

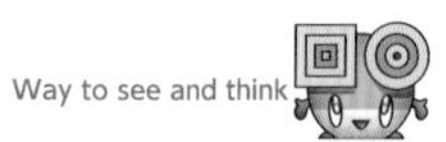

Thinking based on the area diagram.

Nanami's idea

Since $\frac{3}{4}$ dL painted $\frac{2}{5}$ m², then $\frac{1}{4}$ dL can paint $\left(\frac{2}{5} \div 3\right)$ m².

Since 1 dL has four sets of that, $\frac{2}{5} \div 3 \times 4 = \frac{2 \times 4}{5 \times 3}$

$$= \square$$

Daiki's idea

I considered using the division rules.

$$\frac{2}{5} \div \frac{3}{4} = \left(\frac{2}{5} \times \frac{4}{3}\right) \div \left(\frac{3}{4} \times \frac{4}{3}\right) = \frac{2}{5} \times \frac{4}{3} \div 1$$

$$= \frac{2}{5} \times \frac{4}{3} = \frac{2 \times 4}{5 \times 3}$$

$$= \square$$

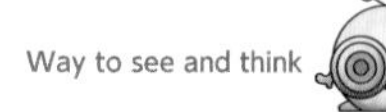

The divisor is converted to a 1 using the reciprocal and division rules.

Summary

When you divide a proper fraction by a proper fraction, calculate by multiplying the dividend by the reciprocal of the divisor.

$$\frac{b}{a} \div \frac{d}{c} = \frac{b}{a} \times \frac{c}{d}$$

Want to confirm

Let's solve the following calculations.

① $\frac{1}{4} \div \frac{1}{3}$ ② $\frac{2}{7} \div \frac{3}{4}$ ③ $\frac{2}{3} \div \frac{7}{8}$ ④ $\frac{3}{5} \div \frac{5}{6}$

Want to explain Operation of improper fractions

3 Let's explain how to solve the following calculations.

① $\frac{8}{3} \div \frac{12}{5} = \frac{8}{3} \times \frac{\square}{\square}$

$= \square$

② $\frac{15}{16} \div \frac{5}{4} = \frac{15}{16} \times \frac{\square}{\square}$

$= \square$

③ $4 \div \frac{2}{5} = \frac{\square}{1} \div \frac{2}{5}$

$= \frac{\square}{1} \times \frac{5}{2}$

$= \square$

④ $\frac{3}{7} \div 6 = \frac{3}{7} \div \frac{\square}{1}$

$= \frac{3}{7} \times \frac{1}{\square}$

$= \square$

If the fraction can be reduced, then let's reduce.

Even when divisors are improper fractions, also calculate by multiplying the dividend by the reciprocal of the divisor. Also, if the whole number is changed into the fraction form, it becomes a operation of fraction÷fraction.

Want to confirm

Let's solve the following calculations.

① $\frac{16}{7} \div \frac{4}{9}$ ② $\frac{4}{3} \div \frac{2}{3}$ ③ $8 \div \frac{2}{3}$ ④ $\frac{8}{9} \div 4$

4

$1\frac{1}{3}$ dL of green paint was used to paint a $\frac{2}{5}$ m^2 wall.
How many square meters can be painted per deciliter of this paint?

① Let's write a math expression.

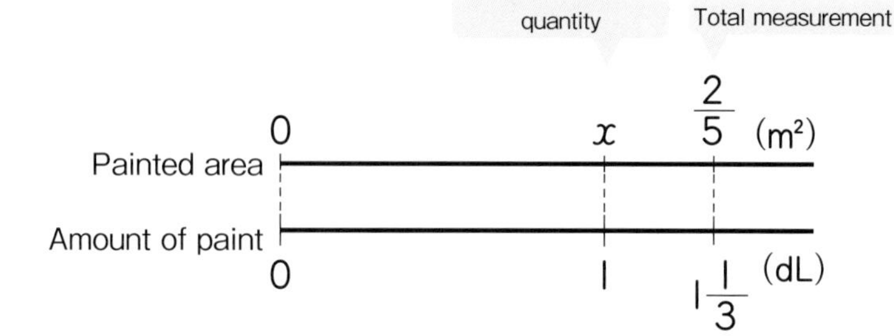

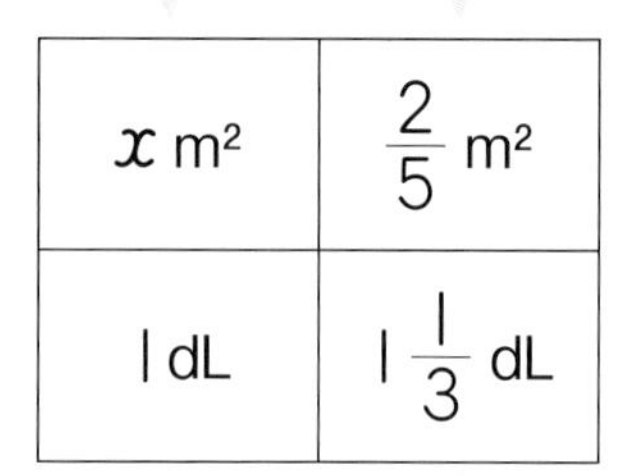

Want to think

② Let's think about how to calculate.

$$\frac{2}{5} \div 1\frac{1}{3} = \frac{2}{5} \div \frac{4}{3}$$
$$= \frac{2}{5} \times \frac{3}{4}$$
$$= \square$$

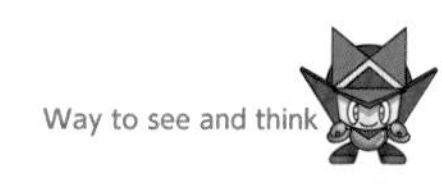

The same as in multiplication, you can think of $1\frac{1}{3}$ as $\frac{4}{3}$.

Even for divisions of fractions, change the mixed fractions into improper fractions and then calculate.

Want to confirm

Let's solve the following calculations.

① $\frac{5}{8} \div 1\frac{4}{5}$　② $\frac{4}{7} \div 1\frac{3}{5}$　③ $\frac{3}{4} \div 2\frac{2}{5}$　④ $\frac{1}{2} \div 1\frac{1}{6}$

Want to solve

5 Let's try to think the various problems shown below.

① There are $1\frac{4}{5}$ L of milk. If a family drinks $\frac{3}{5}$ L of milk at a time, how many times can they drink the milk?

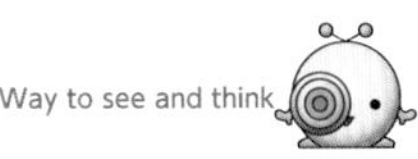
Way to see and think

Since you know the "Total measurement" and "Measure per unit quantity," you can use a division expression.

Amount of milk: $\frac{3}{5}$, $1\frac{4}{5}$ (L)
Number of times: 1, x (times)

Measure per unit quantity	Total measurement
$\frac{3}{5}$ L	$1\frac{4}{5}$ L
1 time	x times

How many units

② There is a wire that weighs $4\frac{1}{2}$ g per meter. If the total weight of the wire is 24 g, how many meters is the length of this wire?

Weight: 0, $4\frac{1}{2}$, 24 (g)
Length: 0, 1, x (m)

$4\frac{1}{2}$ g	24 g
1 m	x m

③ There is a rectangular cloth with an area of $2\frac{2}{3}$ m². If the length is $1\frac{7}{9}$ m, how many meters is the width?

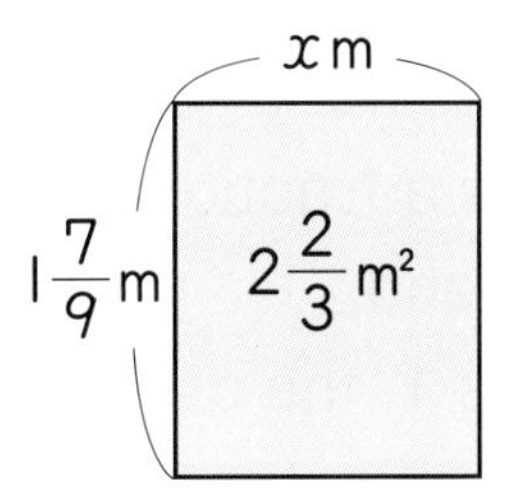

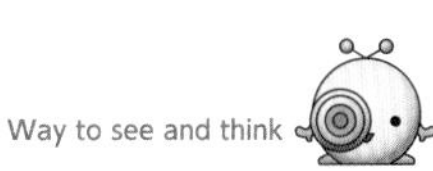
Way to see and think

Use the formula to find the area of a rectangle. Since $1\frac{7}{9} \times x = 2\frac{2}{3}$ then $x = 2\frac{2}{3} \div 1\frac{7}{9}$.

Want to confirm

4 Let's solve the following calculations.

① $1\frac{3}{5} \div \frac{2}{7}$ ② $1\frac{1}{4} \div \frac{5}{8}$ ③ $4\frac{2}{3} \div 1\frac{1}{5}$ ④ $2\frac{1}{3} \div 1\frac{5}{9}$

6 **There is a thin wire with a length of $1\frac{4}{5}$ m that weighs 24 g and a thick wire with a length of $\frac{3}{5}$ m that weighs 24 g. How many grams is the weight per meter of each wire?**

① Let's find the weight per meter of each wire.

Thin wire

Weight: 0, x, 24 (g)
Length: 0, 1, $1\frac{4}{5}$ (m)

x g	24 g
1 m	$1\frac{4}{5}$ m

Thick wire

Weight: 0, 24, x (g)
Length: 0, $\frac{3}{5}$, 1 (m)

x g	24 g
1 m	$\frac{3}{5}$ m

Purpose What kind of relationship exists between the size of the divisor and the quotient?

② When does the quotient become larger than 24?
Also, when does it become smaller than 24?

Summary

When the divisor is a fraction larger than 1, the quotient becomes smaller than the dividend.
When the divisor is a fraction smaller than 1, the quotient becomes larger than the dividend.
When the divisor is 1, the quotient becomes the same as the dividend.

Want to confirm

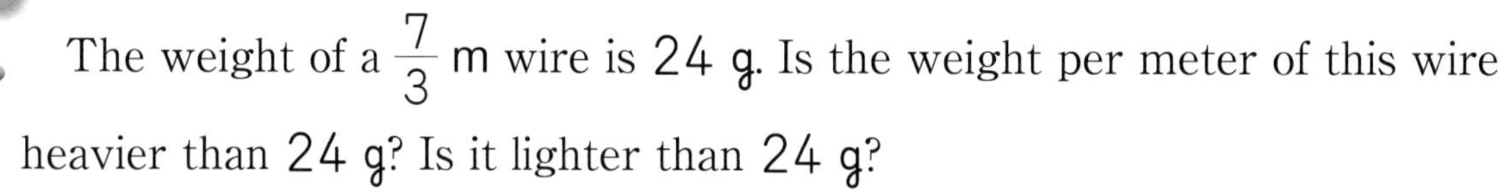

5 The weight of a $\frac{7}{3}$ m wire is 24 g. Is the weight per meter of this wire heavier than 24 g? Is it lighter than 24 g?

Want to try

6 Which of the following calculations has a quotient larger than 7? Let's also explain the reasons.

$7 \div \frac{3}{4}$ $\quad 7 \div 1\frac{2}{3}$ $\quad 7 \div \frac{2}{3}$ $\quad 7 \div 7\frac{7}{8}$

2 Deducing a math expression

Want to solve

1 **Let's read the following problems and write a math expression. Also, let's find the answer.**

① There is a metal bar with a length of $\frac{4}{3}$ m that weighs $\frac{9}{5}$ kg. How many kilograms is the weight for 1 m of this metal bar?

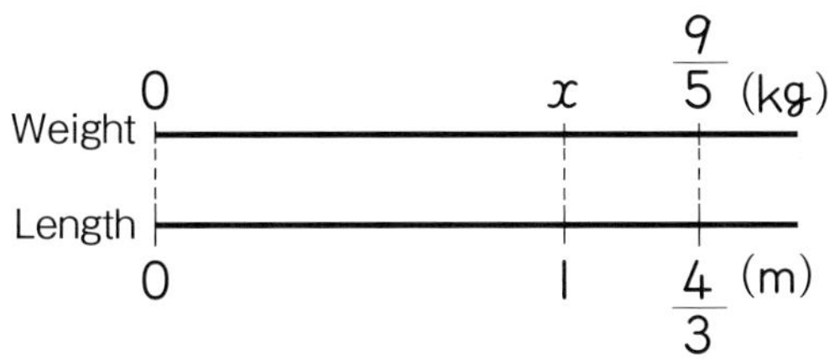

x kg	$\frac{9}{5}$ kg
1 m	$\frac{4}{3}$ m

② There is a metal bar with a length of 1 m that weighs $\frac{5}{3}$ kg. How many kilograms is the weight for $\frac{5}{2}$ m of this metal bar?

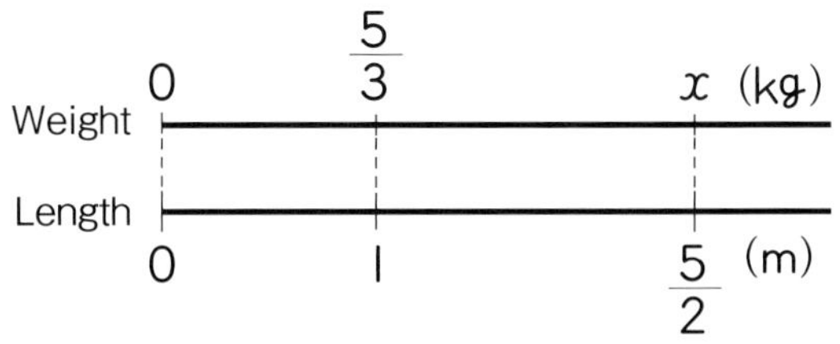

Measure per unit quantity — Total measurement

$\frac{5}{3}$ kg	x kg
1 m	$\frac{5}{2}$ m

How many units

1 Let's think about the problem that was created by Asahi.

> $\frac{6}{7}$ L of water are used to water 1 m² of a field. If $\frac{2}{3}$ m² of a field is watered, x L of water are needed. Let's find the number that applies for x.

① Let's answer the problem created by Asahi.

② Let's change the number inside each ☐ and create a multiplication or division problem.

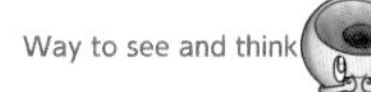

Just replace the sentence into a table or diagram and decide which values you want for the problem.

What you can do now

☐ **Understanding how to solve divisions of fractions.**

1 Let's fill in each ☐ with a number.

① $\frac{7}{14} \div \frac{3}{5} = \frac{7}{14} \times \square$　　② $3 \div \frac{4}{7} = 3 \times \square$

☐ **Can solve calculations that have fractions as divisors.**

2 Let's solve the following calculations.

① $\frac{2}{5} \div \frac{3}{7}$　② $\frac{1}{5} \div \frac{9}{10}$　③ $\frac{4}{9} \div \frac{2}{3}$

④ $\frac{3}{4} \div \frac{15}{16}$　⑤ $9 \div \frac{5}{6}$　⑥ $4 \div \frac{8}{9}$

⑦ $2\frac{2}{9} \div \frac{2}{7}$　⑧ $5\frac{1}{4} \div \frac{3}{8}$　⑨ $\frac{1}{6} \div 1\frac{1}{18}$

⑩ $3\frac{1}{3} \div 1\frac{5}{7}$　⑪ $4\frac{1}{6} \div 2\frac{1}{2}$　⑫ $1\frac{1}{14} \div 1\frac{2}{7}$

☐ **Understanding the relationship between the quotient and the divisor.**

3 Which calculation has a quotient larger than 5?

Ⓐ $5 \div \frac{2}{3}$　Ⓑ $5 \div 1\frac{1}{2}$　Ⓒ $5 \div \frac{5}{4}$　Ⓓ $5 \div \frac{7}{9}$

☐ **Can create a division expression and find the answer.**

4 Let's answer the following questions.

① A tape that is $1\frac{4}{5}$ m long is cut into pieces with a length of $\frac{3}{10}$ m. How many pieces with a length of $\frac{3}{10}$ m can be cut?

② Today, Ryunosuke read 16 pages of a book. This represents $\frac{2}{29}$ of the book he is reading. As for this book, how many pages are there in total?

Supplementary problems p.227

Usefulness and efficiency of learning

1 There is a parallelogram with an area of 12 cm². If the length of the base is $4\frac{4}{5}$ cm, how many centimeters is the length of the height?

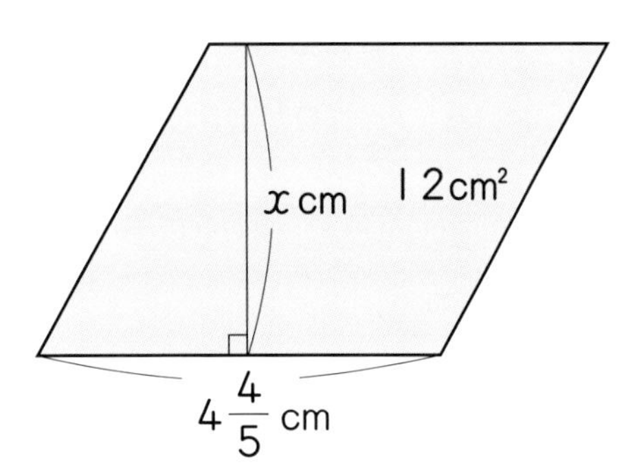

- Understanding how to solve divisions of fractions.
- Can solve calculations that have fractions as divisors.
- Understanding the relationship between the quotient and the divisor.
- Can create a division expression and find the answer.

2 Let's answer the following questions.

① There are $\frac{2}{3}$ L of paint that weighs $\frac{3}{4}$ kg.
How many kilograms is the weight for 1 L of this paint?

② There is a $2\frac{1}{4}$ m checkered cloth. This length is $\frac{3}{8}$ times of a polka dot cloth.
How many meters is the length of the polka dot cloth?

3 Let's choose the ones that become a division expression from Ⓐ～Ⓔ.

- Can understand a sentence and represent it by a math expression.

Ⓐ A $\frac{4}{5}$ m aluminum bar weighed $\frac{2}{3}$ kg.
How many kilograms is the weight per meter of this aluminum bar?

Ⓑ If a $12\frac{1}{2}$ m rope is cut into pieces with a length of $1\frac{1}{4}$ m, in how many pieces can it be cut?

Ⓒ There is an oil that weighs $\frac{6}{7}$ kg per liter. How many kilograms is the weight for $\frac{1}{3}$ L of this oil?

Ⓓ A $\frac{2}{3}$ m² wall can be painted with $\frac{4}{5}$ L of paint. How many square meters can be painted with 1 L of this paint?

Ⓔ I bought $\frac{4}{5}$ kg of rice that had a price of 540 yen per kilogram. How many yen was the cost?

Want to connect

I can solve calculations of fraction × fraction and fraction ÷ fraction. Now, I want to solve calculations that incorporate not only fractions but also decimal numbers and whole numbers.

6 Calculations with Decimal Numbers and Fractions

Let's think about how to operate mixed calculations of decimal numbers and fractions.

1 Mixed calculations of decimal numbers and fractions

Want to think Addition and subtraction

1 **Let's think about how to solve the following calculation.**

$\frac{2}{5} + 0.5$

① Let's calculate by aligning to decimal numbers.

$\frac{2}{5} = 2 \div \square = \square$ $\square + 0.5 = \square$

② Let's calculate by aligning to fractions.

$0.5 = \frac{5}{\square} = \square$ $\frac{2}{5} + \square = \square$

Want to confirm

1 Let's calculate $0.9 - \frac{1}{6}$.

① Let's calculate by aligning to decimal numbers.

$\frac{1}{6} = 1 \div \square = 0.1666...$ $0.9 - 0.167 = \square$

↓

0.167

Since it's not divisible, it's not possible to calculate exactly using decimal numbers .

Nanami

② Let's calculate by aligning to fractions.

$0.9 = \frac{\square}{\square}$ $\frac{\square}{\square} - \frac{1}{6} = \square$

As for an addition or subtraction that includes decimal numbers and fractions, the calculation is done after aligning to decimal numbers or fractions. When the numbers after the decimal point continue infinitely, the calculation should be performed by aligning to fractions.

Want to try

Let's solve the following calculations.

① $0.6+\frac{4}{9}$　② $0.7+\frac{4}{5}$　③ $\frac{3}{7}+0.4$　④ $\frac{2}{3}+0.45$

⑤ $\frac{7}{8}-0.3$　⑥ $1\frac{4}{7}-0.4$　⑦ $\frac{8}{7}-0.25$　⑧ $\frac{1}{5}-0.12$

Want to think　Multiplication and division

Let's think about how to solve the following calculations.

① $\frac{5}{9}\div\frac{3}{4}\times\frac{7}{10}$　② $7\times\frac{1}{6}\div 1.4$

① $\frac{5}{9}\div\frac{3}{4}\times\frac{7}{10}=\frac{5}{9}\times\frac{\square}{\square}\times\frac{7}{10}$

$=\frac{5\times\square\times 7}{9\times\square\times 10}$

$=\square$

② $7\times\frac{1}{6}\div 1.4=\frac{7}{\square}\times\frac{1}{6}\div\frac{\square}{10}$

$=\frac{7}{\square}\times\frac{1}{6}\times\frac{\square}{\square}$

$=\frac{7\times 1\times\square}{\square\times 6\times\square}$

$=\square$

As for math expressions that incorporate division and multiplication of fractions, if the divisor is changed for the multiplication of its reciprocal, then the math expression will only include multiplications.

Want to confirm

Let's find the area of the triangle shown on the right.

① Let's write a math expression.

② Let's calculate.

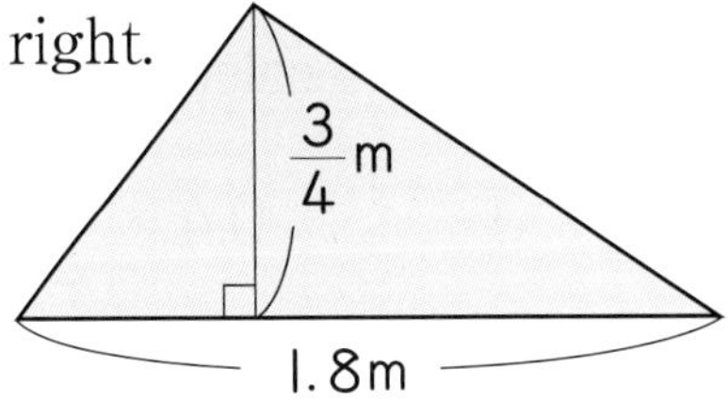

4 Let's solve the following calculations.

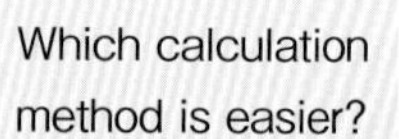

① $0.3 \times 0.48 \div 0.45$

Ⓐ Let's calculate by decimal numbers.

Ⓑ Let's calculate using fractions.

$$0.3 \times 0.48 \div 0.45 = \frac{3}{\square} \times \frac{48}{\square} \div \frac{45}{\square}$$

$$= \frac{3}{\square} \times \frac{48}{\square} \times \frac{\square}{45}$$

$$= \frac{3 \times 48 \times \square}{\square \times \square \times 45}$$

$$= \square$$

② Let's calculate $14 \div 6 \times 3$ using fractions.

$$14 \div 6 \times 3 = \frac{14}{\square} \div \frac{6}{\square} \times \frac{3}{\square}$$

$$= \frac{14}{\square} \times \frac{\square}{6} \times \frac{3}{\square}$$

$$= \frac{\square \times \square}{\square}$$

$$= \square$$

14÷6=2.333...
Ah... it is not divisible.

Hiroto

5 Let's calculate using fractions.

① $\frac{1}{3} \div 0.4 \times \frac{5}{3}$ ② $27 \div 48 \times 32$ ③ $0.8 \times \frac{3}{5} \div 0.36$

④ $\frac{3}{7} \div 0.75 \times \frac{9}{14}$ ⑤ $0.7 \times 0.35 \div 0.25$ ⑥ $0.5 \div 0.21 \times 0.7$

2 Various problems

Want to think

1

Car A traveled 270 km using 15 L of gasoline.

Car B traveled 372 km using 24.8 L of gasoline.

How many liters of gasoline did each car need to travel 100km?

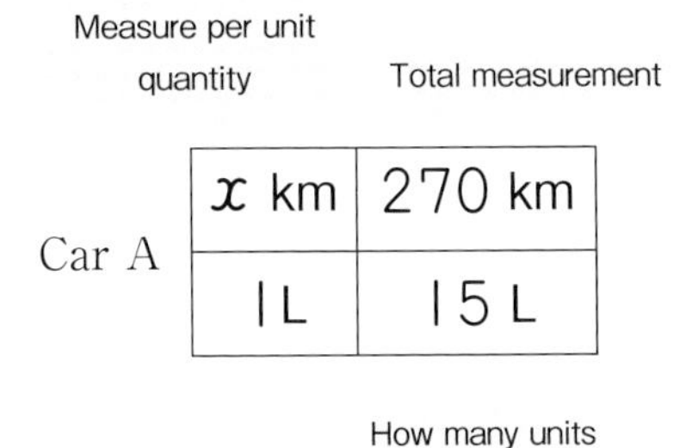

Measure per unit quantity / Total measurement

Car A

x km	270 km
1 L	15 L

How many units

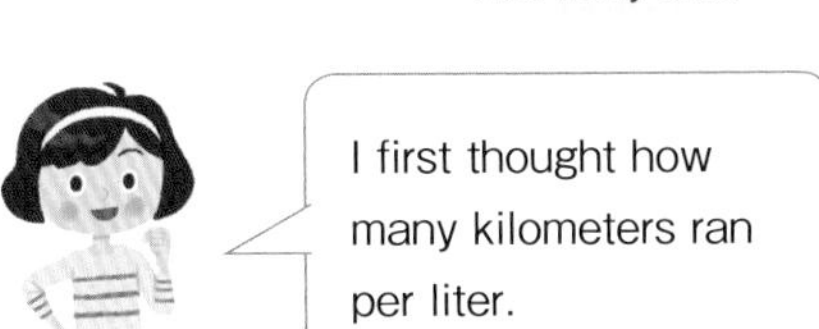

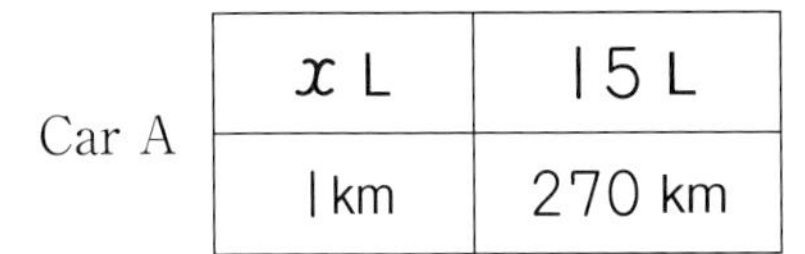

Car A

x L	15 L
1 km	270 km

I thought how many liters of gasoline were used per kilometer.

Want to confirm

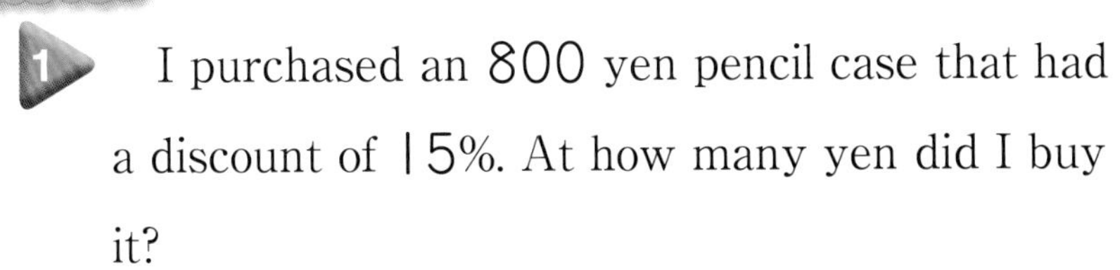

1. I purchased an 800 yen pencil case that had a discount of 15%. At how many yen did I buy it?

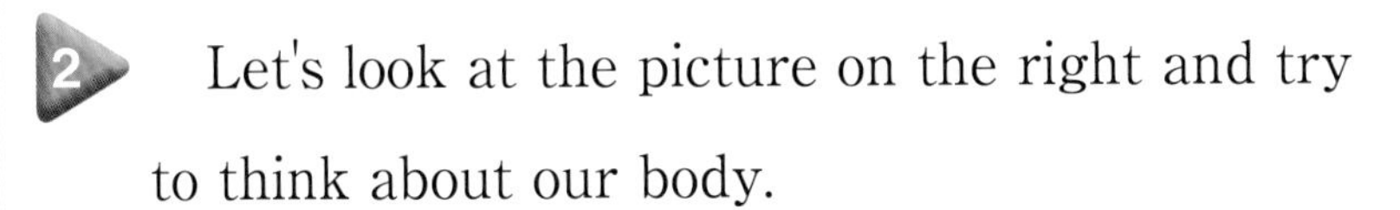

2. Let's look at the picture on the right and try to think about our body.

① About how many kilograms is the weight of the brain of a person that weighs 36 kg?

② About $\frac{1}{7}$ of the bones are in the head. About how many bones are there in the whole body?

③ About how many kilograms is the water inside the body of a person that weighs 45 kg?

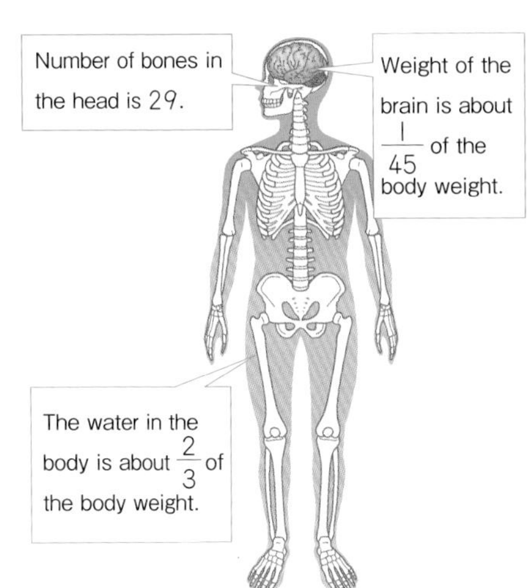

What you can do now

☐ **Can find the sum, difference, product, and quotient of a mixed calculation of decimal numbers and fractions.**

1 Let's find the sum, difference, product, and quotient of the following pairs composed by a decimal number and a fraction. As for the quotient, divide the number on the left by the number on the right.

① $\frac{1}{3}$, 0.2　　② 3.5, $2\frac{1}{3}$

☐ **Can solve the calculations that incorporate the addition and subtraction of decimal numbers and fractions.**

2 Let's solve the following calculations.

① $0.8+\frac{1}{3}$　　② $\frac{2}{5}+2.6$

③ $\frac{5}{7}-0.09$　　④ $0.32-\frac{1}{5}$

☐ **Understanding how to solve the calculations that incorporate the multiplication and division of decimal numbers and fractions.**

3 Let's solve the following calculations using fractions.

① $\frac{1}{5}\div 0.6\times\frac{2}{3}$　　② $36\div 27\times 16$

③ $0.9\times\frac{2}{7}\div 0.18$　　④ $\frac{5}{12}\div 0.25\div\frac{3}{10}$

⑤ $0.2\div 0.16\div 0.35$　　⑥ $0.7\div 0.35\div 0.5$

☐ **Can solve using calculations with fractions.**

4 There is a rhombus, like the one shown on the right, that has an area of 4 cm².

How many centimeters is the length of the other diagonal?

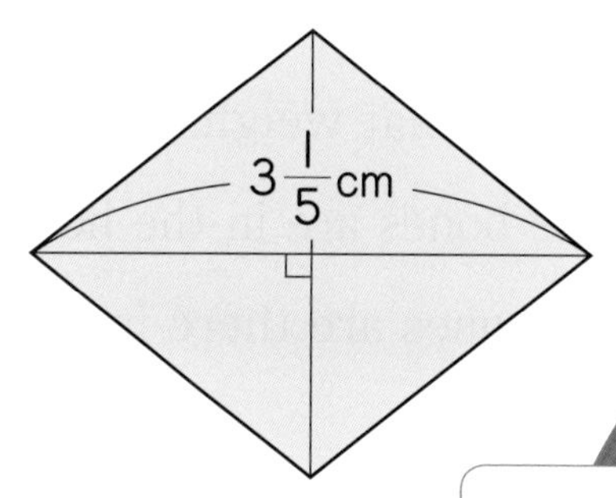

Let's deepen.

Is there anything useful that can be represented by fractions?

Supplementary problems p.229

Utilize in life.

Deepen.

How to represent time
—Let's try to represent by fractions—

Want to know

As for the divisors of 60, there are many numbers as shown below.

Divisors of 60: 1, 2, 3, 4, 5, 6, 10, 12, 15, 20, 30, 60.

Since 1 hour is in units of 60 minutes, it is convenient to use fractions.

Let's think about a method to represent time by fractions.

① How long is 20 minutes in hour?

$\frac{20}{60} = \frac{\square}{\square}$　　$\frac{\square}{\square}$ of 1 hour

If you divide by 60, you can represent time by fractions.

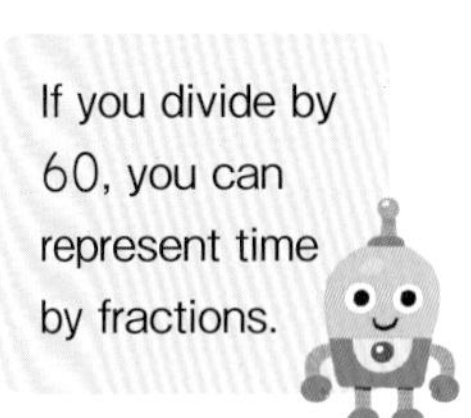

② How long are the following Ⓐ~Ⓓ periods of time in hour?

Ⓐ 18 minutes　Ⓑ 35 minutes　Ⓒ 90 minutes　Ⓓ 100 minutes

Want to deepen

There is a machine that makes 92 accessories in 1 hour.

This machine has been creating accessories for 45 minutes.

① **How long is 45 minutes in hour?**

② **How many accessories has the machine created in 45 minutes?**

Measure per unit quantity　　Total measurement

92 accessories	x accessories
1 hour	$\square$ of 1 hour

How many units

Develop in Junior High School

Problem

Let's represent math sentences created with 3, 5 and +, −, ×, ÷ on the number line.

Addition

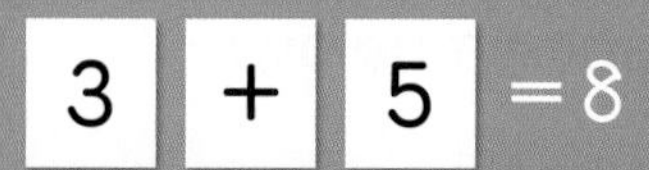

3 + 5 = 8

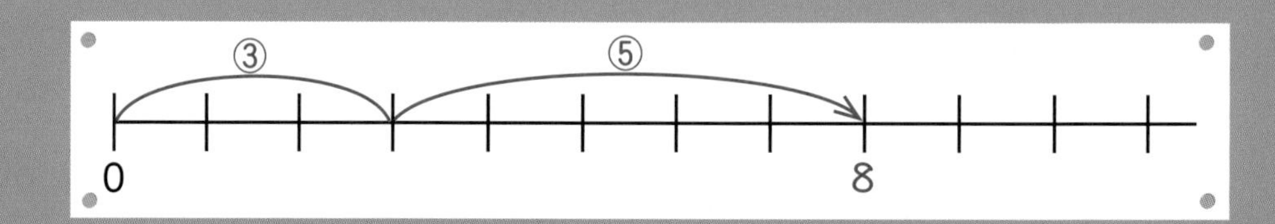

5 + 3 = 8

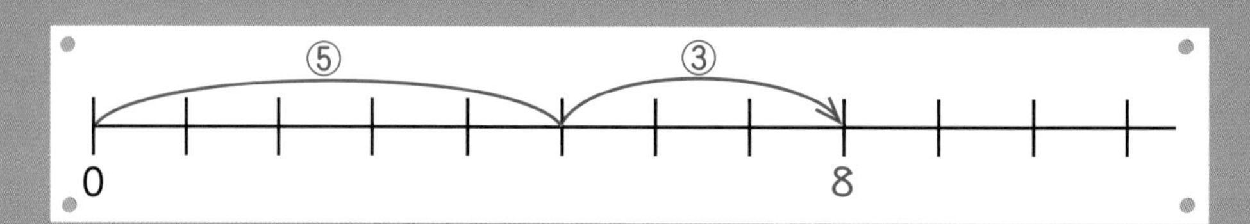

Subtraction

5 − 3 = 2

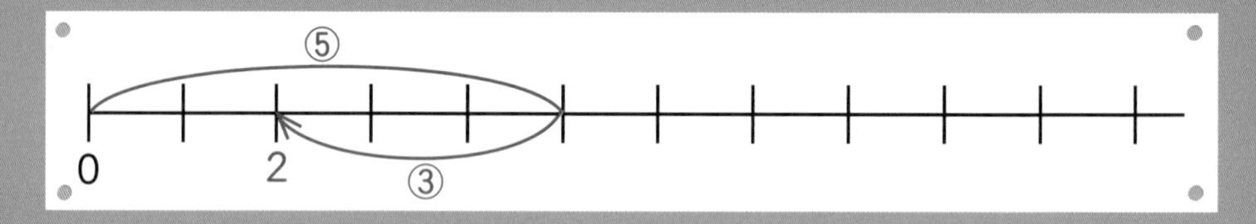

3 − 5 = ×

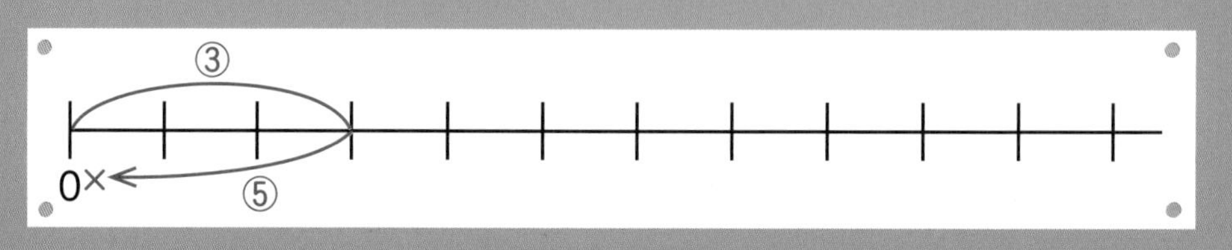

Except for subtraction, both math sentences are valid.

For both, addition and multiplication, the math sentences have the same answer.

Multiplication

$3 \times 5 = 15$

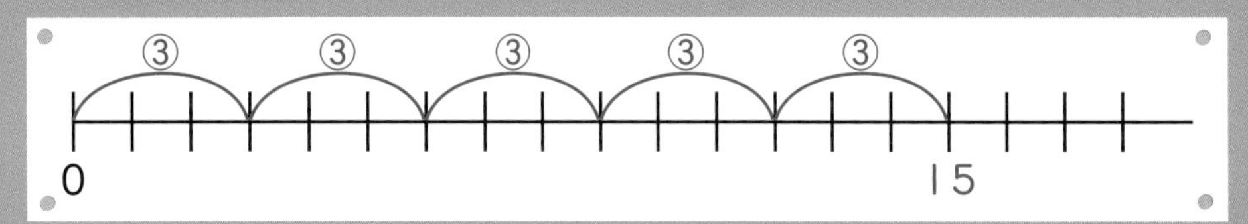

$5 \times 3 = 15$

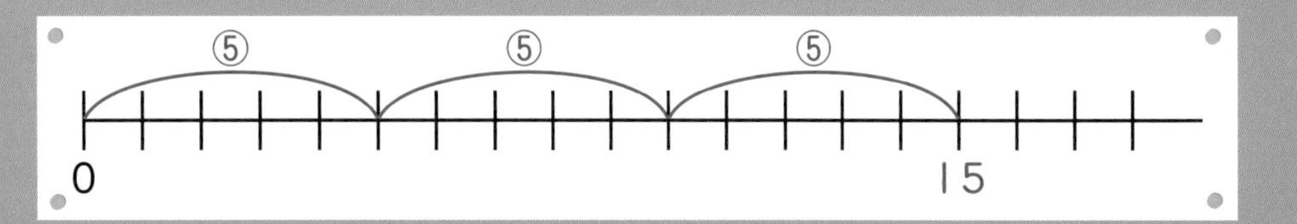

Division

$3 \div 5 = 0.6 = \frac{6}{10} = \frac{3}{5}$

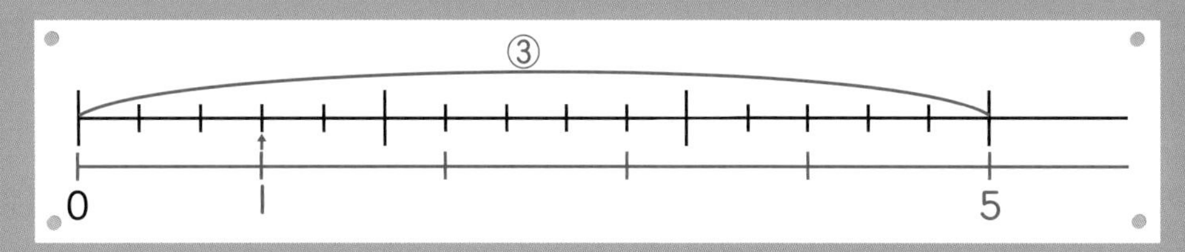

$5 \div 3 = \frac{5}{3}$

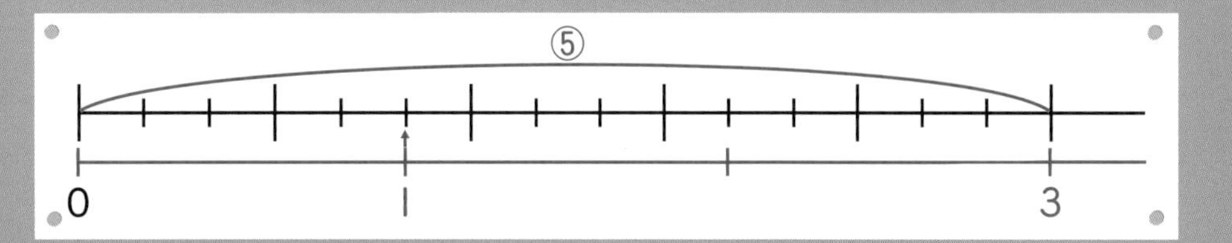

Summary

Answers can be found for the addition, multiplication, and division of whole numbers. However, as for subtraction, there is a case when the answer cannot be found.

Want to connect

How many times: Softball throw

Want to think　Times a fraction

1 **The throwing records for Yukie's softball team were taken and the mean was 18 m. Let's compare the throwing distance and the mean.**

① Yukie's record was 24 m. How many times of the mean is 24 m?

Let's represent using fractions.

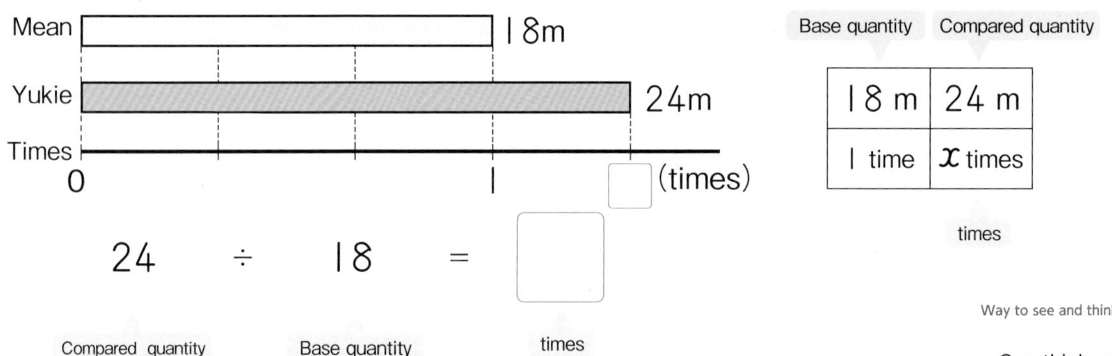

18 m	24 m
1 time	x times

Way to see and think

Can think as compared quantity÷base quantity=times.

In occasions, times are represented by fractions.

Want to represent

③ Mari's record was 15 m. How many times of the mean is 15 m?

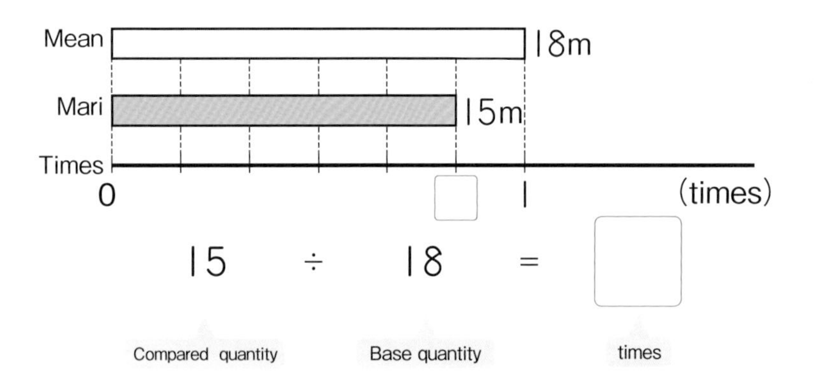

18m	15m
1 time	x times

Sometimes the ratio represented by the fraction is smaller than 1.

Want to confirm

1 Let's find the number that applies inside the ☐ using fractions.

① 15 m is ☐ times of 9 m.　② 35 kg is ☐ times of 42 kg.

Want to think

2 The throwing records for Sota's softball team were taken and the mean was 30 m. Let's compare the throwing distance based on the mean.

① Sota's record was $\frac{7}{5}$ times of the mean. How many meters was Sota's throwing distance?

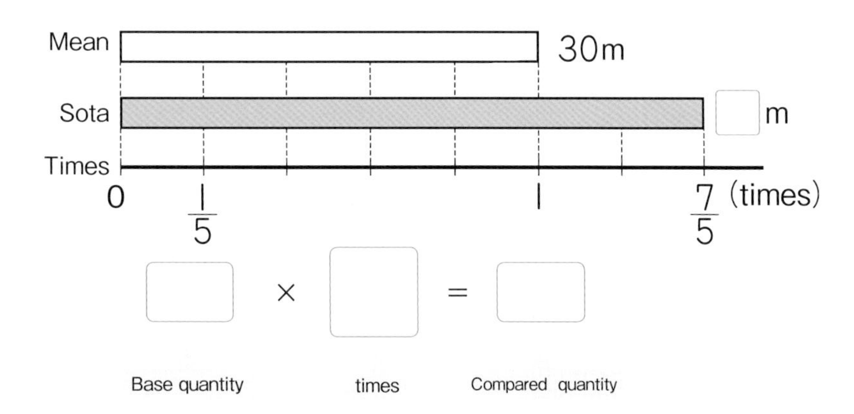

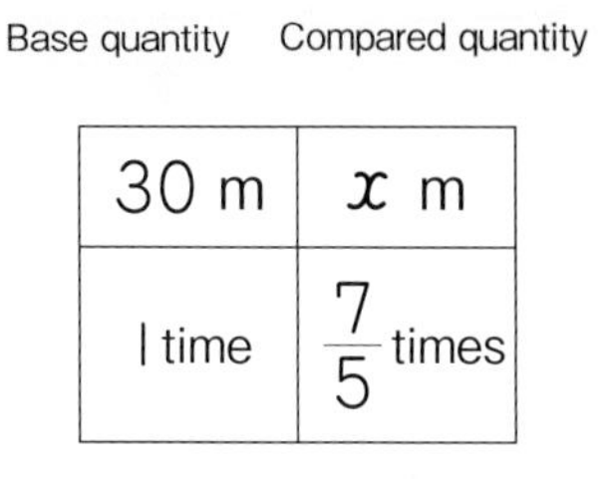

Base quantity	Compared quantity
30 m	x m
1 time	$\frac{7}{5}$ times

times

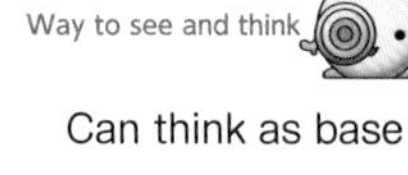

Can think as base quantity × times = compared quantity.

② Sota's throwing distance was $\frac{7}{6}$ times of Haruto's distance. How many meters was Haruto's throwing distance? Let's think and represent through a table and diagram considering x m as Haruto's record.

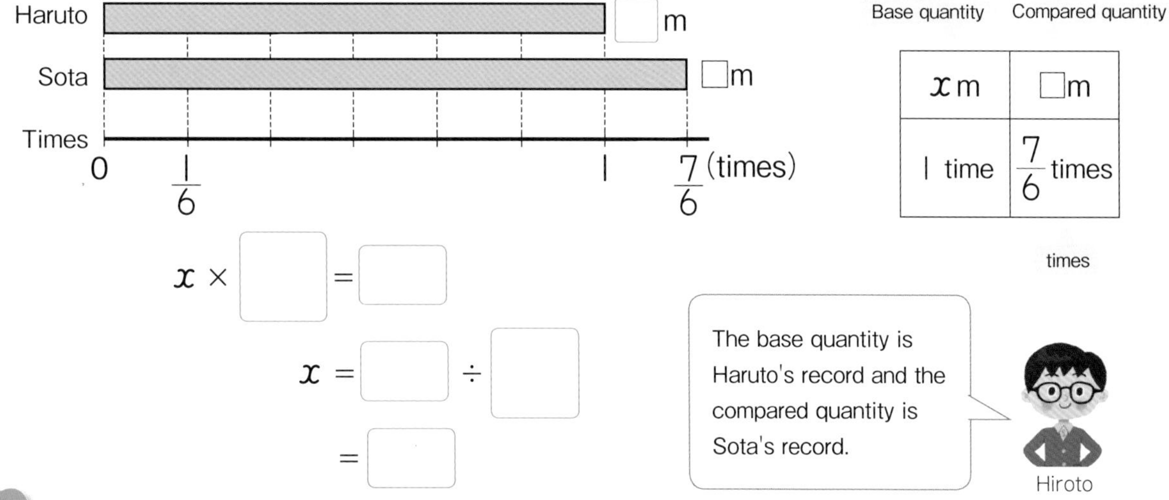

Base quantity	Compared quantity
x m	□ m
1 time	$\frac{7}{6}$ times

Want to confirm

3 Let's find the number that applies inside each □.

① $\frac{8}{7}$ times of 14 kg is □ kg. ② $\frac{5}{6}$ times of □ kg is 50 kg.

The difference between the problem where the overlap is not erased and the problem where the overlap is erased...

Want to explore How should we handle the overlap when counting the combinations?

There are cards with written questions that ask to find combinations and the number of ways to order.

Ⓐ There is one card for each number from 1 ~ 4. From these cards, 3 cards are chosen. How many ways to choose three cards are there?

Ⓑ There is one card for each number from 1 ~ 4. From these cards, 3 cards are chosen to create a 3-digit number. How many ways to create 3-digit numbers are there?

Ⓒ There is one icecream for each of the 4 types available. From these types, 2 types will be bought. How many ways to choose two types of icecreams are there?

Ⓓ There is one icecream for each of the 4 types available. From these types, 2 types are chosen and the eating order is defined. How many ways to choose two types of icecreams and eating order are there?

Ⓔ From a group of 5 children, the team leader and assistant team leader will be decided. How many ways to decide are there?

Ⓕ From a group of 5 children, two representatives will be decided. How many ways to decide two representatives are there?

There are similar problems.

What about the answers?

Think by myself

1 Let's solve each of the problems written in the cards.

Hiroto's idea

I thought problem Ⓐ and Ⓑ.

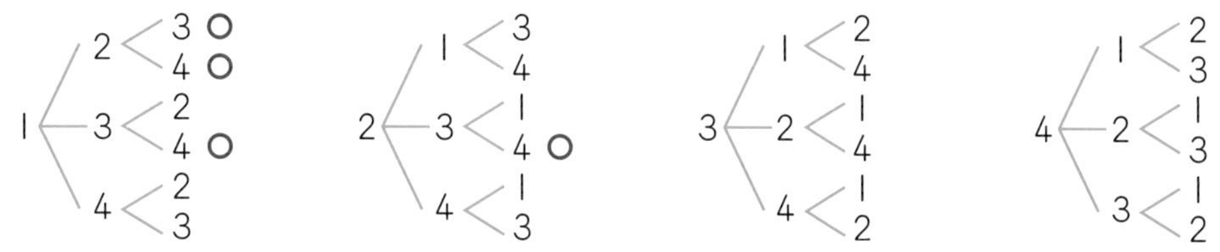

First, I considered all the ways of ordering, in total 24 ways. (answer for Ⓑ)
After, from these ways I erased the overlap, in total 4 ways. (answer for Ⓐ)

Nanami

In problem Ⓐ, the overlap was erased. In problem Ⓑ, the overlap was not erased. What about for Ⓒ, Ⓓ, Ⓔ, and Ⓕ?

Think in groups

2 Let's discuss how to think and solve the problems by separating the class into groups that will discuss problems Ⓐ and Ⓑ, problems Ⓒ and Ⓓ, and problems Ⓔ and Ⓕ. Also, let's discuss the differences between each pair of problems.

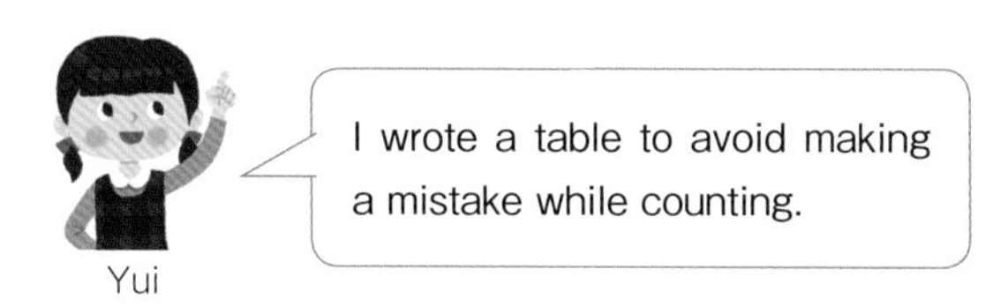

Think as a class

3 Let's present what each group discussed. Also, listen to the presentation of the other groups and try to explain the differences with your group.

01602

Utilizing rule of three on a 4-cell table

1 There is a 1 m metal bar that weighs $\frac{5}{3}$ kg. How many kilograms is the weight for $\frac{5}{2}$ m of this metal bar?

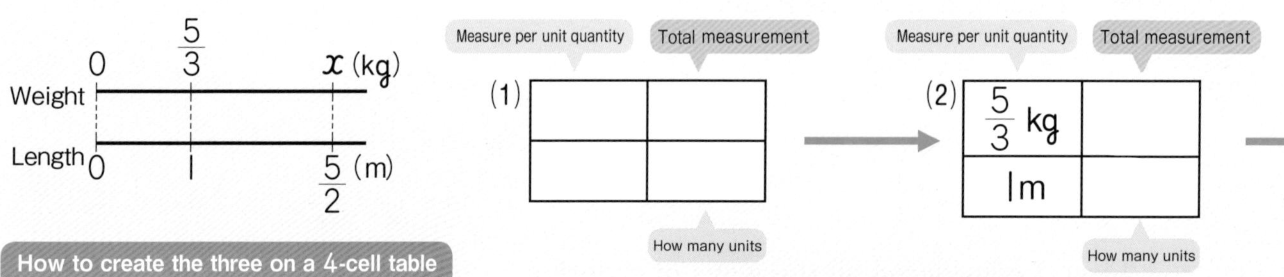

How to create the three on a 4-cell table

(1) Write a table with four entries.

(2) Since a 1 m metal bar weighs $\frac{5}{3}$ kg, write "1 m" and "$\frac{5}{3}$ kg" in the left column.

(3) Since you don't know the weight for a metal bar that is $\frac{5}{2}$ m, consider it as x kg and write "$\frac{5}{2}$ m" and "x kg" in the right column.

Even if you write it in line with the first figure, you can also write as shown on the right.

1m	$\frac{5}{2}$ m
$\frac{5}{3}$ kg	x kg

2 There is a $\frac{3}{4}$ m metal bar that weighs $\frac{9}{5}$ kg. How many kilograms is the weight for 1 m of this metal bar?

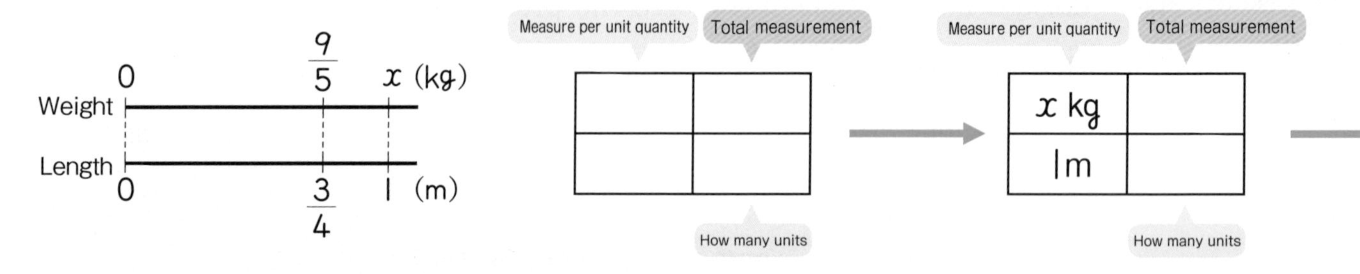

3 There is a 1 m metal bar that weighs $\frac{3}{5}$ kg. How many meters is the length of this metal bar when it weighs $\frac{1}{3}$ kg?

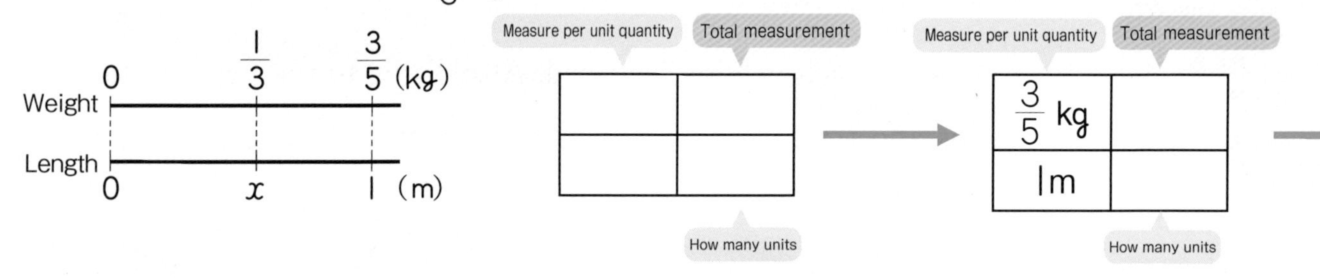

If you think with a word formula... | **If you think using "times"...**

Measure per unit quantity | Total measurement

(3)

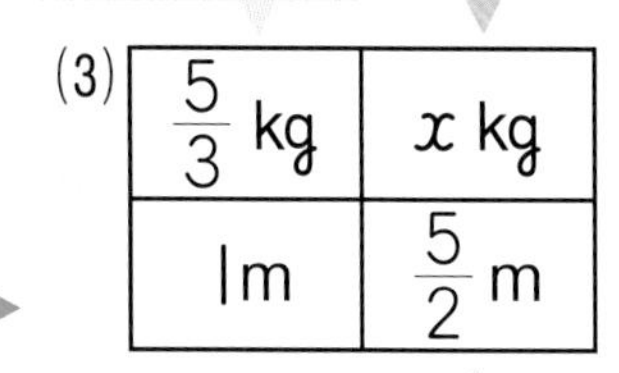

$\frac{5}{3}$ kg	x kg
1m	$\frac{5}{2}$ m

How many units

1

Since,

Measure per unit quantity × How many units = Total measurement

$$\frac{5}{3} \times \frac{5}{2} = \frac{25}{6}$$

Answer: $\frac{25}{6}$ kg $\left(4\frac{1}{6}\text{ kg}\right)$

$\times\frac{5}{2}$

$\frac{5}{3}$ kg	x kg
1m	$\frac{5}{2}$ m

$\times\frac{5}{2}$

$$\frac{5}{3} \times \frac{5}{2} = \frac{25}{6}$$

Answer: $\frac{25}{6}$ kg $\left(4\frac{1}{6}\text{ kg}\right)$

The unit in each row is the same.

Measure per unit quantity | Total measurement

x kg	$\frac{9}{5}$ kg
1m	$\frac{3}{4}$ m

How many units

2

Since,

Total measurement ÷ How many units = Measure per unit quantity

$$\frac{9}{5} \div \frac{3}{4} = \frac{12}{5}$$

Answer: $\frac{12}{5}$ kg $\left(2\frac{2}{5}\text{ kg}\right)$

$\times\frac{3}{4}$

x kg	$\frac{9}{5}$ kg
1m	$\frac{3}{4}$ m

$\times\frac{3}{4}$

$$x \times \frac{3}{4} = \frac{9}{5}$$

$$\frac{9}{5} \div \frac{3}{4} = \frac{12}{5}$$

Answer: $\frac{12}{5}$ kg $\left(2\frac{2}{5}\text{ kg}\right)$

Measure per unit quantity | Total measurement

$\frac{3}{5}$ kg	$\frac{1}{3}$ kg
1m	x m

How many units

3

Since,

Total measurement ÷ Measure per unit quantity = How many units

$$\frac{1}{3} \div \frac{3}{5} = \frac{5}{9}$$

Answer: $\frac{5}{9}$ m

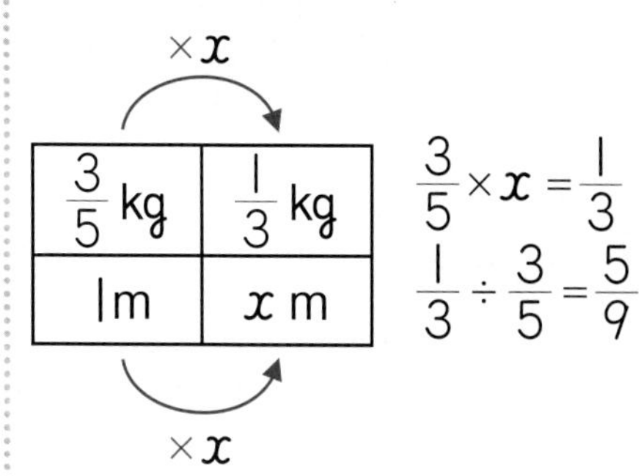

$\times x$

$\frac{3}{5}$ kg	$\frac{1}{3}$ kg
1m	x m

$\times x$

$$\frac{3}{5} \times x = \frac{1}{3}$$

$$\frac{1}{3} \div \frac{3}{5} = \frac{5}{9}$$

Answer: $\frac{5}{9}$ m

Find the ?

Which figure can you make?

Problem

Various figures were folded using origami sheets.
What are the characteristics of each figure?

Symmetry

Let's explore the properties of balanced figures and its categorization.

1 Line symmetric figures

1 **Let's try to separate the figures, created by Hiroto and friends, into members of the category.**

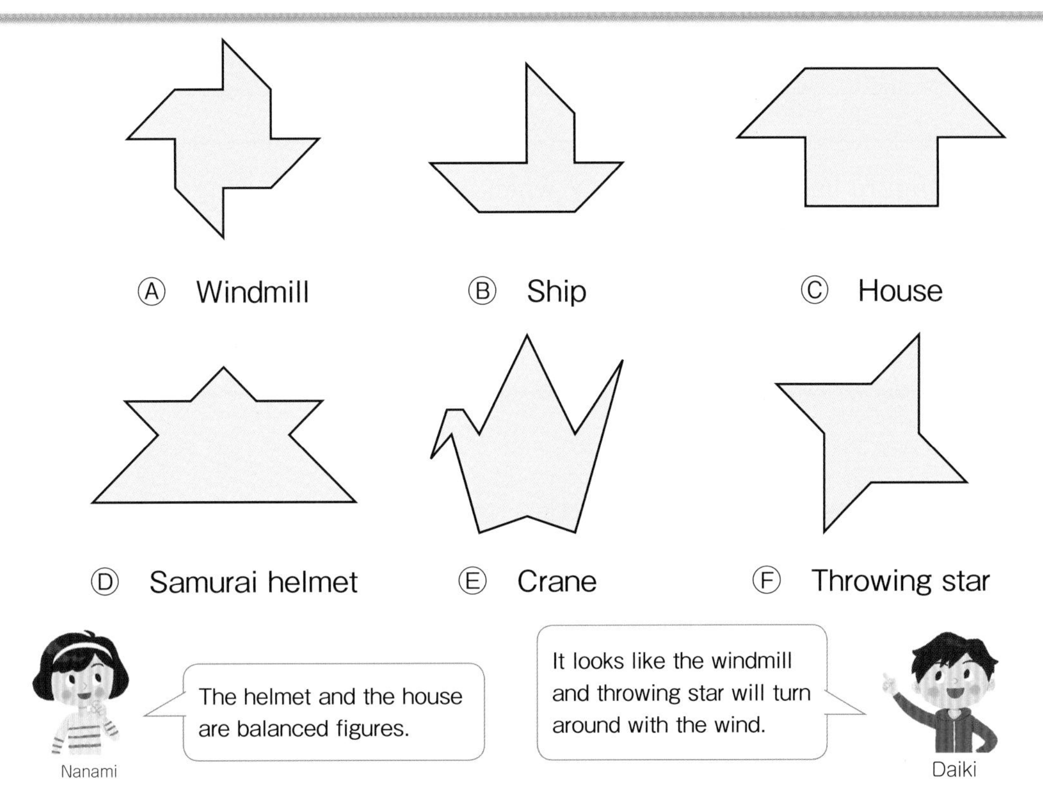

From the above figures Ⓐ~Ⓕ, which discription from ①~③ correctly applies?

① If it is folded by the middle, it will overlap exactly.

② If it is rotated 180°, it looks exactly as the original figure.

③ Neither of the two mentioned above.

Let's try to explore by cutting the figures in p.245.

Want to find

2 **The following diagram shows figures that overlap exactly when folded in two. Where should the figures be folded? Let's draw the folding line.**

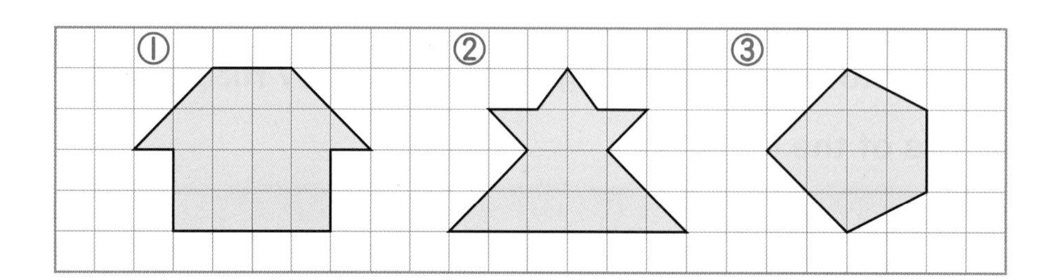

A figure has **line symmetry** when it can be folded along a straight line and both sides overlap exactly. The folding line is called the **axis of symmetry** or **line of symmetry.**

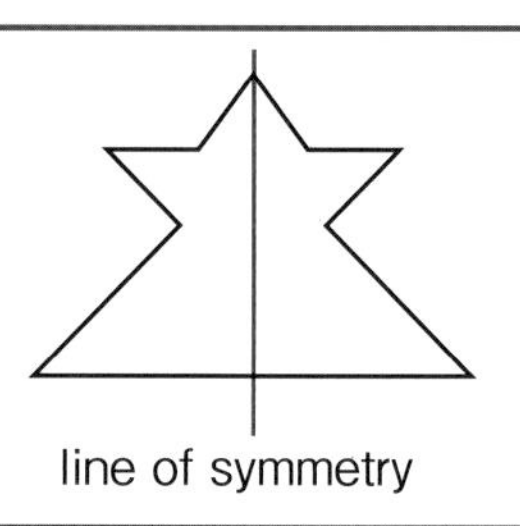

Want to confirm

Let's choose the line symmetric figures from the following Ⓐ～Ⓔ and draw the line of symmetry.

Line symmetric figures are (   ).

 Isosceles Triangle  Parallelogram  Rhombus

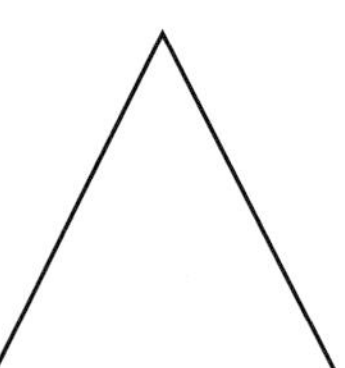

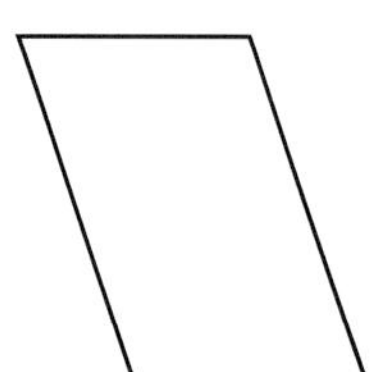

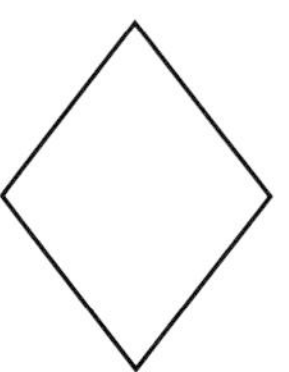

Ⓓ Trapezoid Ⓔ Equilateral Triangle

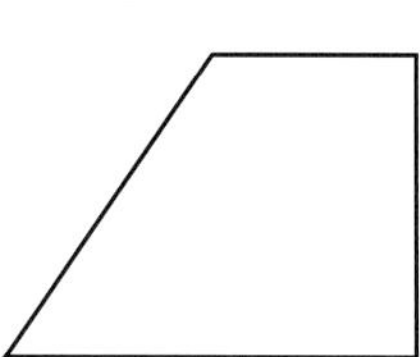

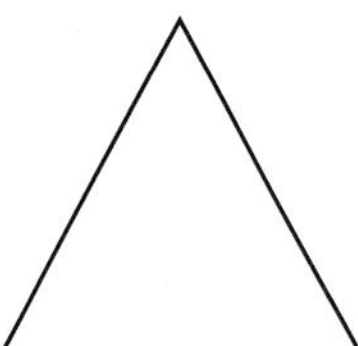

Let's try to fold thin paper to actually reproduce.

Way to see and think

Classifying the figures you have learned so far into line symmetric figures and not line symmetric figures.

Want to find in our life

From your surroundings, let's look for line symmetric figures.

Want to explore

3 The following diagram is a line symmetric figure. Let's explore the points, sides, and angles when the figure is folded by the line of symmetry.

X, A, B, H, C, G, D, E, F, Y

line of symmetry

Purpose In a line symmetric figure, what happens with the length of the overlapping sides and the size of the overlapping angles?

① Which are the respective overlapping points for point B and point C?

② Which are the respective overlapping sides for side AB and side CD?

③ Which are the respective overlapping angles for angle B and angle F?

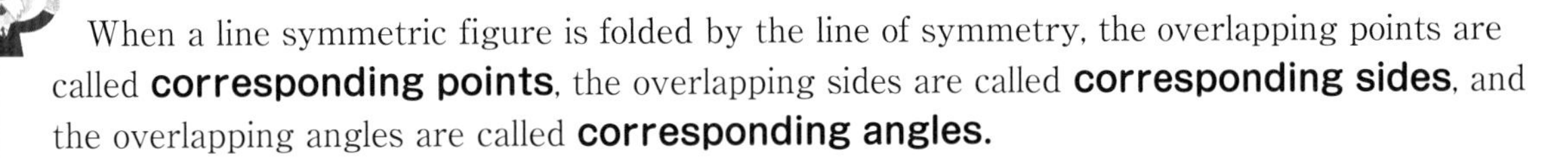
When a line symmetric figure is folded by the line of symmetry, the overlapping points are called **corresponding points**, the overlapping sides are called **corresponding sides**, and the overlapping angles are called **corresponding angles.**

④ Let's explore the relationship between the length of the corresponding sides and the size of the corresponding angles.

Summary

In line symmetric figures, the length of the corresponding sides and the size of the corresponding angles are respectively equal.

Want to confirm

3 The diagram on the right is a line symmetric figure that has straight line XY as the line of symmetry. Let's answer the following questions.

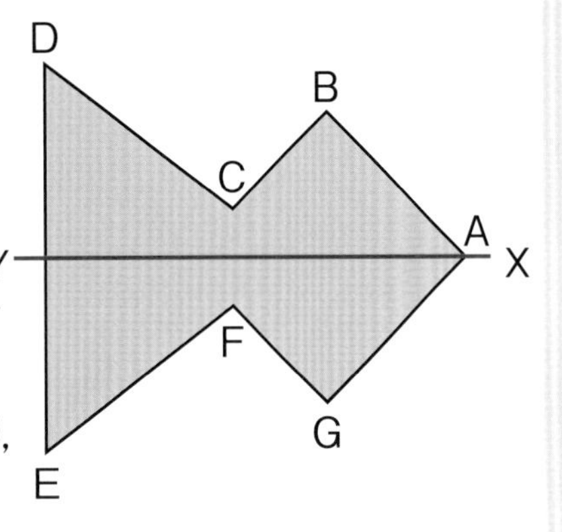

① Which are the respective corresponding points for points D, C, and B?

② Which are the respective corresponding sides for sides AB, BC, and CD?

③ Which are the respective corresponding angles for angles B, C, and D?

Let's imagine you actually fold along the line of symmetry.

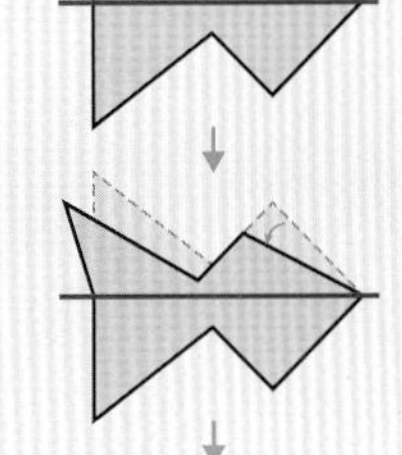

Want to explore

4 Let's explore the following things about the line symmetric figure shown on the right.

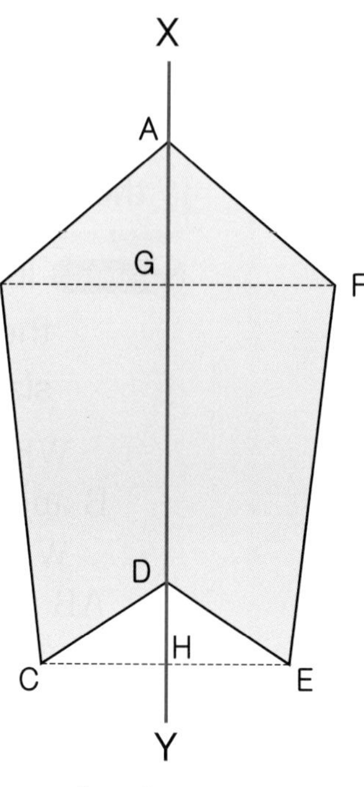

① How does straight line BF, connecting the corresponding points B and F, intersect with the line of symmetry?

② How does straight line CE, connecting the corresponding points C and E, intersect with the line of symmetry?

③ Let's compare the length of straight lines BG and FG. Also, let's compare the length of straight lines CH and EH.

For line symmetric figures, the intersection between the straight line that connects two corresponding points and the line of symmetry is perpendicular. Also, the length from the line of symmetry to each of the corresponding points is equal.

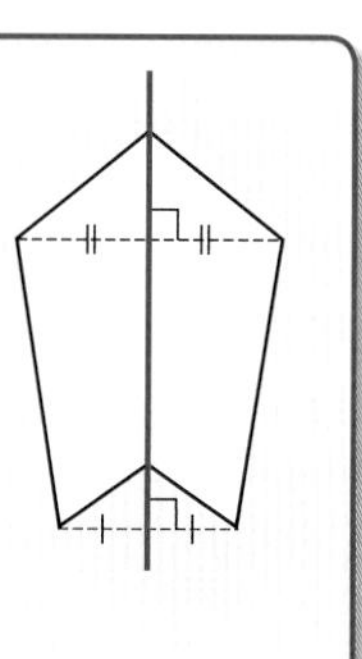

Want to try

 The diagram on the right is a line symmetric figure that has straight line XY as the line of symmetry. Let's answer the following questions.

① How does straight line CE intersect the line of symmetry?

② The length of straight line BJ is 25mm. How many millimeters is the length of straight line FJ?

③ Let's draw in the diagram the corresponding point M to point L.

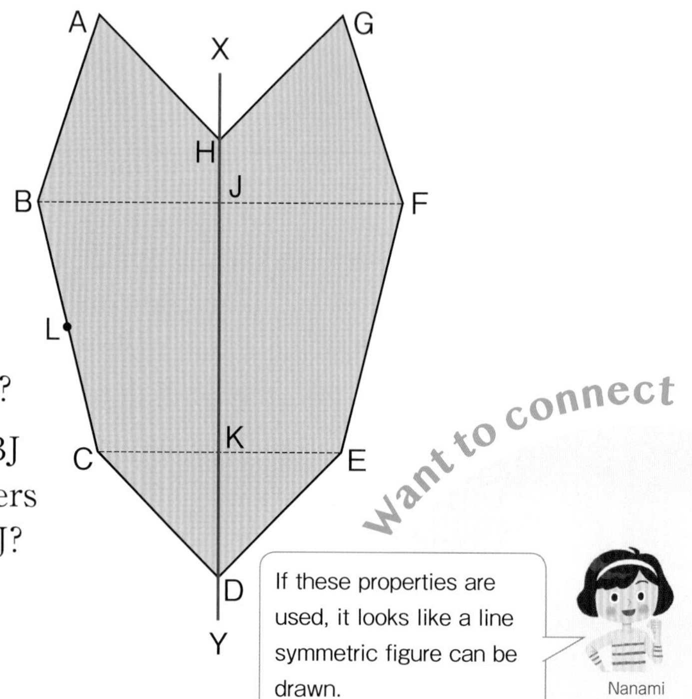

5 The following diagram represents half of a line symmetric figure that has the straight line XY as the line of symmetry. Let's draw the remaining half.

①

X

Y

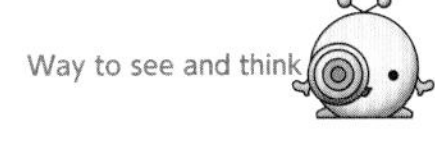

The length of corresponding sides is equal.

Way to see and think

For line symmetric figures, the length from the line of symmetry to each of the corresponding points is equal.

②

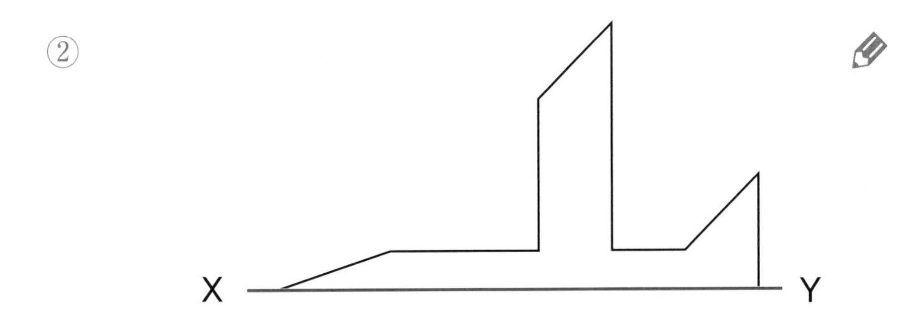

Want to explain

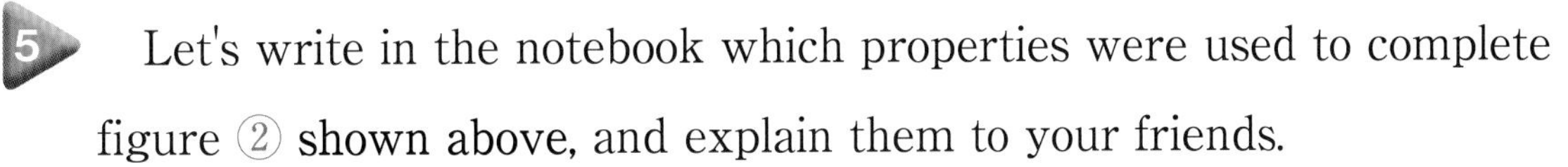

5 Let's write in the notebook which properties were used to complete figure ② shown above, and explain them to your friends.

Become a writing master.

Summary notebook

Let's summarize what you learned that day.

Write today's date.

September 10

Let's write the problem.

The diagram shown on the right is a line symmetric figure. Let's explore the points, sides, and angles when the figure is folded by the line of symmetry.

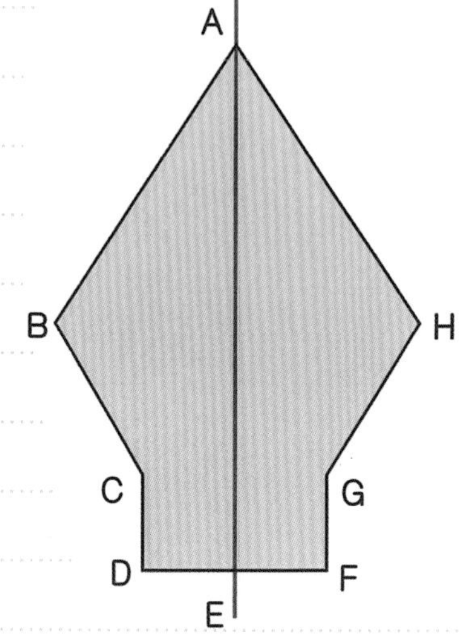

Let's learn with purpose.

Purpose

In a line symmetric figure, what happens with the length of the overlapping sides and size of the overlapping angles?

Organize and write your thoughts.

〈My ideas〉

AE is the line of symmetry. If the figure is folded by AE, both sides overlap exactly.

① Point B overlaps with point H.
   Point C overlaps with point G.

② Side AB overlaps with side AH.
   Side CD overlaps with side GF.

③ Angle B overlaps with angle H
   Angle F overlaps with angle D.

Summary

A line symmetric figure is folded by the line of symmetry.

- Overlapping points ··· corresponding points
- Overlapping sides ··· corresponding sides

  the length is equal

- Overlapping angles ··· corresponding angles

  the size is equal

As shown in the diagram, the summary can be later read and better understood by using colors, drawing lines and surrounding it.

〈Kaori's discovery〉

If overlapping points are connected with straight lines after the figure is folded in two, the straight lines become parallel.

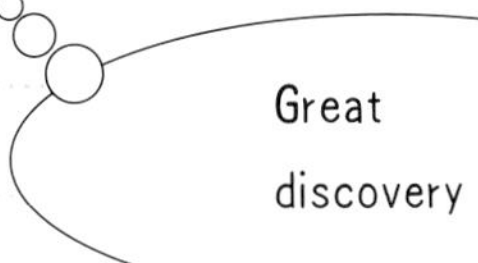

Let's write down what you think is good from your friend's ideas.

〈Reflect〉

- Since the overlap is exact when the figure is folded by the line of symmetry, the length of the corresponding sides and the size of the corresponding angles are equal. ← I was able to find this by myself.
- Kaori discovered that the straight lines connecting corresponding points are parallel. It's an amazing discovery. I want to confirm it by myself.

As for reflection, the following must be written:
- understood things,
- not understood things,
- noticed things,
- what you want to do more,
- and what you definitely achieved.

〈Kaori's notebook〉

The straight lines that connect the overlapping points B and H, and points C and G are parallel.

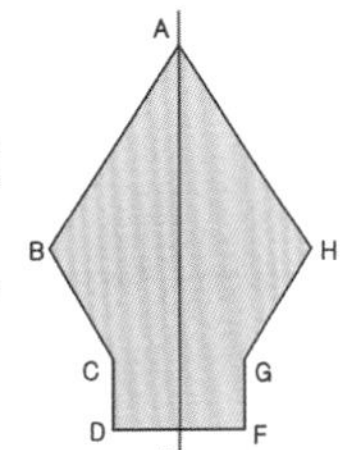

Kaori's discovery.

2 Point symmetric figures

Want to try

1 **In the following diagram, if a figure is rotated 180° around the central point "・", it matches the original figure. Let's confirm.**

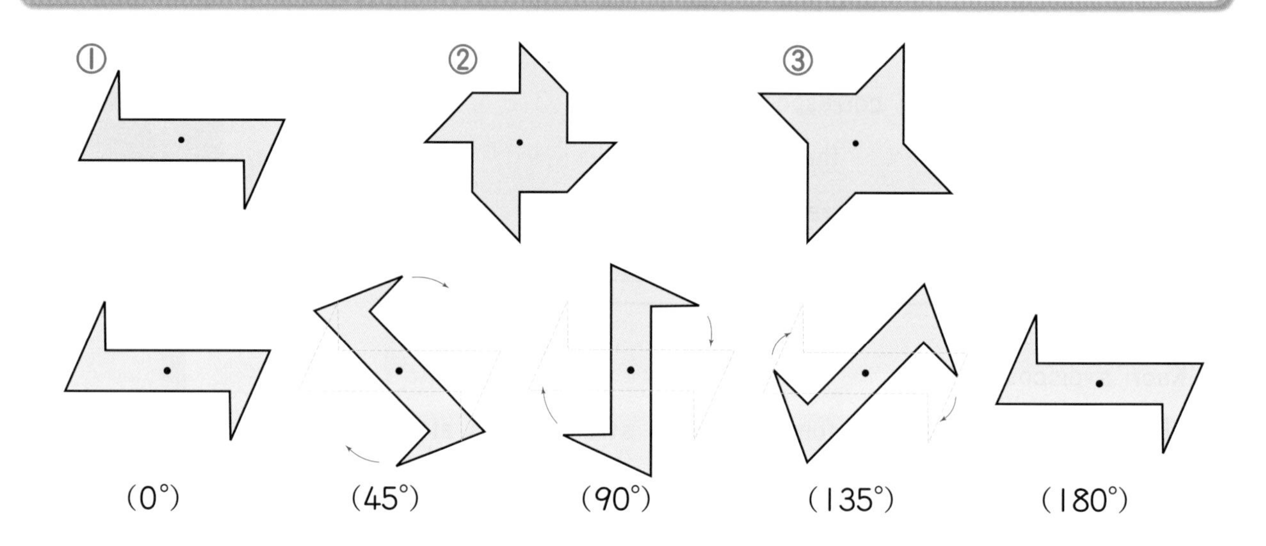

Let's confirm by rotating 180° the figures that were cut from p.245 and used in p.87. Only use figures Ⓐ and Ⓕ.

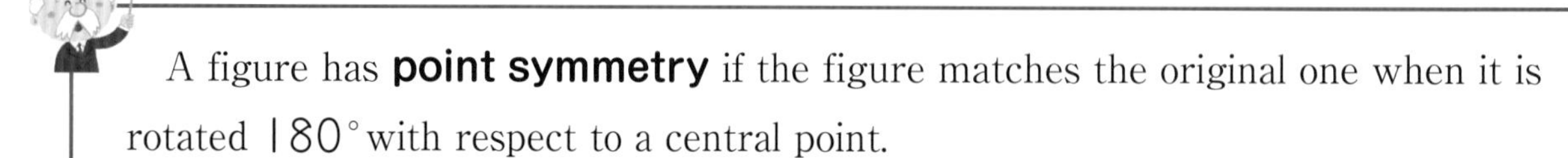

A figure has **point symmetry** if the figure matches the original one when it is rotated 180° with respect to a central point.

Also, this central point is called **point of symmetry** or **center of symmetry**.

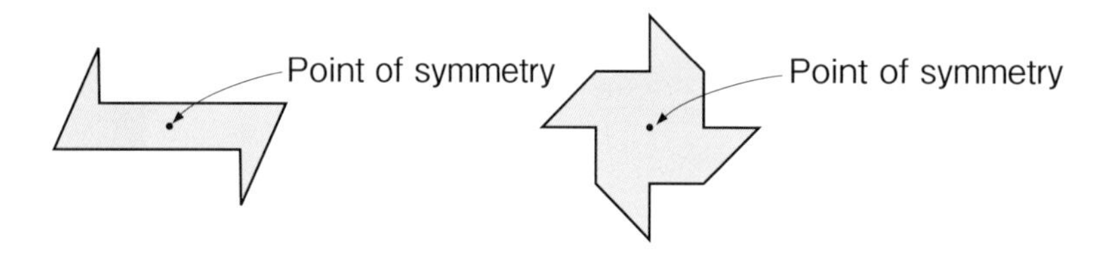

Want to find in our life

In your surroundings, let's look for point symmetric figures.

Want to explore | Properties of point symmetric figures

2 **The diagram on the right is a point symmetric figure considering point O as the point of symmetry. Let's reproduce the figure in thin paper and explore the points, sides, and angles when the figure is rotated 180° around the point of symmetry.**

Purpose: In a point symmetric figure, what happens with the length of the matching sides and the size of the matching angles?

① Which are the respective matching points for point B and point C?

② Which are the respective matching sides for side AB and side BC?

③ Which are the respective matching angles for angle B and angle D?

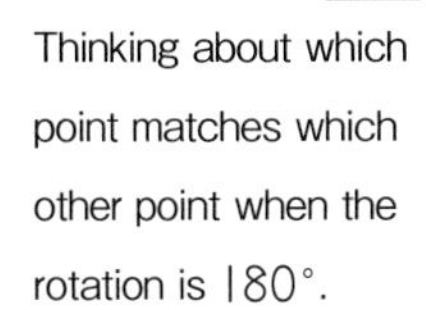

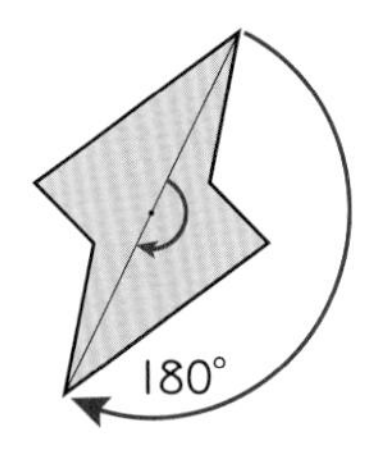

When a point symmetric figure is rotated 180° around the point of symmetry, the matching points are called **corresponding points,** the matching sides are called **corresponding sides,** and the matching angles are called **corresponding angles.**

④ Let's explore the relationship between the length of the corresponding sides and the size of the corresponding angles.

Summary

In point symmetric figures, the length of the corresponding sides and the size of the corresponding angles are respectively equal.

Want to confirm

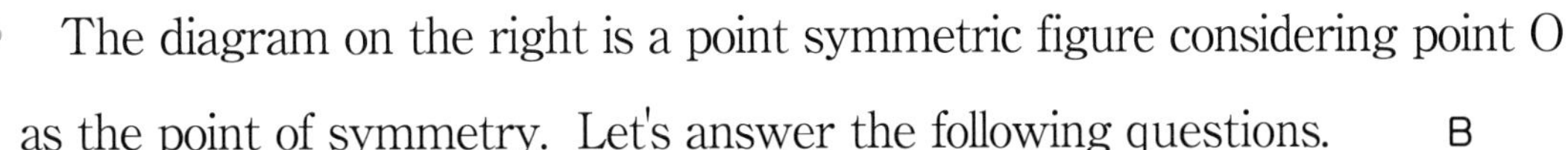

2 The diagram on the right is a point symmetric figure considering point O as the point of symmetry. Let's answer the following questions.

① Which are the respective corresponding points for points A, B, and C?

② Which are the respective corresponding sides for sides AB, BC, and CD?

③ Which are the respective corresponding angles for angles A, D, and F?

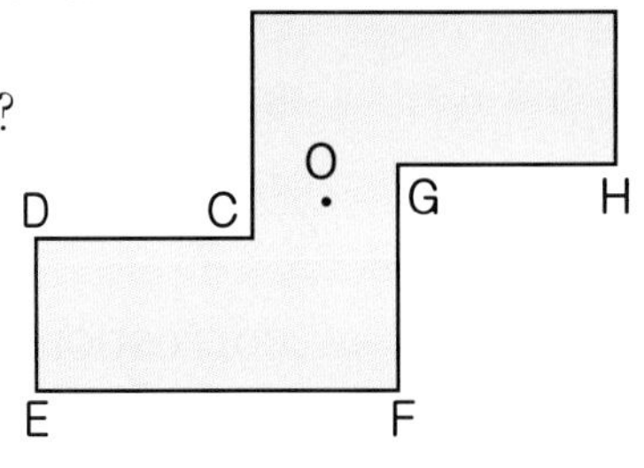

Want to explore

3 Let's explore the following things about the point symmetric figure.

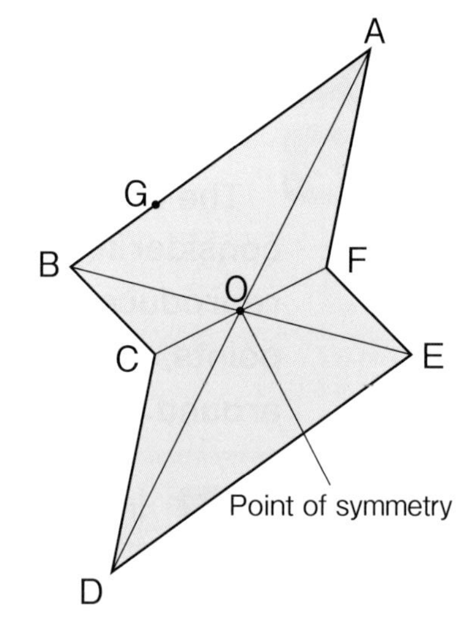

① The straight lines AD, BE, and CF are connecting corresponding points. Where do these straight lines intersect?

② Let's draw in the figure the corresponding point H to point G, which is on side AB.

③ Let's compare the lengths of straight lines AO and DO. Also, let's compare the lengths of straight lines GO and HO.

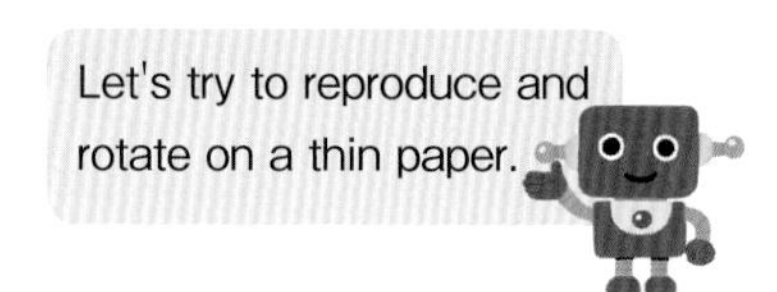

For point symmetric figures, the straight line that connects two corresponding points always passes through the point of symmetry. Also, the length from the point of symmetry to each of the corresponding points is equal.

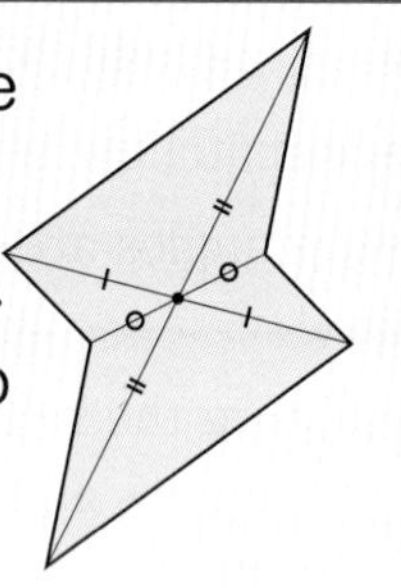

Way to see and think

The point of symmetry is on the straight line that connects two corresponding points.

Want to try

3 The diagram on the right is a point symmetric figure. Let's answer the following questions.

① Let's draw the point of symmetry. Also, let's explain how you found the point of symmetry.

② Let's draw in the figure the corresponding point B to point A.

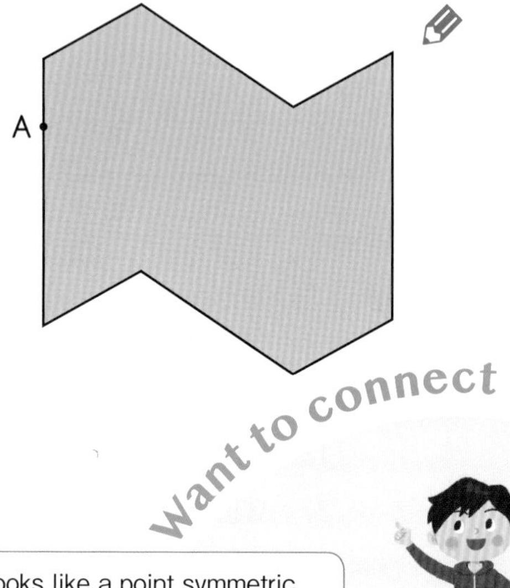

4 **The following diagram represents half of a point symmetric figure that has the point O as the point of symmetry. Let's draw the remaining half.**

①

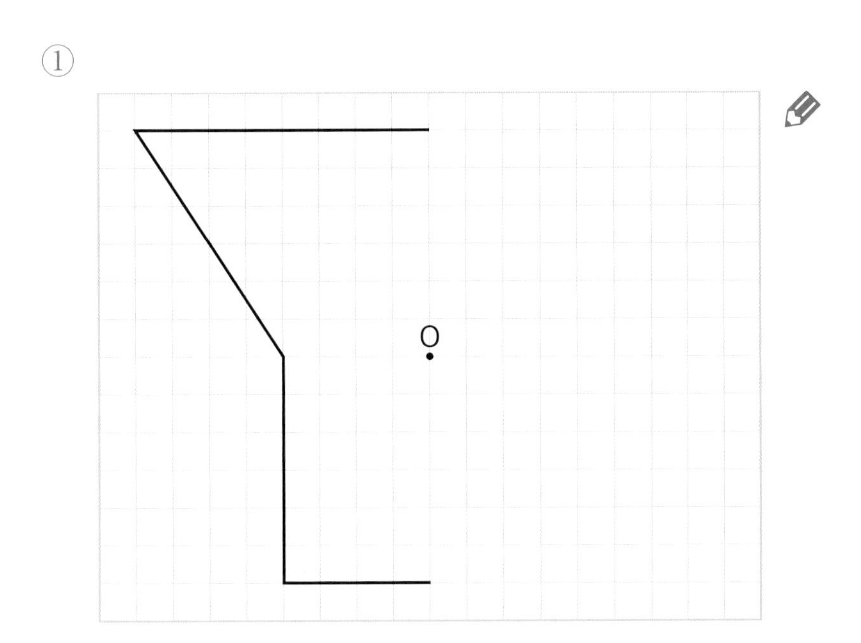

The length of the corresponding sides of a point symmetric figure is equal.

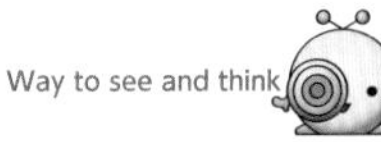

In a point symmetric figure, the length from the point of symmetry to each of the corresponding points is equal.

②

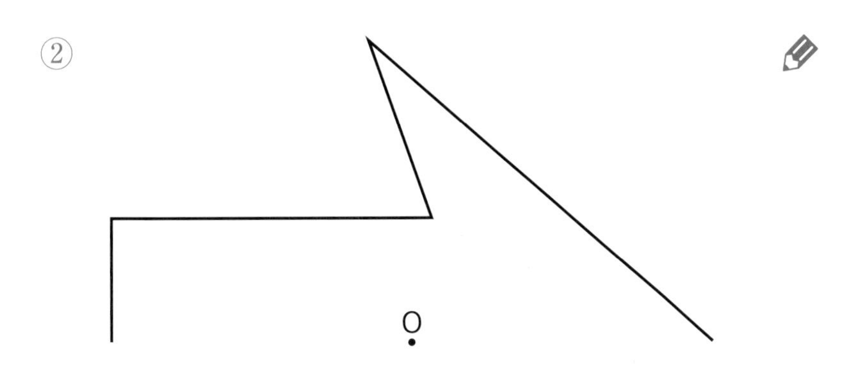

Want to explain

Let's write in the notebook which properties were used to complete figure ② shown above, and explain them to your friends.

Want to explore

1 Let's explore the things shown below about the following quadrilaterals.

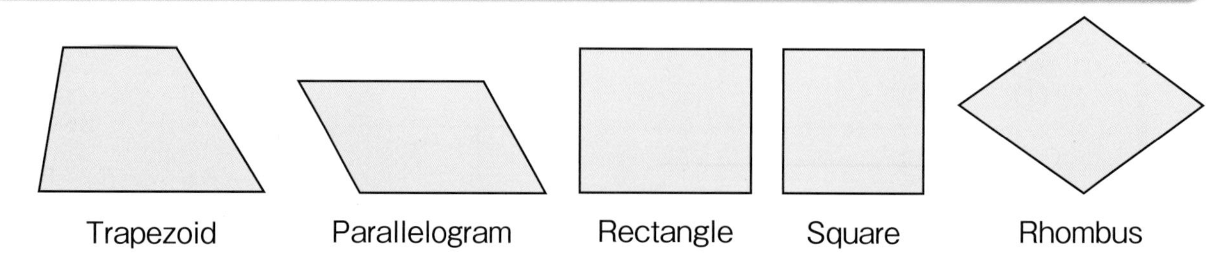

① Let's write ○ and × in the table to separate line symmetric figures and point symmetric figures. Also, as for line symmetric figures, how many lines of symmetry does each have?

② Let's draw the point of symmetry on each point symmetric figure.

	Line symmetry	Number of lines of symmetry	Point symmetry
Trapezoid			
Parallelogram			
Rectangle			
Square			
Rhombus			

③ Let's make a presentation of what you notice from the table.

Want to try

1 Let's explore the things shown below about the following triangles.

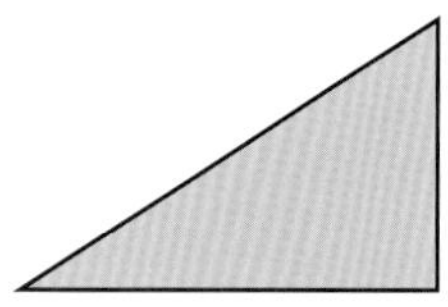

Right triangle

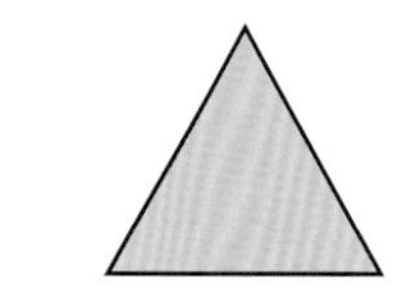

Equilateral triangle

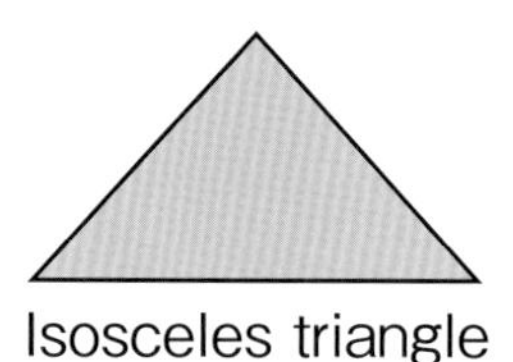

Isosceles triangle

① Which are line symmetric triangles? Also, as for line symmetric triangles, how many lines of symmetry does each have?

② Is there any point symmetric triangle?

2 **Let's explore the things shown below about the following regular polygons.**

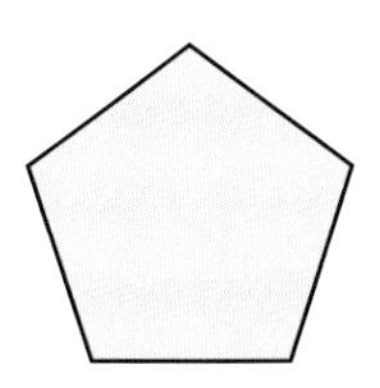
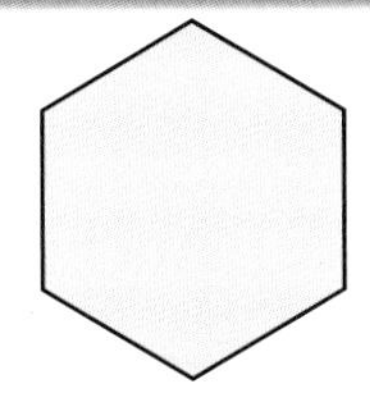
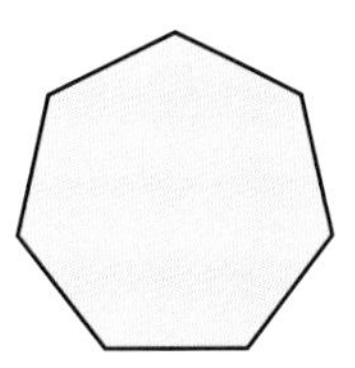
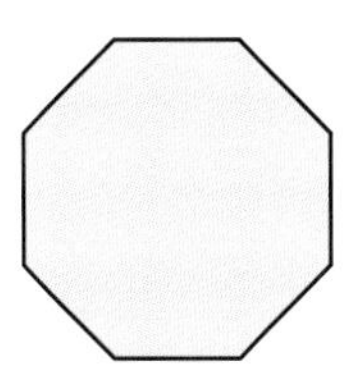
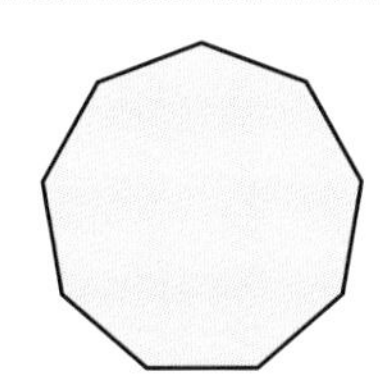

Regular pentagon Regular hexagon Regular heptagon Regular octagon Regular nonagon

① Let's write ○ and × in the table to separate line symmetric figures and point symmetric figures. Also, as for line symmetric figures, how many lines of symmetry does each have?

② Let's draw the point of symmetry on each point symmetric figure.

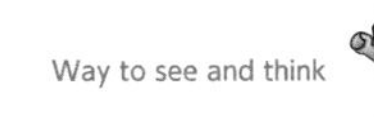

You can see the rules by representing it in a table.

	Line symmetry	Number of lines of symmetry	Point symmetry
Regular pentagon			
Regular hexagon			
Regular heptagon			
Regular octagon			
Regular nonagon			

Let's try to add equilateral triangles and squares to the table as regular polygons.

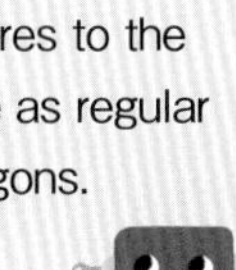

③ Let's write in your notebook and discuss with your classmates what you have noticed while exploring.

Want to confirm

Let's explore about the circle.

① Is the circle a line symmetric figure?

Also, how many lines of symmetry does it have?

② Is the circle a point symmetric figure?

Also, where is the point of symmetry?

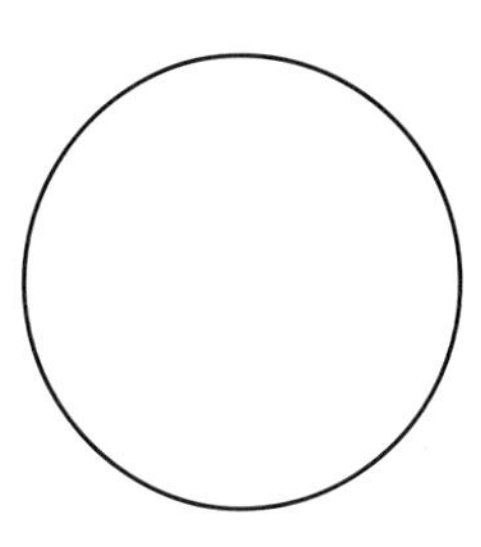

What you can do now

☐ **Can draw line symmetric figures and point symmetric figures based on their properties.**

1 Let's complete the line symmetric figure in ① and the point symmetric figure in ②.

① Straight line XY is the line of symmetry. ② Point O is the point of symmetry.

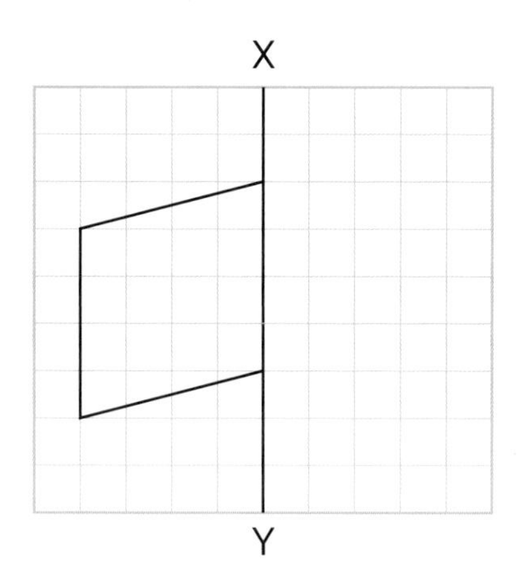

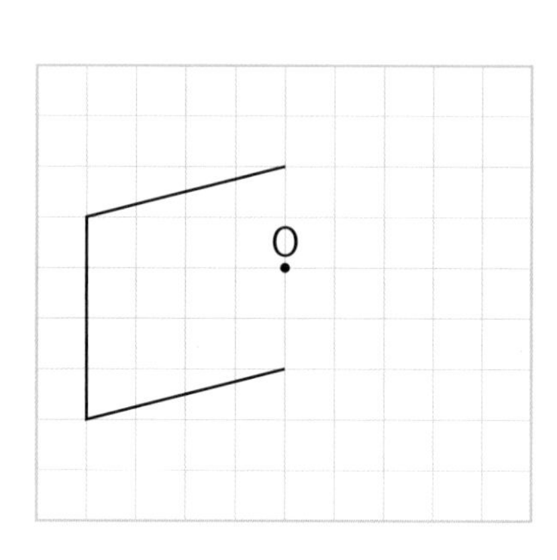

☐ **Can categorize and organize figures in a symmetrical way.**

2 What type of quadrilateral are the following 6 quadrilaterals? Let's summarize in the table below.

Ⓐ 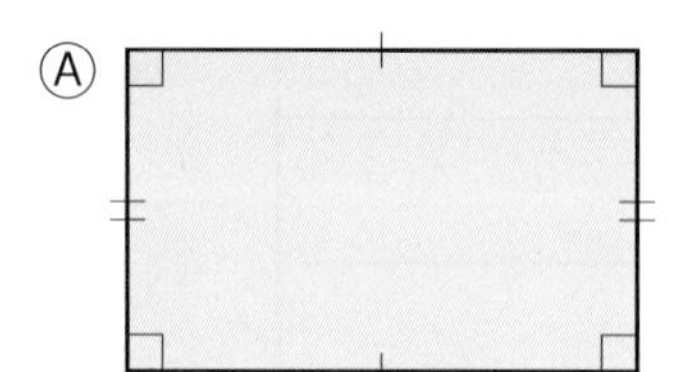 Ⓑ 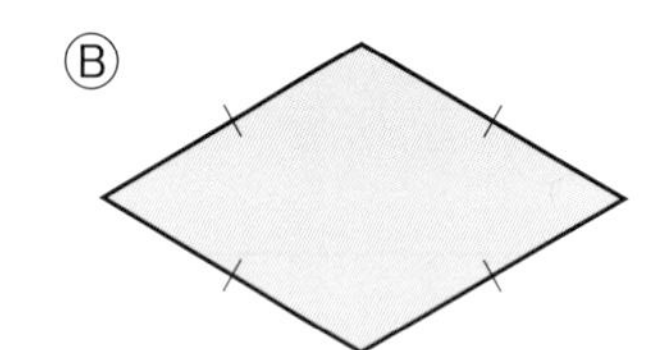 Ⓒ 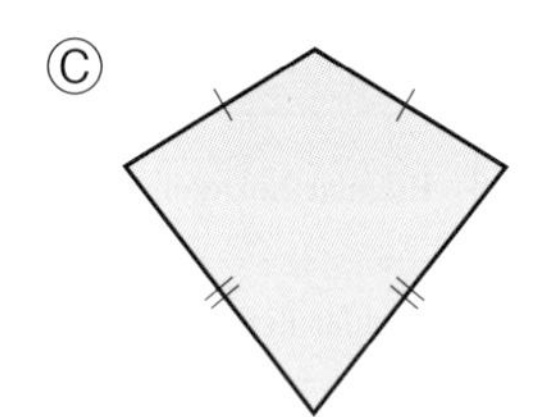

Ⓓ 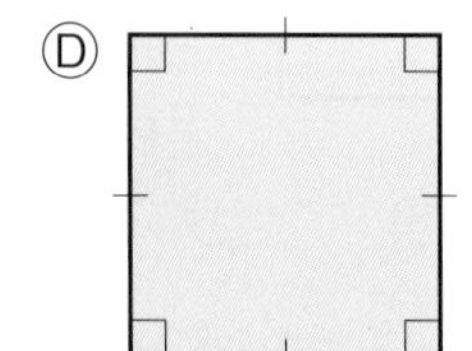 Ⓔ 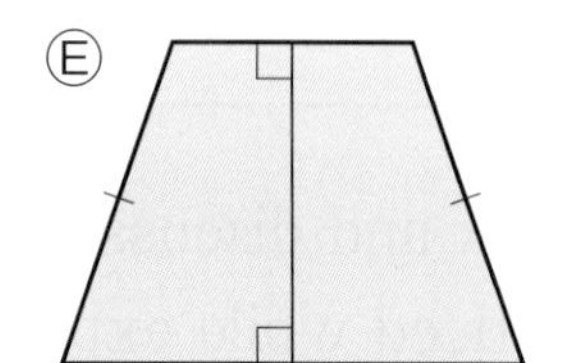 Ⓕ 

	Ⓐ	Ⓑ	Ⓒ	Ⓓ	Ⓔ	Ⓕ
Line symmetric figure	○					
Number of lines of symmetry	2					
Point symmetric figure	○					

Supplementary problems p.230

Usefulness and efficiency of learning

1 Which are line symmetric figures? Also, which are point symmetric figures?

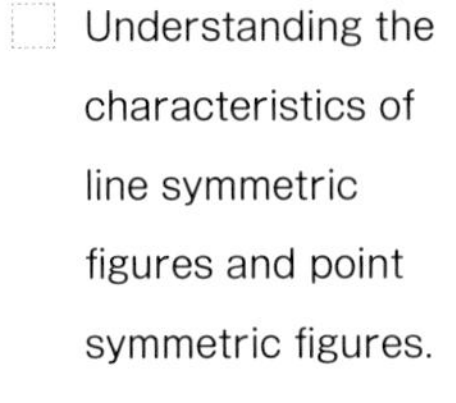
Understanding the characteristics of line symmetric figures and point symmetric figures.

①  ② 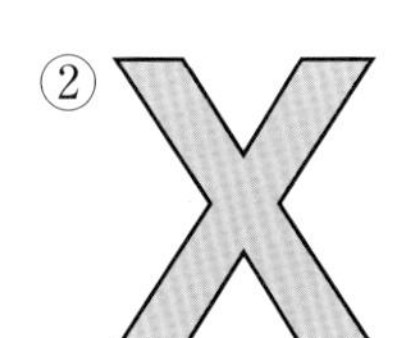 ③  ④ 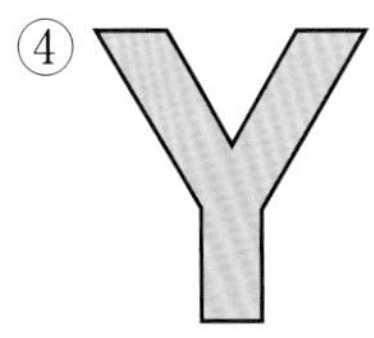

2 ① is a line symmetric figure. Let's draw the line of symmetry.
Also, ② is a point symmetric figure. Let's draw the point of symmetry.

Understanding the properties of the line of symmetry and the point of symmetry.

①

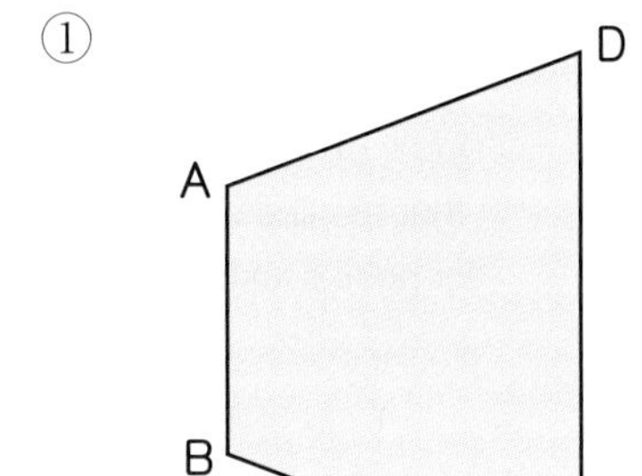

②

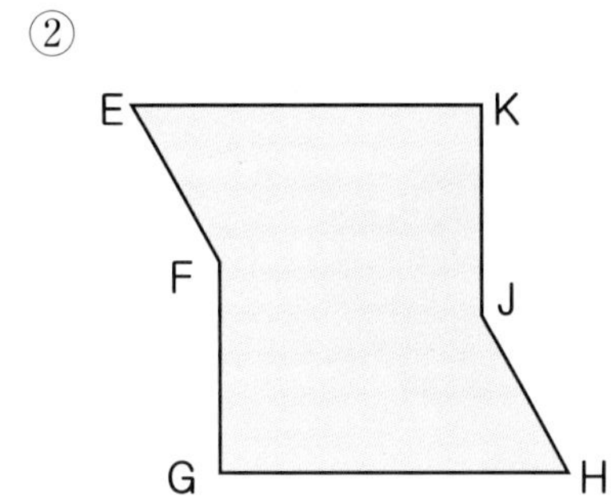

3 A regular dodecagon is a line symmetric figure. How many lines of symmetry does it have? Let's draw all the lines of symmetry in the figure shown on the right.

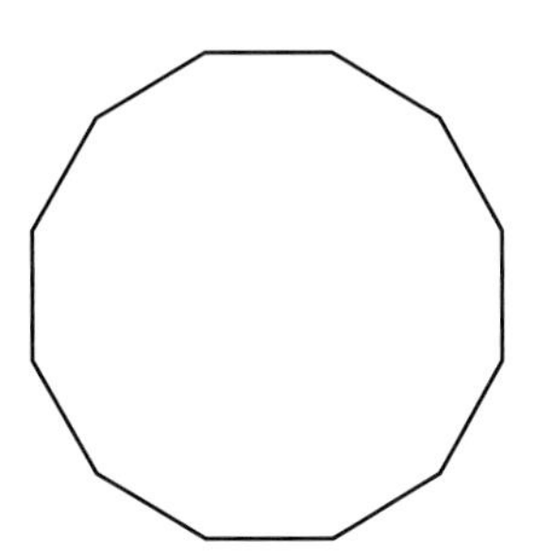

Understanding the relationship between regular polygons and symmetric figures.

Let's deepen.

Let's find line symmetric figures and point symmetric figures in our surroundings.

Deepen.

Let's look for symmetrical shapes.

What about the emblem of the prefecture where you live?

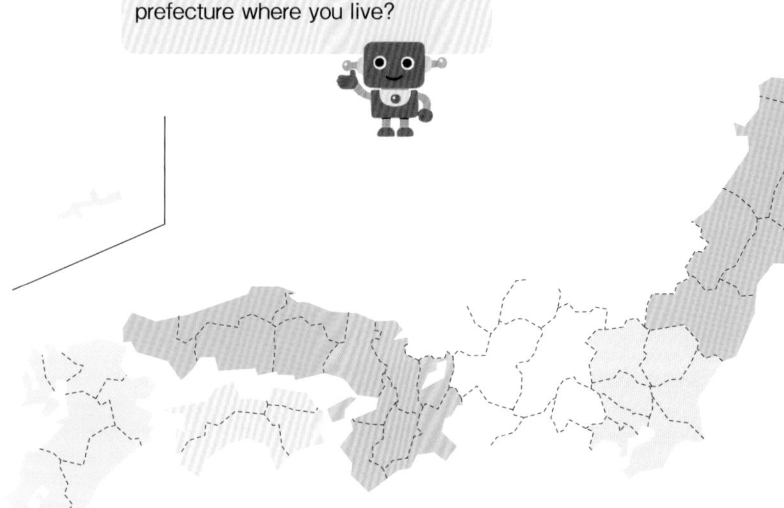

Let's explore emblems that are line symmetric or point symmetric.

Prefecture Emblems

① Hokkaido ② Aomori ③ Iwate ④ Miyagi ⑤ Akita ⑥ Yamagata ⑦ Fukushima ⑧ Ibaraki

  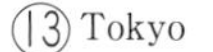  

⑨ Tochigi ⑩ Gunma ⑪ Saitama ⑫ Chiba ⑬ Tokyo ⑭ Kanagawa ⑮ Niigata ⑯ Toyama

 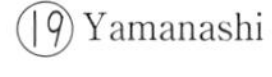 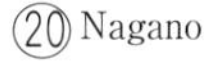  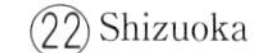 

⑰ Ishikawa ⑱ Fukui ⑲ Yamanashi ⑳ Nagano ㉑ Gifu ㉒ Shizuoka ㉓ Aichi ㉔ Mie

㉕ Shiga ㉖ Kyoto ㉗ Osaka

㉘ Hyogo ㉙ Nara ㉚ Wakayama

㉛ Tottori ㉜ Shimane ㉝ Okayama

㉞ Hiroshima ㉟ Yamaguchi ㊱ Tokushima

㊲ Kagawa ㊳ Ehime ㊴ Kochi

㊵ Fukuoka ㊶ Saga ㊷ Nagasaki

㊸ Kumamoto ㊹ Oita ㊺ Miyazaki

㊻ Kagoshima ㊼ Okinawa

A pictogram is a simple drawing symbol of what you want to represent.

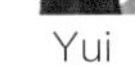

Traffic Signs

Ⓐ Ⓑ Ⓒ Ⓓ Ⓔ

Ⓕ Ⓖ Ⓗ Ⓘ Ⓙ

Ⓚ Ⓛ Ⓜ Ⓝ

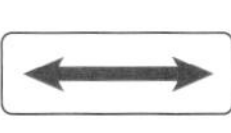

I've seen those in town.

Map symbols

Ⓐ Ⓑ Ⓒ Ⓓ Ⓔ Ⓕ

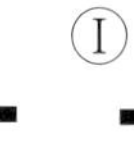

Ⓖ Ⓗ Ⓘ Ⓙ Ⓚ Ⓛ

Pictograms

Ⓐ Toilets Ⓑ Men Ⓒ Women Ⓓ Nursery

Ⓔ Escalator Ⓕ Trash box Ⓖ Collection facility for the recycling products Ⓗ Telephone

Ⓘ Stairs Ⓙ Bus stop Ⓚ Coin locker Ⓛ Railway station

Which pizza is the biggest?

How can we find the area of a circle?

Area of a Circle

Let's think about how to find the area of a circle.

1 Area of a circle

Want to solve

1 Let's think about how to find the area of a circle with a radius of 10 cm.

① Let's look at the following diagrams and estimate the area.

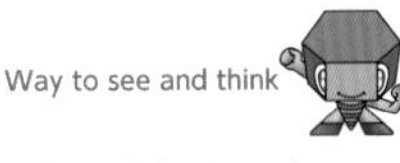

Considering the circle that fits inside a square and the square that fits inside a circle.

Ⓐ 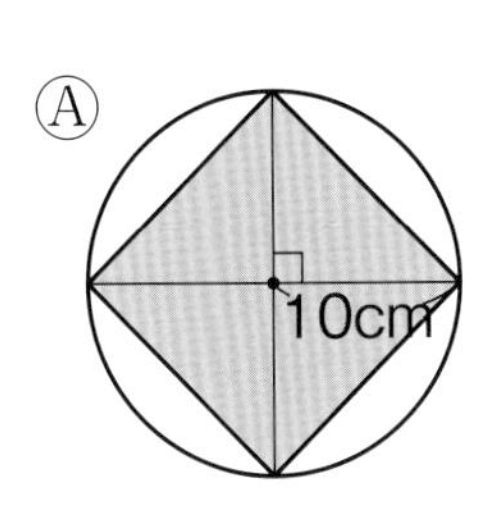

Ⓑ 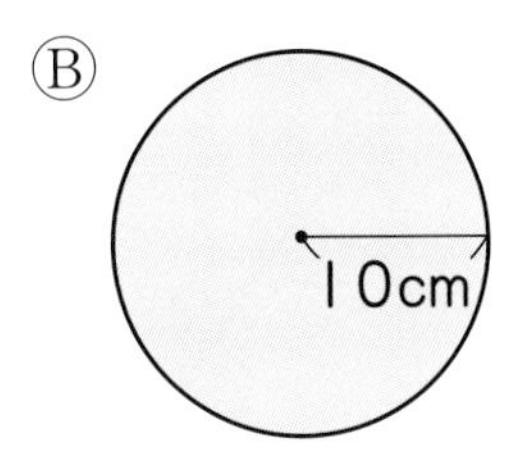

Ⓒ 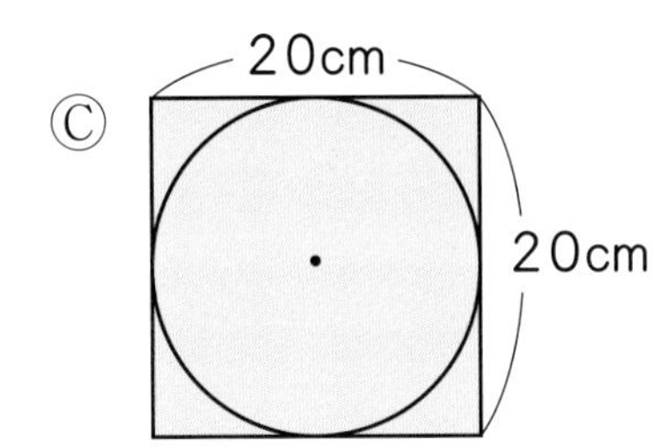

The area of square Ⓐ is ☐ cm². The area of square Ⓒ is ☐ cm².

The area of circle Ⓑ is larger than ☐ cm² and smaller than ☐ cm².

Want to explore

② Let's explore the area of circle Ⓑ using graph paper with 1 cm side units.

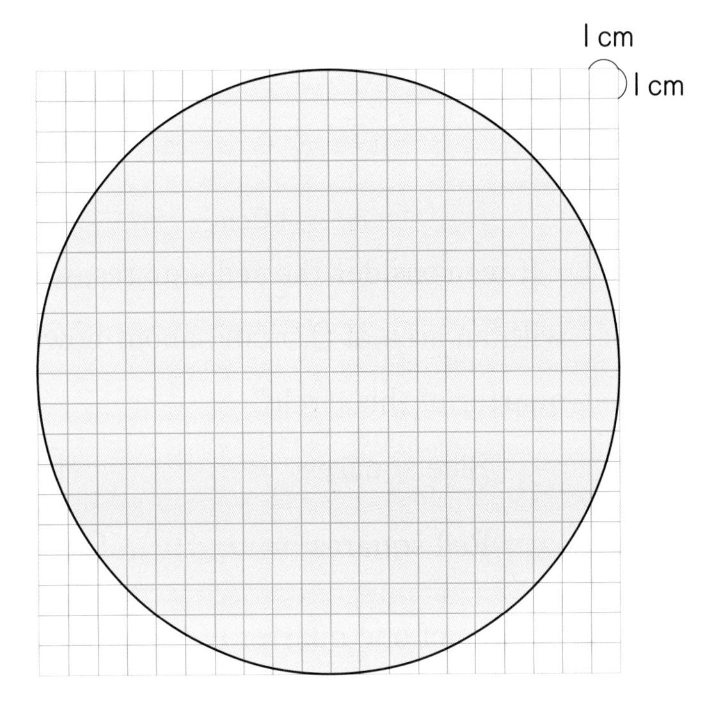

Want to think

③ Let's separate the circle into 4 equal parts and consider one part.

Ⓐ In the diagram below, how many blue and red squares are there?

blue squares …… ☐ red squares …… ☐

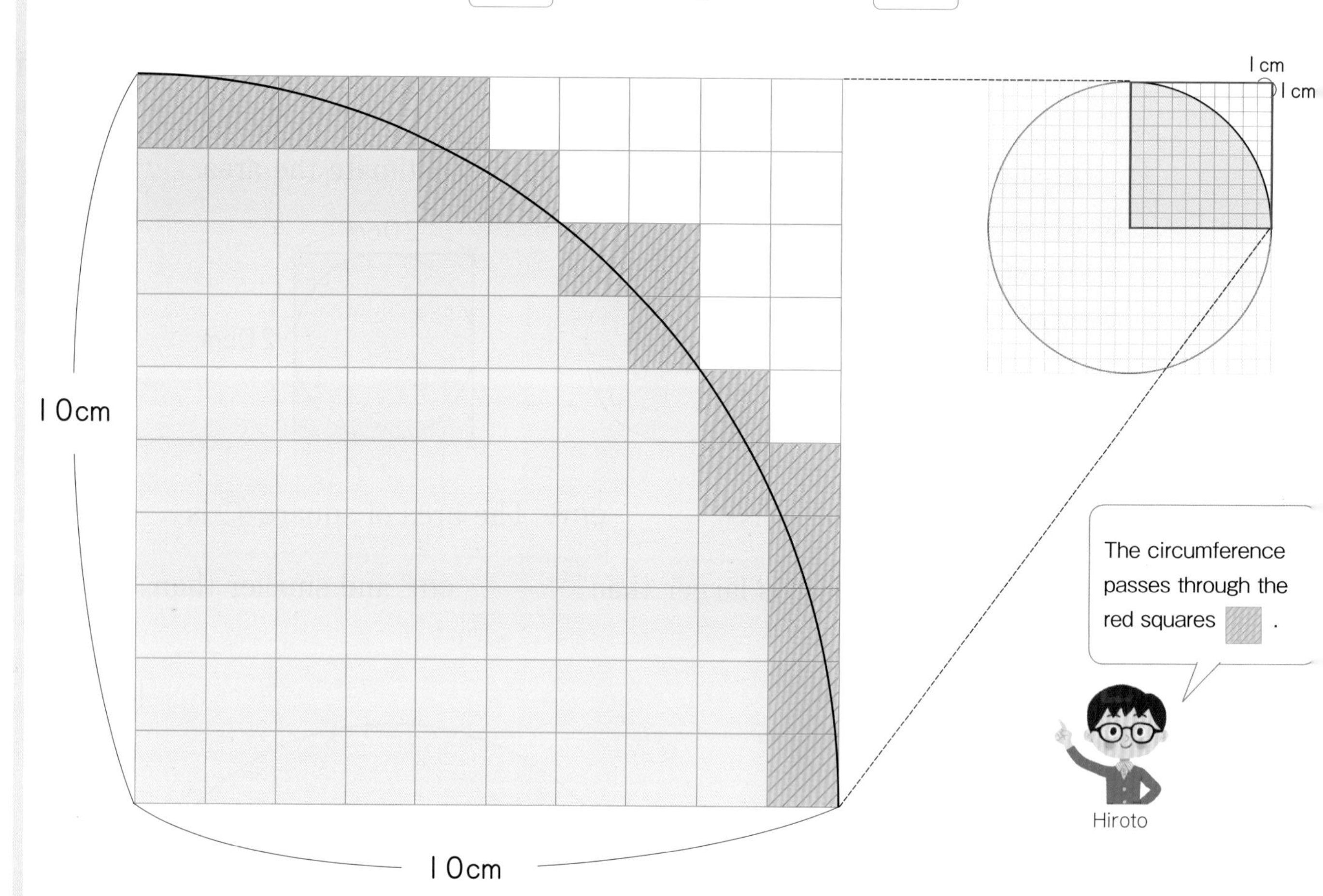

Ⓑ If we consider the red squares, where the circumference passes, as squares with an area of 0.5 cm², about how many square centimeters is the area of a quarter of this circle?

Blue squares…………… $1 \times$ ☐ $=$ ☐ (cm²)

Red squares…………… $0.5 \times$ ☐ $=$ ☐ (cm²)

The area of one quarter of the circle is about ☐ (cm²)

④ About how many square centimeters is the area of the entire circle?

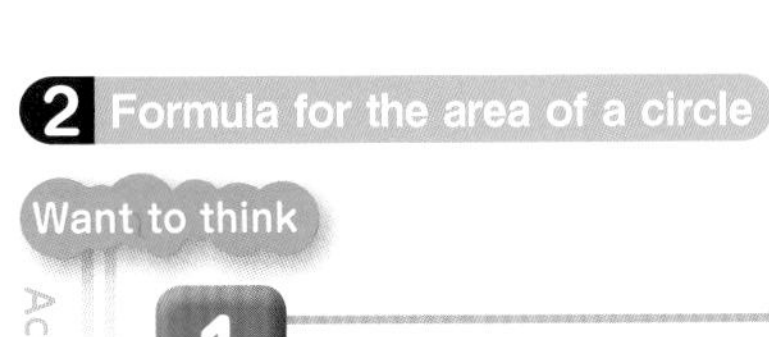

2 Formula for the area of a circle

Want to think

Activity

1 **Let's think about how to find the area of a circle with a radius of 5 cm.**

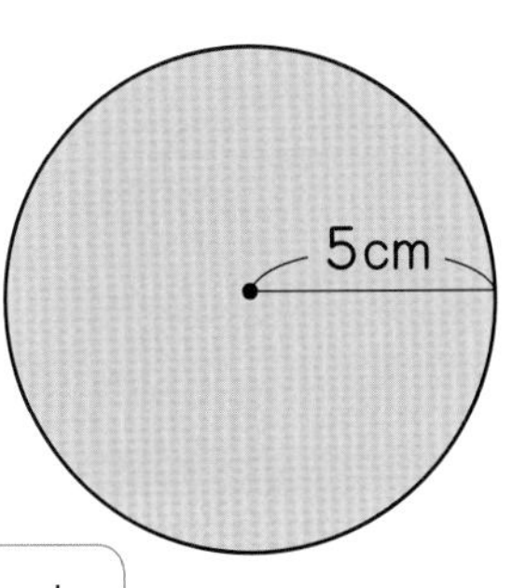

As for parallelograms and triangles, the area was changed to a known figure.

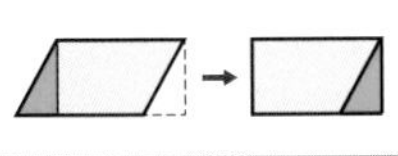

There is a formula for rectangles and triangles. Is there one for circles?

Purpose Is there a formula to find the area of a circle?

Want to explain

① Daiki and Yui rearranged the circle using small equal parts. Let's explain the ideas of the two children.

Let's confirm using the figures from p.245.

Daiki's idea

The circle was separated into ☐ parts and rearranged as a ☐.

circumference÷4

4 times of the radius

circumference÷4

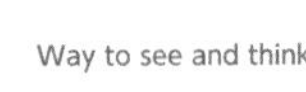
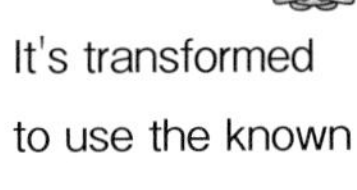

It's transformed to use the known formula of a triangle.

Yui's idea

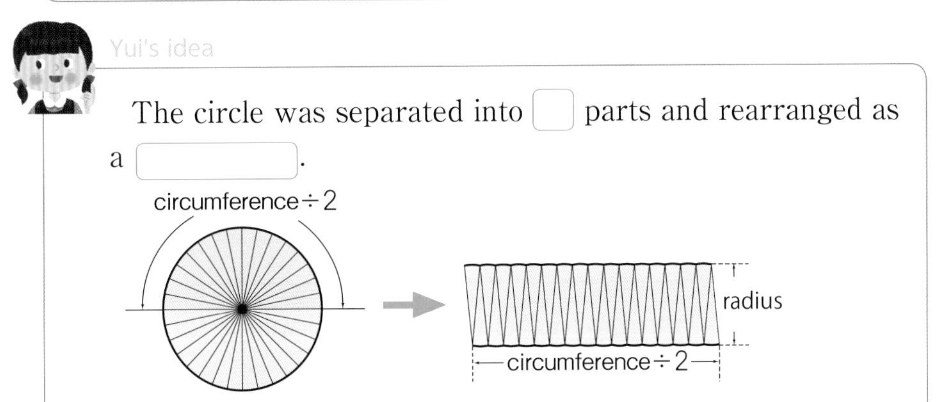

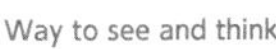
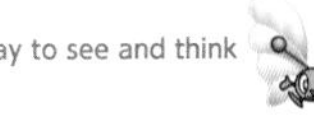

If you divide the circle into small parts, you can get closer to a rectangular figure.

② Let's try to find the area of the circle based on the ideas shown in ①.

Want to explain

③ Based on the idea presented by Yui in the previous page, let's explain the formu

to find the area of a circle.

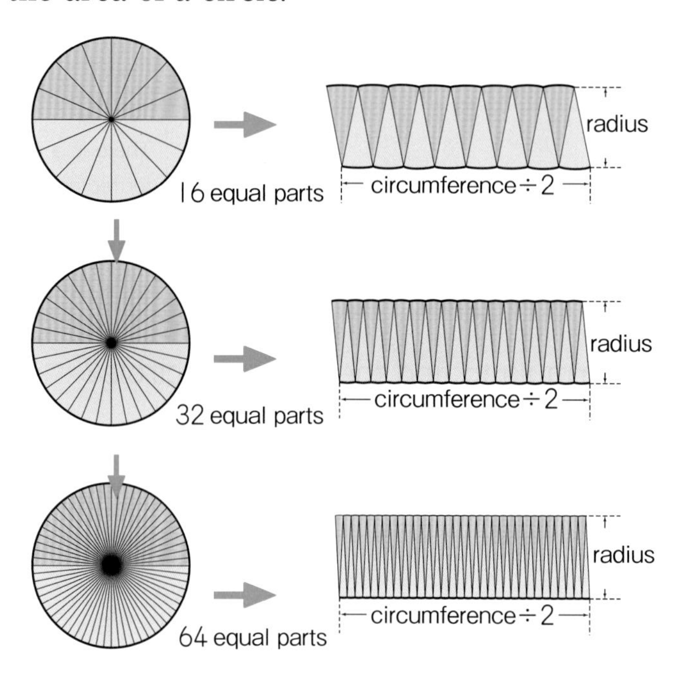

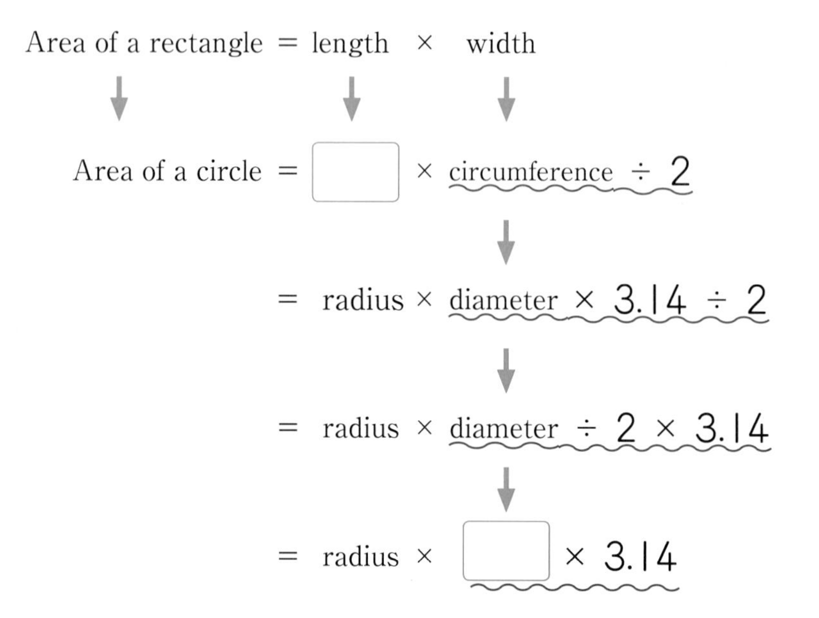

As we divide the circle finely into small sections of equal size, it approaches the shape of a rectangle.

Summary

The area of a circle can be found with the following formula:

Area of a circle = radius × radius × 3.14

Want to confirm

1 Let's find the area of a circle with the following radius.

① 8cm　　② 10cm　　③ 12cm

Want to try

2 Let's find the area of the following figures.

①

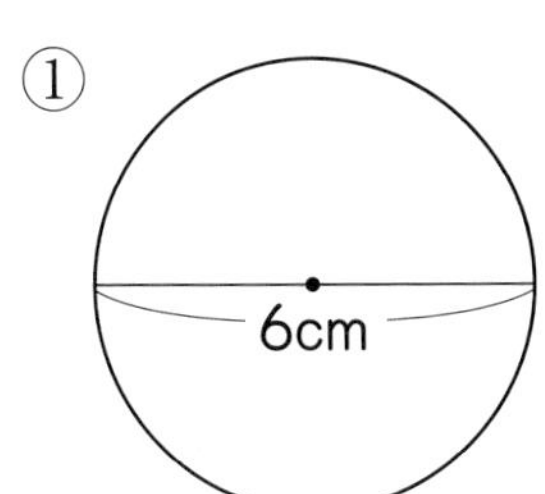

②

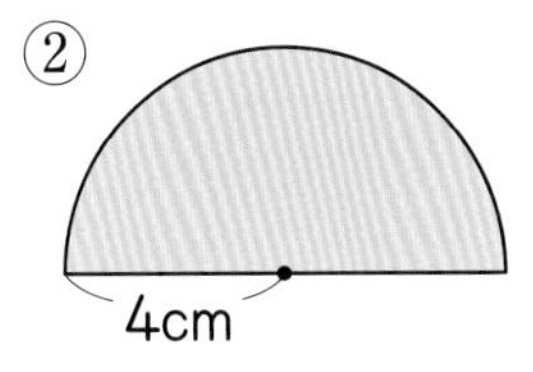

③

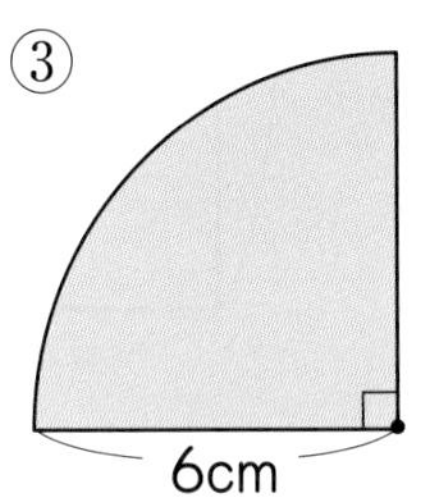

3 The following ①~③ are the circumferences of circles. Let's find the length of the radius and area of each circle.

① 62.8cm　　② 18.84cm　　③ 15.7cm

Want to deepen

4 There are two circles with diameters 2 cm (Ⓐ) and 4 cm (Ⓑ).

Let's think about these two circles.

① Let's find the circumference and area of each circle.

② The diameter of Ⓑ is 2 times the diameter of Ⓐ. How many times the circumference and the area of Ⓐ is that of Ⓑ?

Ⓐ

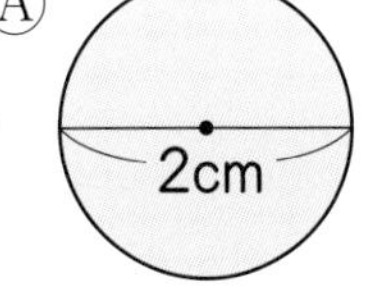

Ⓑ

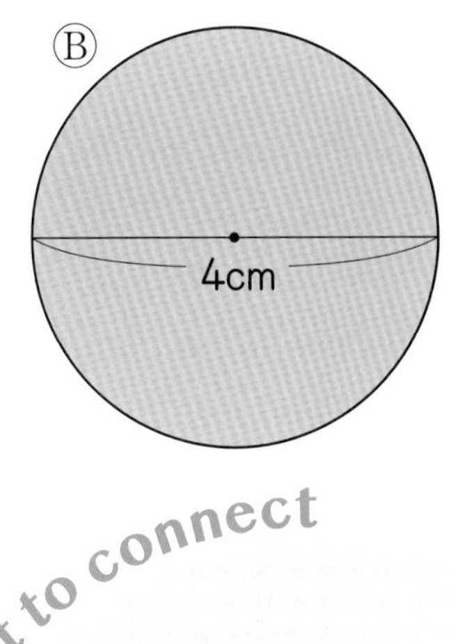

Want to connect

I want to try with other figures. For example, what happens when it's a square?

Want to improve

Activity

1 For the following diagrams, let's find the area of the colored parts.

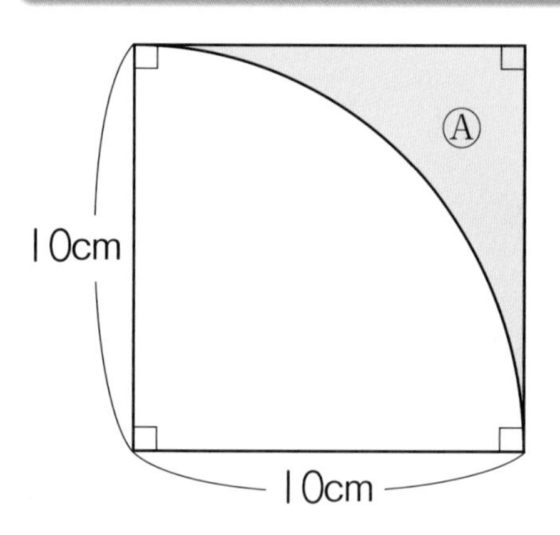

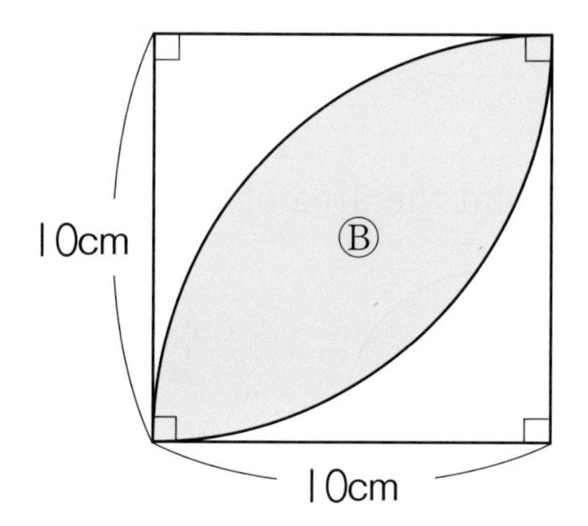

Searching for a figure that you know how to find the area.

① Let's write a math expression to find the area of Ⓐ and its corresponding answer.

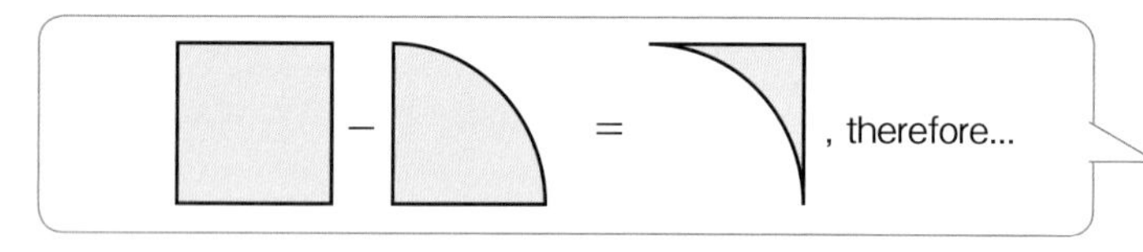

Daiki

Want to explain

② Let's explain the methods to find the area of Ⓑ followed by the following children.

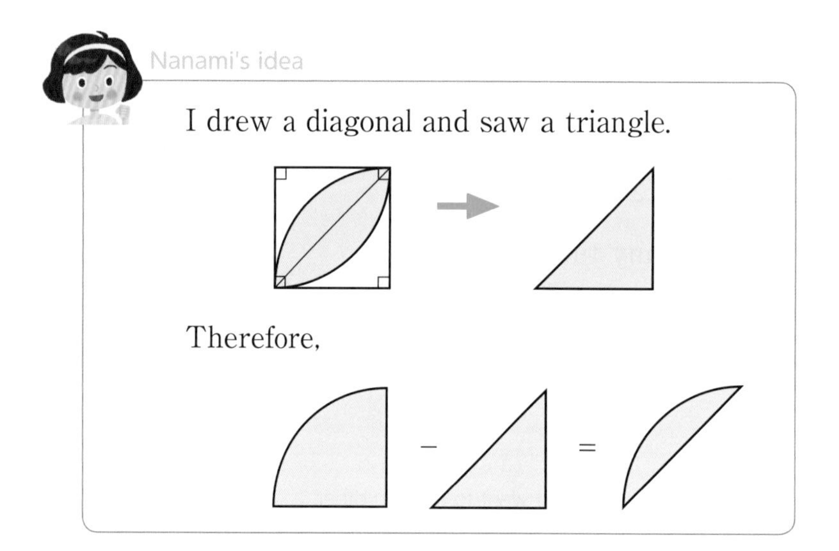

Based on the area of a quarter of a circle and a right triangle.

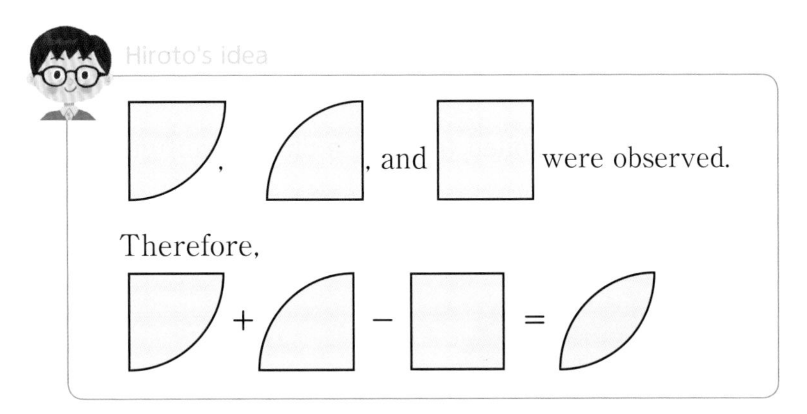

Based on the area of a quarter of a circle and a square.

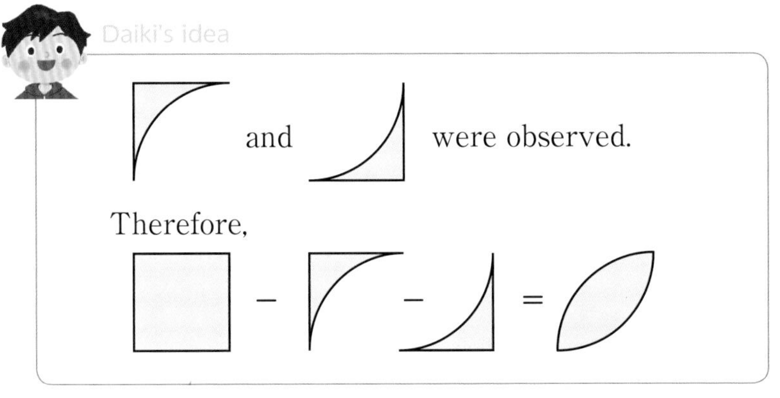

Way to see and think

Based on the area of Ⓐ and a square.

③ Let's find the area of Ⓑ using the ideas from ②.

Want to confirm

1 For the following diagrams, let's find the area of the colored parts.

①

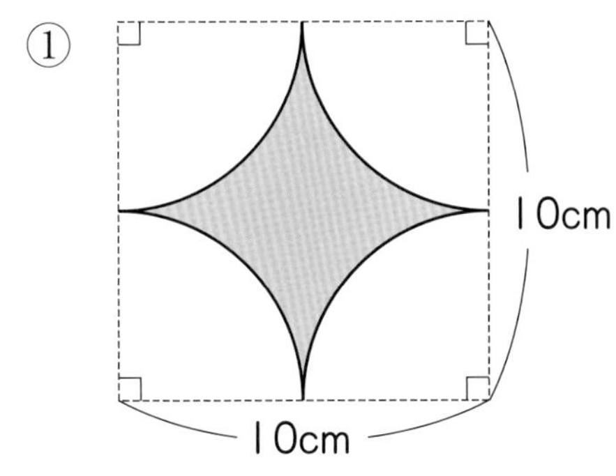

②

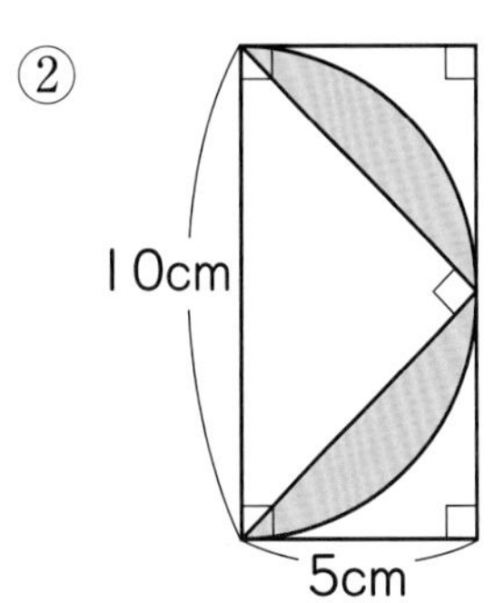

③

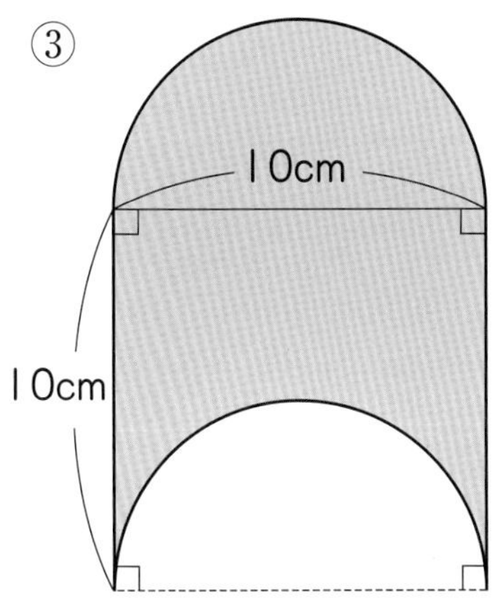

④

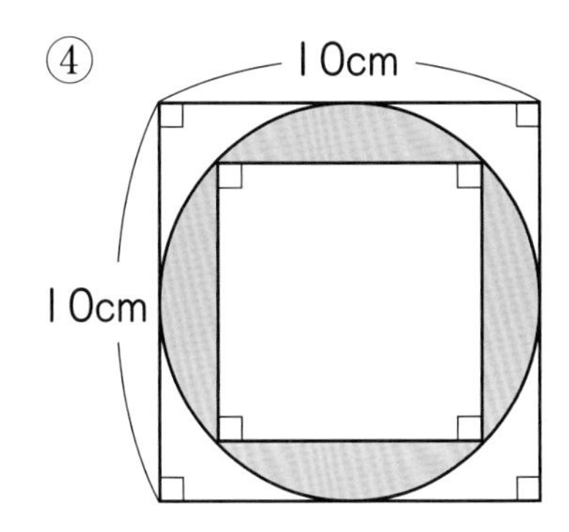

Let's confirm the area of a circle using a rope.

As shown in the picture below, a rope is coiled to make a circle-like figure that has a radius of 5 cm.

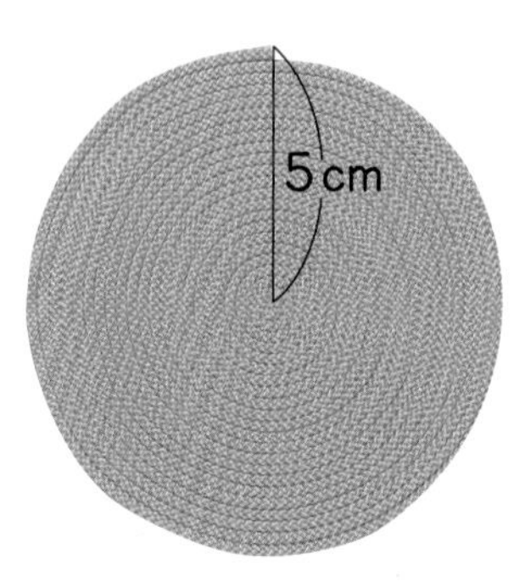

① The following figure was created by spreading the rope after cutting through the radius. Which lengths of the above circle correspond to the lengths of AB and CD?

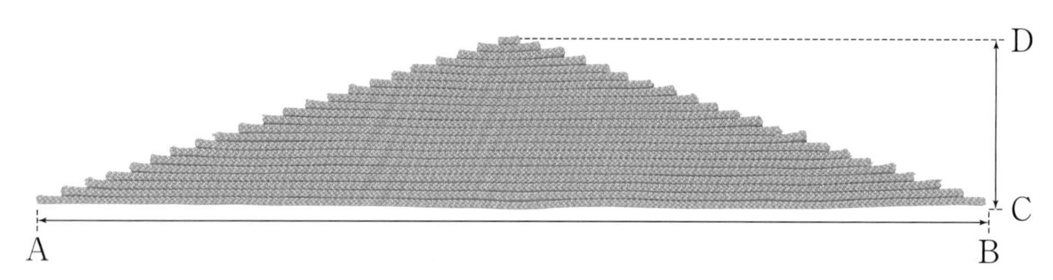

② Manami thought of a formula to find the area of a circle based on the figure in ①. Let's write the words or numbers that apply inside each ☐ to complete Manami's idea. Let's explain it to your classmates.

Area of a triangle = ☐ × ☐ ÷ ☐

↓

Area of a circle = ☐ × 3.14 × ☐ ÷ ☐

↓

= ☐ × 2 × 3.14 × ☐ ÷ ☐

= ☐ × ☐ × 3.14

4 Approximate area

Want to think

1 **As shown on the right, there is a field that lies between rivers. Let's find the approximate area of the field surrounded by the black line.**

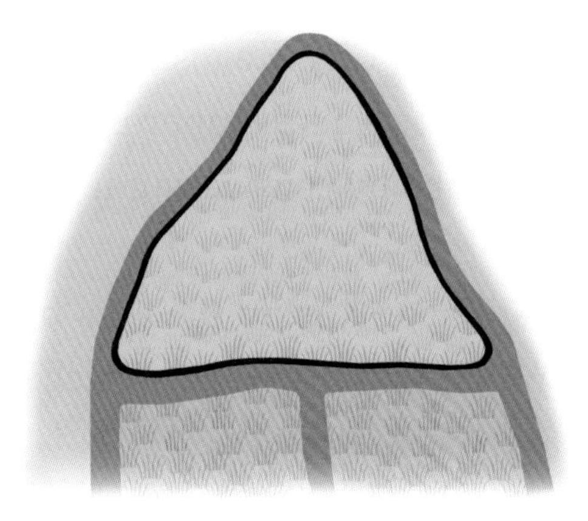

① In the diagram on the right, how many blue and red squares are there?

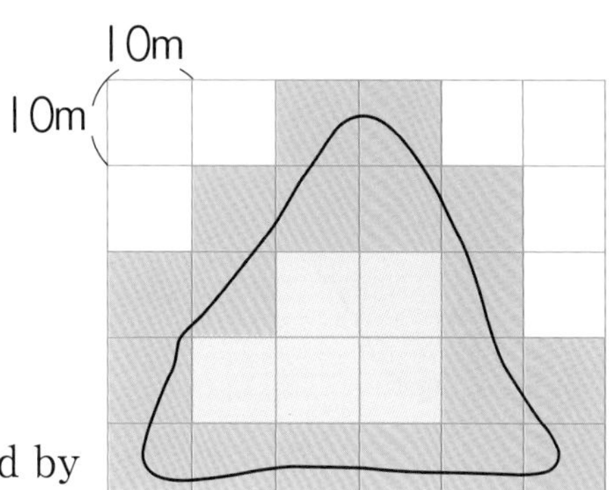

② Let's find the area of the field by considering the area of any 2 red squares where the perimeter passes through as 100 m².

③ Let's find the area of the field shown on the right by considering it as a triangle.

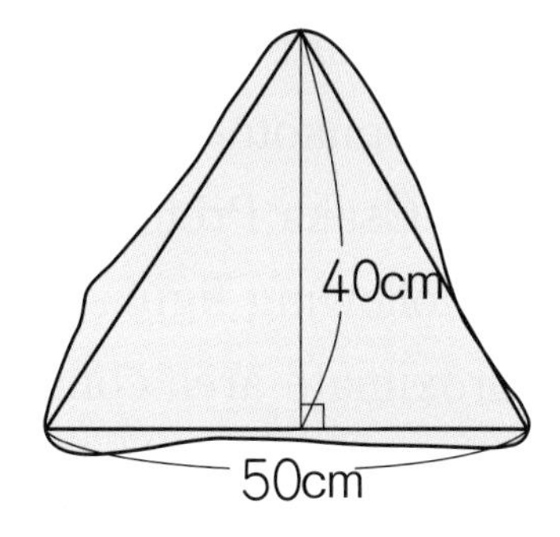

Way to see and think

Can we do the same as when we calculated the approximate area of a circle?

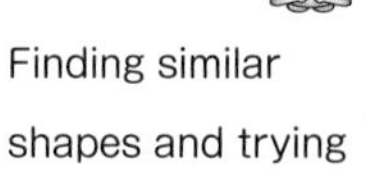

Way to see and think

Finding similar shapes and trying to apply them.

Want to confirm

1 Let's use graph paper to find the approximate area of various leaves.

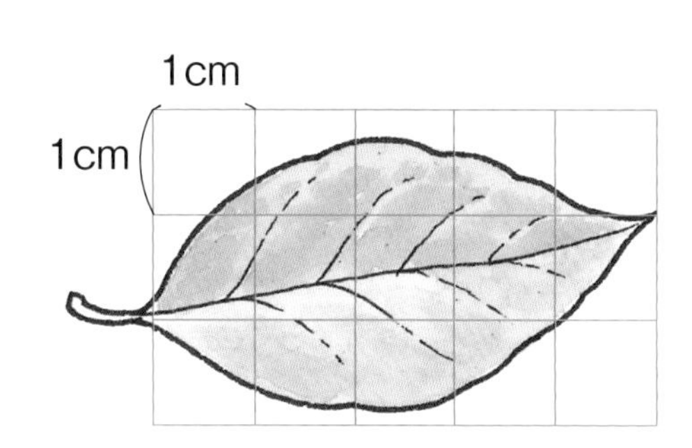

Want to try

2 The lake shown on the right is Lake Ikeda in Ibusuki City, Kagoshima Prefecture. Let's find the approximate area of the lake by considering it as a circle.

Also, let's approximate the area by considering it as a trapezoid. Which is closer to the actual area?

Lake Ikeda (Ibusuki City, Kagoshima Prefecture)

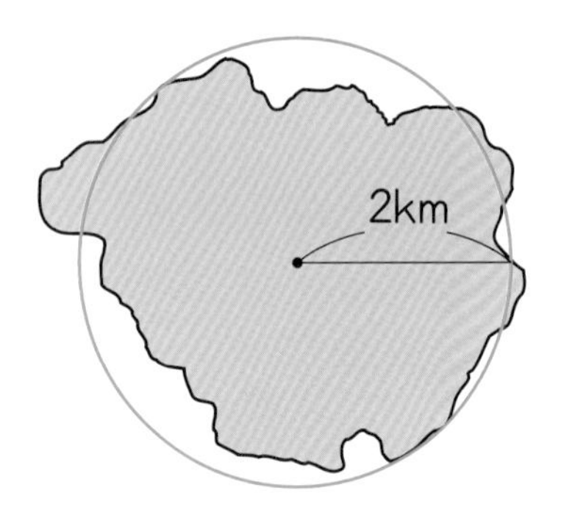

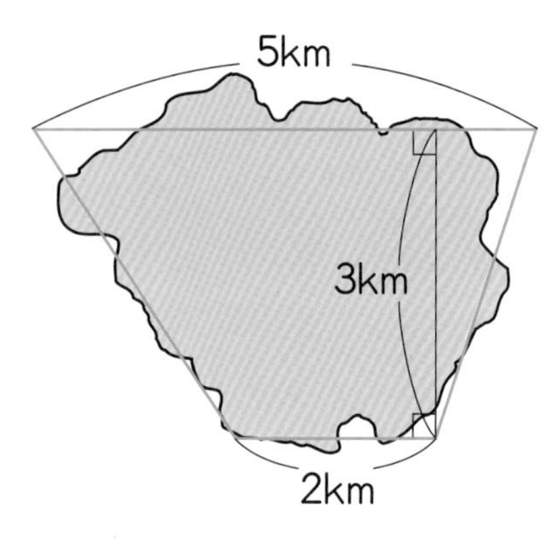

Let's explore the actual area on an encyclopedia or the internet.

3 The mounded tomb on the right is Nakatsuhime-no-mikoto-ryo Kofun in Fujiidera City, Osaka Prefecture. As shown in the following figure, let's find the approximate area considering it a combination of half a circle and a trapezoid.

Nakatsuhime-no-mikoto-ryo Kofun (Fujiidera City, Osaka Prefe

What you can do now

☐ **Can use the area formula of a circle.**

1 Let's find the area of the following diagrams.

①

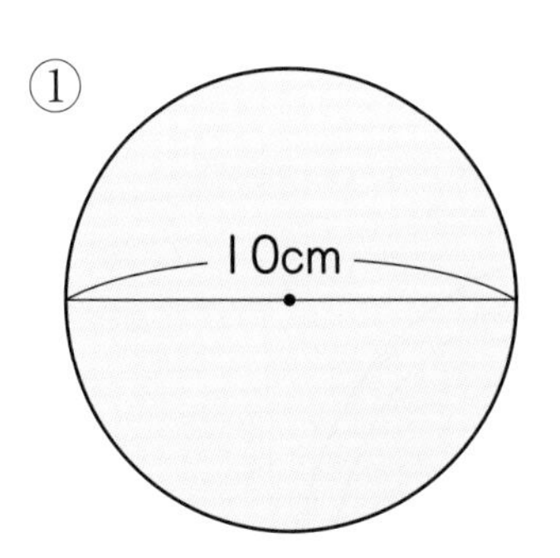

②

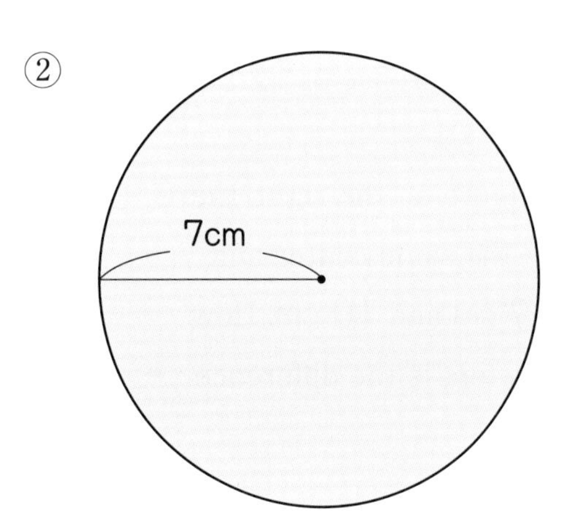

③

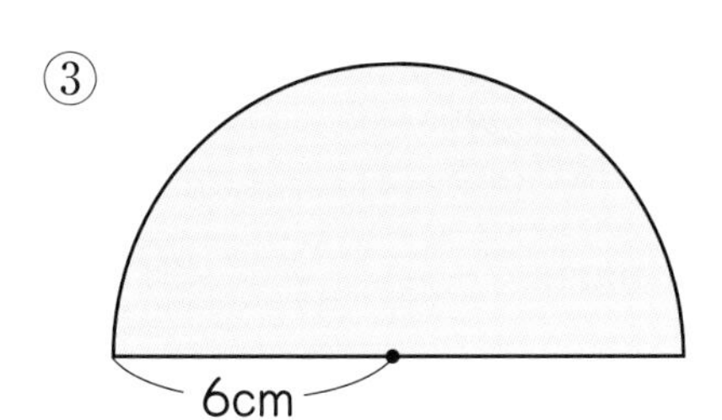

④

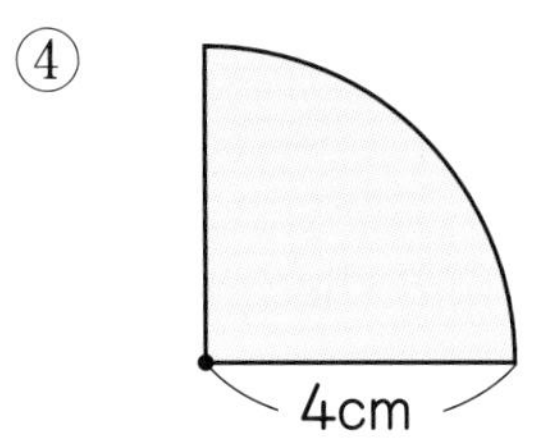

☐ **Can find the area of figures using new ideas.**

2 For the following diagrams, let's find the area of the colored parts.

①

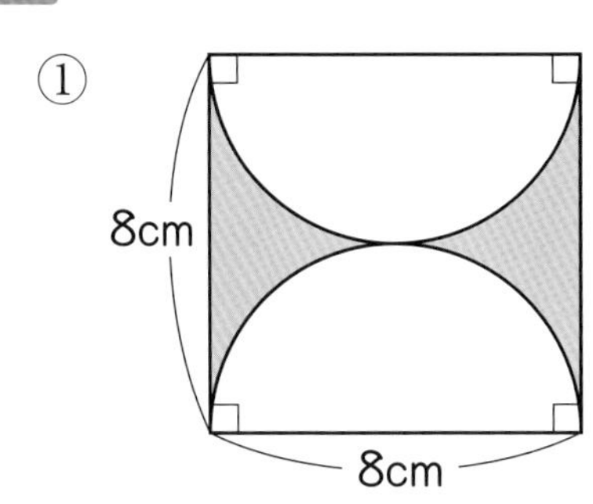

②

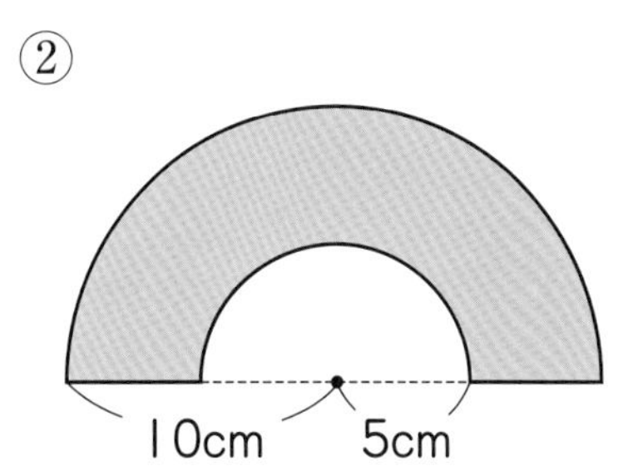

Supplementary problems p.231

Usefulness and efficiency of learning

1 Let's find the diameter and area of the following circles.

① Circle with a circumference of 6.28 cm.

② Circle with a circumference of 12.56 cm.

☐ Can use the area formula of a circle.

2 As shown in the figure on the right, there are two circles with the same center and radius with lenghs 9 cm and 10 cm. How many square centimeters is the difference between the areas?

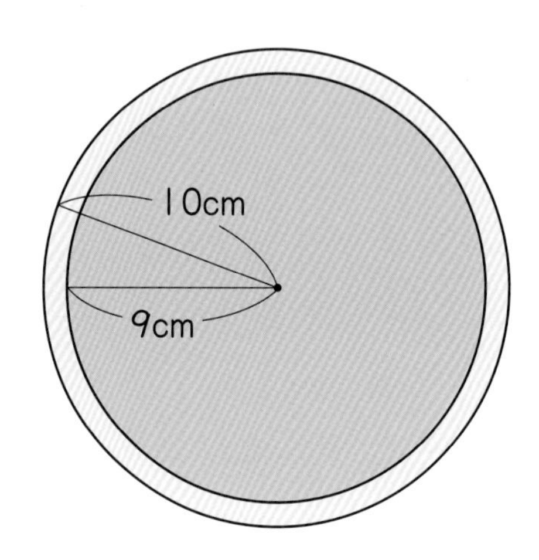

☐ Can use the area formula of a circle.

3 For the following diagrams, let's find the area of the colored parts.

☐ Can find the area of figures using new ideas.

①

10cm
10cm

②

20cm
10cm

③

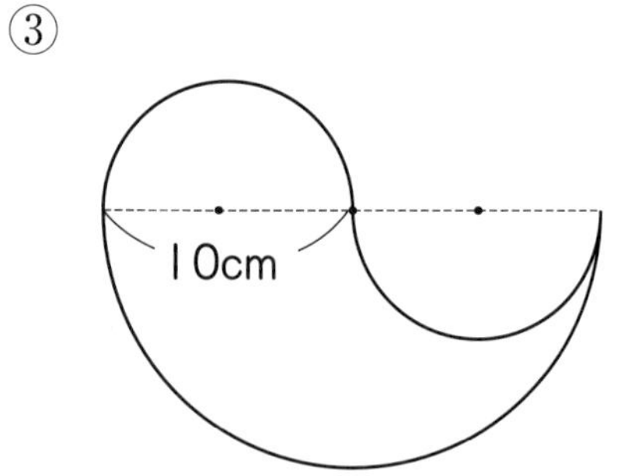

④

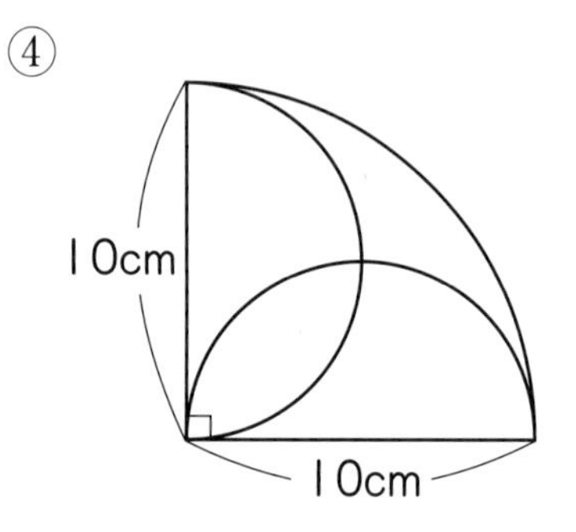

Let's deepen.

I can find the area of figures like ◜ or ◠, but can I find the area of figures like ▽?

Develop in Junior High School

Utilize in life.

Deepen.

Let's find the area of a figure that looks like a hand fan.

Want to express

Let's draw a figure like the one shown on the right.

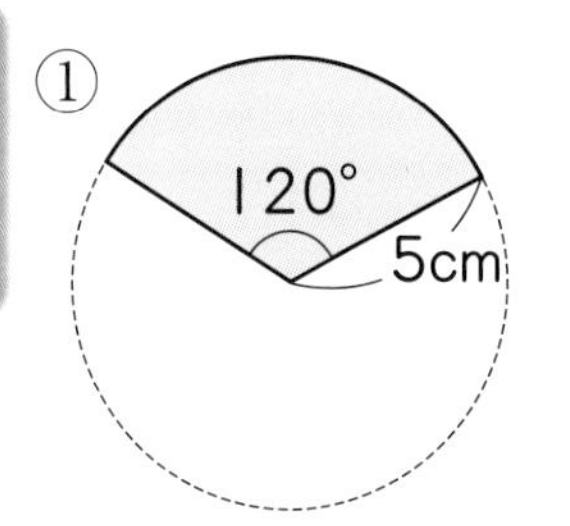

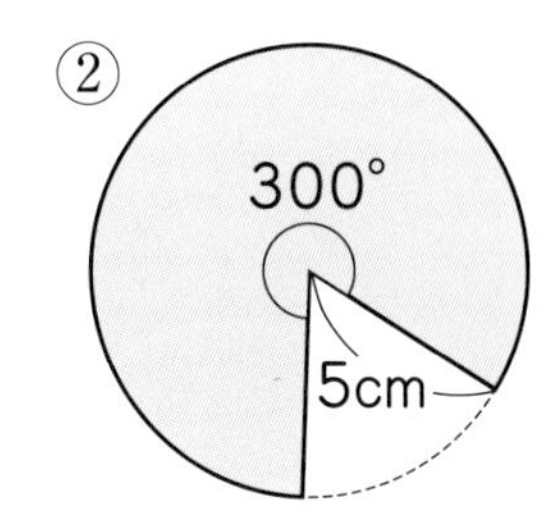

The area of a circle separated by two radii is called a **circular sector**. The angle formed by the two radii is called a **center angle**.

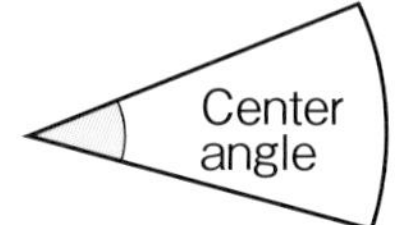

Want to explore

As shown in the figure on the right, the center angles are doubled, tripled, ..., without changing the radius of the circular sector.

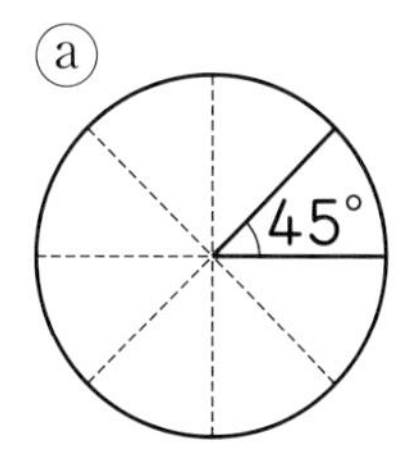

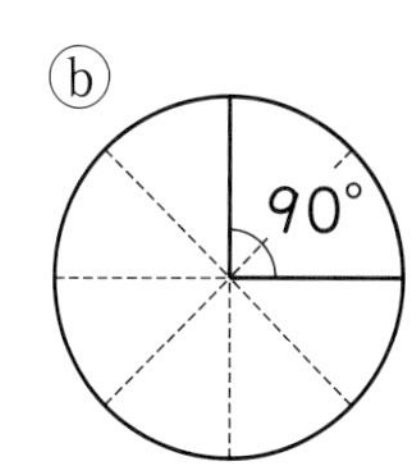

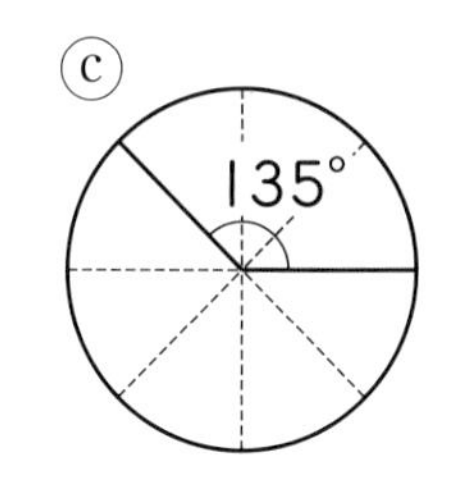

① Let's write in the following table the relationship between the center angle and the area when the area of a circular sector with center angle 45° is x cm².

Center angle (℃)	45	90	135	180
Area (cm²)	x	$x \times 2$		

The area of the circular sector is proportional to the center angle.

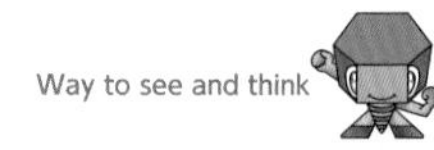

The area of a circular sector with center angle 90° has two parts of the area of a circular sector with center angle 45°.

Want to confirm

② Let's find the area of a circular sector with center angle 45° and radius 4 cm.

Problem

Let's draw the regular dodecagon.

◎ How should you draw it ?

Use a circle to divide the center angle into 12 equal parts.

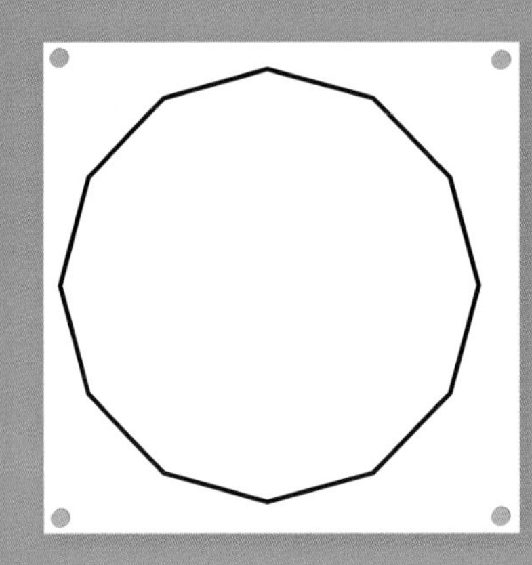

You can think based on one angle of a regular dodecagon.

◎ If you divide the center angle of a circle in 12 equal parts ...

$360 \div 12 = 30$ $\underline{30^\circ}$

It's drawn!

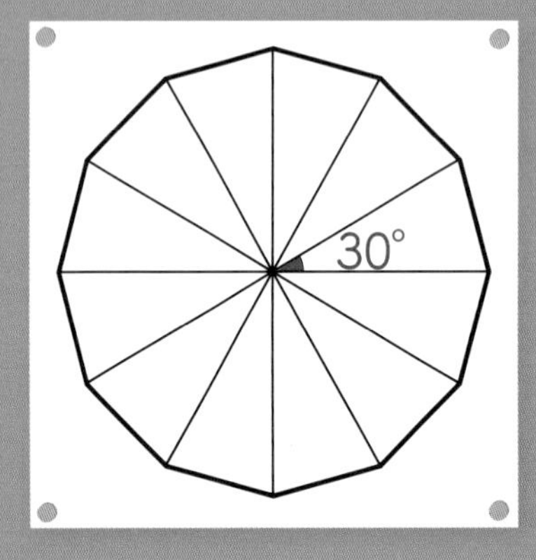

There are a lot of congruent figures.

How many degrees is one angle of the regular dodecagon?

The area is?

How should you draw a regular dodecagon? After drawing it, let's try to think various things about the regular dodecagon.

Using a circle, the center angle can be divided into 12 equal parts.

It can also be drawn considering one angle.

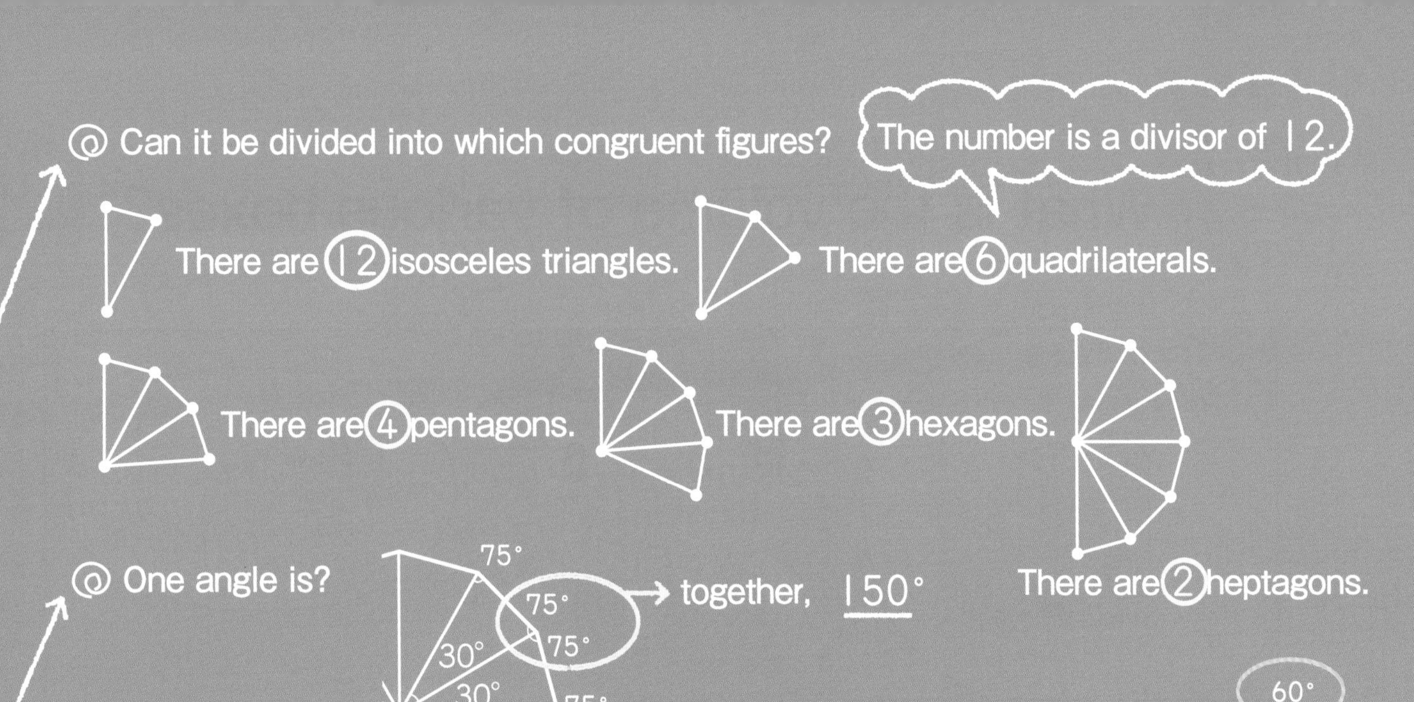

@ What about the area?

If you gather a regular hexagon as shown on the right, you can see the quadrilateral ABCD.
Let's try to think about this quadrilateral ABCD.

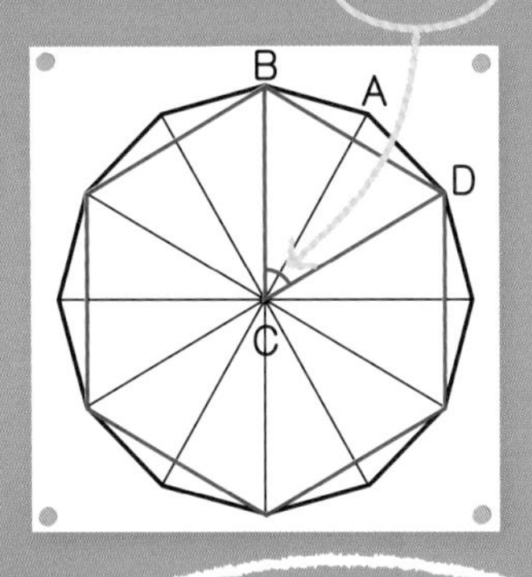

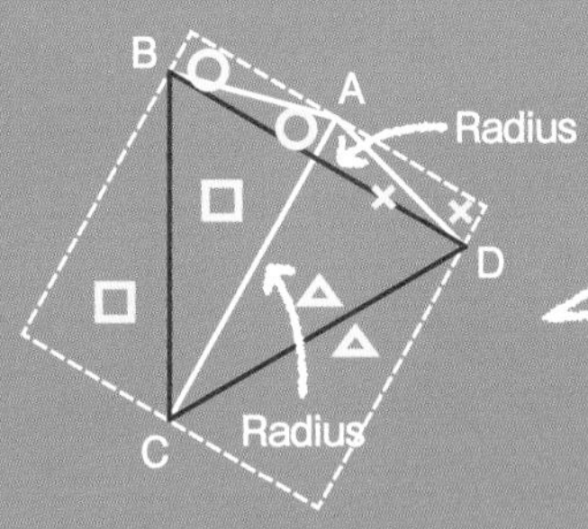

A square can be made using two times the quadrilateral ABCD.

A hexagon is two times the quadrilateral ABCD.

The area of two times the quadrilateral ABCD is radius × radius → The area of the hexagon is radius × radius.

Since it's an equilateral triangle, the length of BD is the same as the radius.

Since there are three hexagons, radius × radius × 3.

Daiki: It can be divided into which congruent figures? How many degrees is one angle?

Nanami: If you know the radius of the original circle, you can easily find the area of the regular dodecagon.

Continue at Junior High School.

Yui: Can I draw a regular dodecagon using a circle, a triangle ruler, and a compass?

Problem Can we find the volume of a figure that is not a cuboid or cube?

Volume of Solids

Let's think about how to find the volume of a solid and its formula.

1 Volume of prisms

Want to know Volume of quadrangular prisms

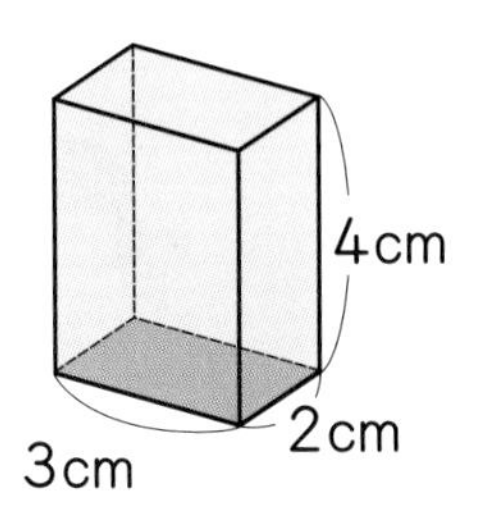

1 **Let's find the volume of a quadrangular prism with a rectangular base as shown on the right.**

Purpose How can we find the volume of a prism?

① Let's find the volume of the quadrangular prism using the volume formula of a cuboid.

Length Width Height Volume

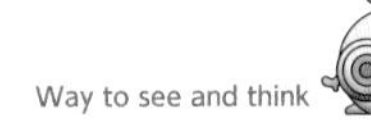

A quadrangular prism with a rectangular base can be seen as a cuboid.

② Let's find the volume of the quadrangular prism with a height of 1 cm.

4th layer
3rd layer
2nd layer
1st layer
1 cm

The bottom area is called the **area of the base.**

③ Let's find the area of the base and compare it to the number that represents the volume of the quadrangular prism with a height of 1 cm.

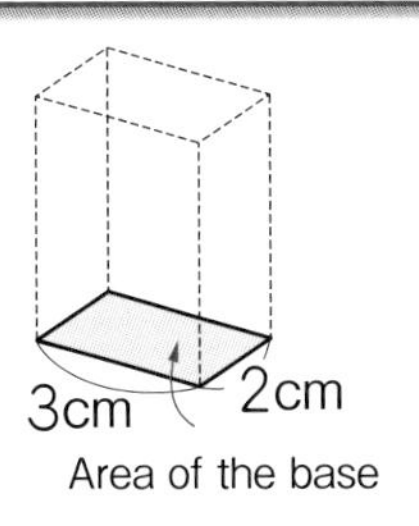

Area of the base

④ Let's find the volume of a quadrangular prism by using the area of the base.

Length Width Height Volume

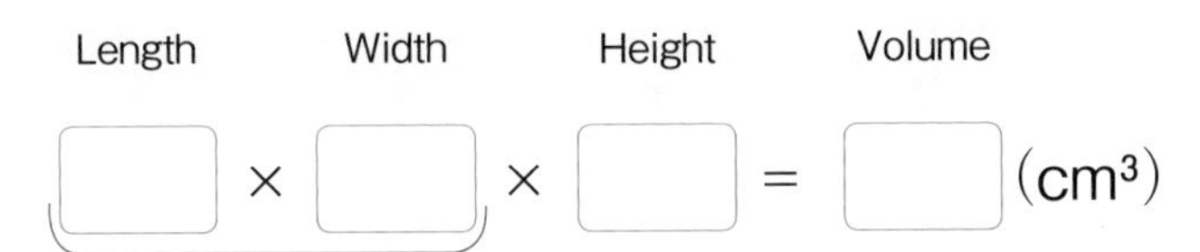

Area of the base

Hiroto

It seems that the volume can be calculated by (area of the base) x (height).

It feels like a quadrangular prism gathers a lot of bases.

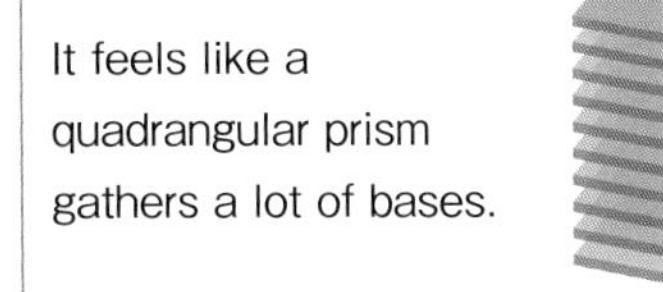

Daiki

Want to think Volume of a triangular prism

Let's think about how to find the volume of a triangular prism as the one shown on the right.

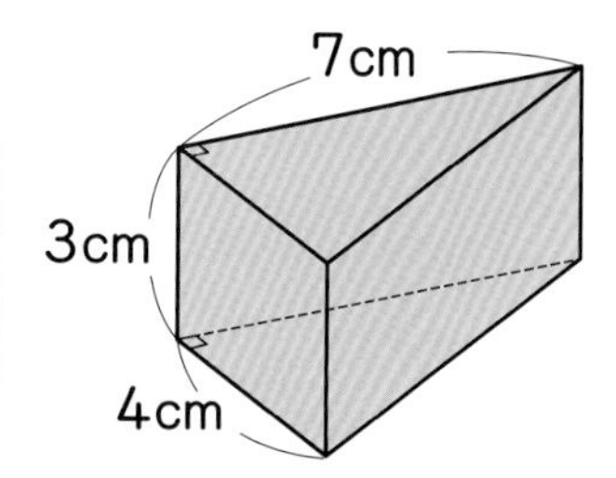

① Let's find the volume of the triangular prism considering it as half of a cuboid.

$7 \times 4 \times 3 \div 2 = \square$ (cm³)

Volume of a cuboid ($7 \times 4 \times 3$)

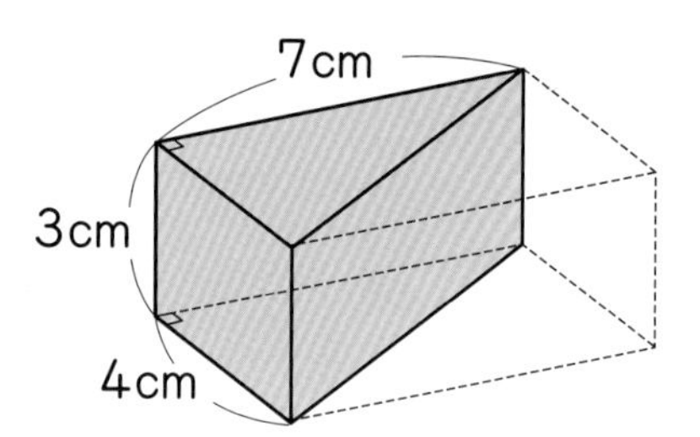

② Let's find the volume of the triangular prism using the area of the base.

$7 \times 4 \div 2 \times 3 = \square$ (cm³)

Area of the base ($7 \times 4 \div 2$)

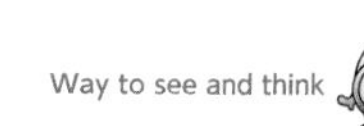

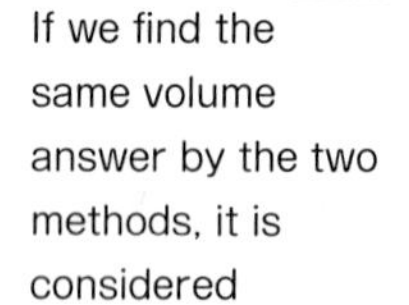

Answer ① is equivalent to answer ②, so the volume of the triangular prism can also be found by the formula (area of the base) × (height).

Summary

The volume of all prisms can be calculated as follows:

Volume of a prism = area of the base × height

Want to confirm

There is a prism with a rhomboidal base as shown on the right. Let's find its volume.

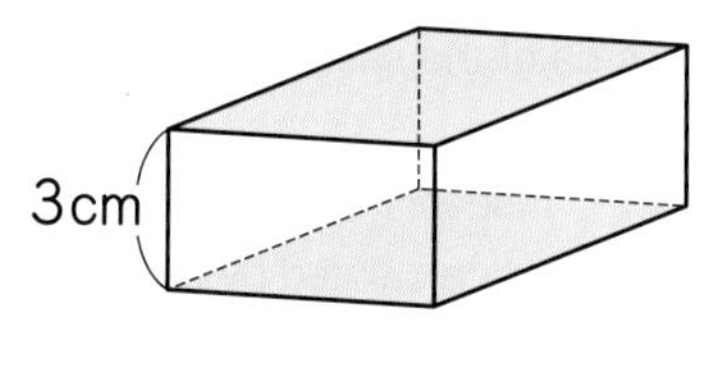

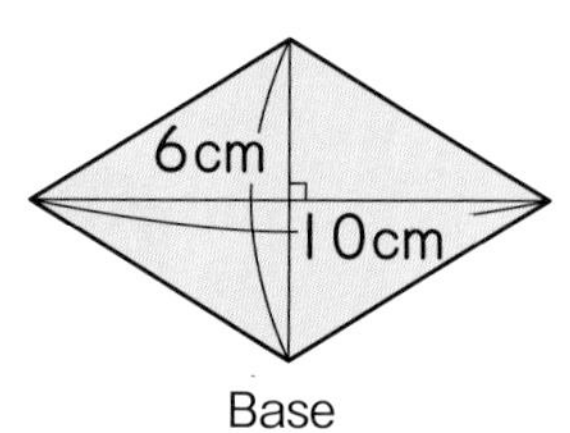

Base

2 Volume of cylinders

Want to know Volume of cylinders

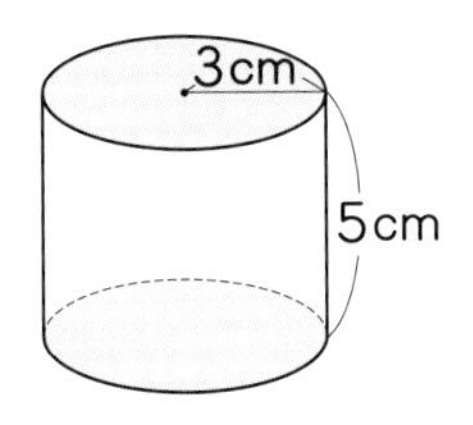

1 **Let's think about how to find the volume of a cylinder as the one shown on the right.**

Area of a circle= radius × radius × 3.14

Purpose Can the volume of a cylinder also be calculated by the formula (area of the base) × (height)?

① Let's think about how to find the volume of a cylinder by looking at the following diagrams.

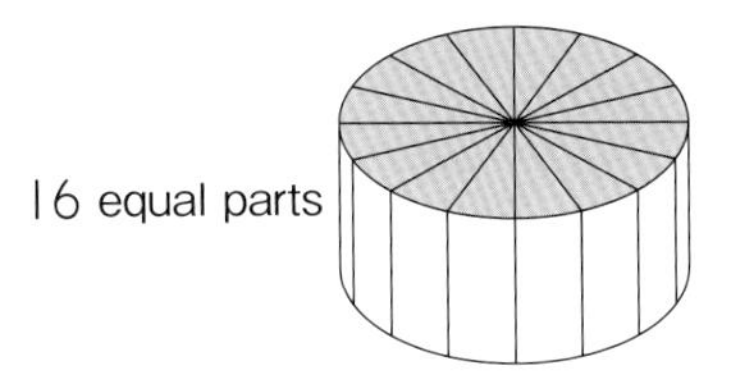

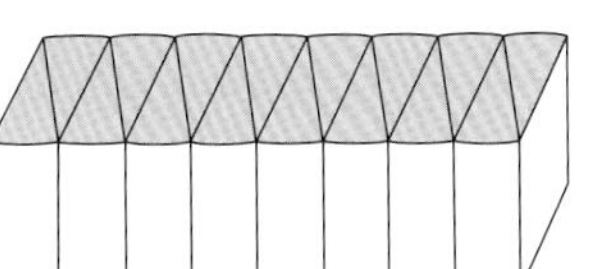

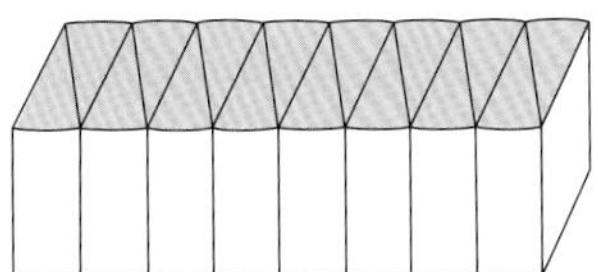

Way to see and think

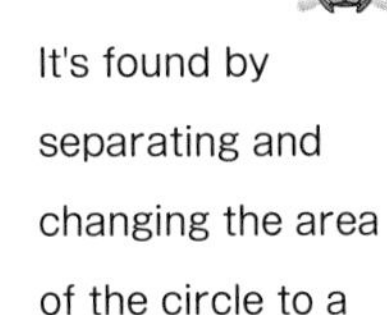

It's found by separating and changing the area of the circle to a rectangular area.

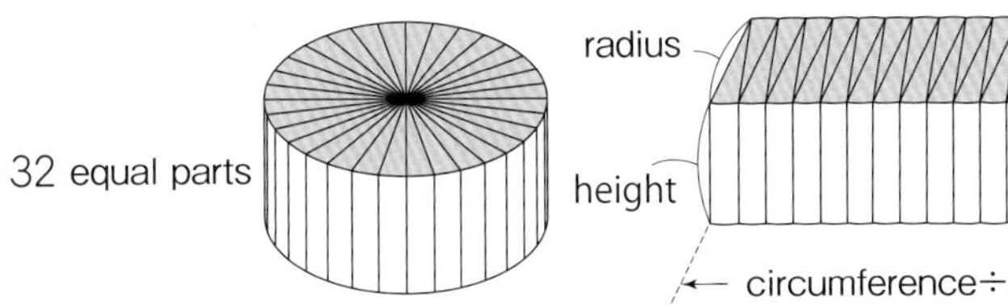

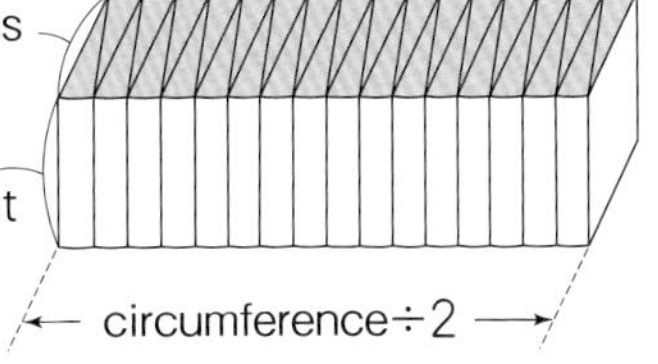

$(3 \times 3 \times 2 \times 3.14 \div 2) \times 5 = \square$ (cm³)

Radius | Half the length of the circumference | Height | Volume

② Let's find the area of the base of the cylinder.

③ Let's find the volume of the cylinder in the same way as the prism.

$(3 \times 3 \times 3.14) \times 5 = \square$ (cm³)

Area of the base | Height | Volume

Want to compare

Way to see and think

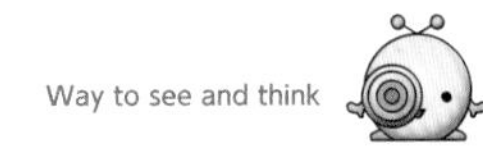

Tried to compare two methods to find the volume.

④ Let's explain what became equal in math sentence ① and ③. Also, what can be said from this?

Summary

The volume of a cylinder can be calculated as follows:

Volume of a cylinder = area of the base × height

Want to confirm

Let's find the volume of the following solids.

From what we have learned so far, the volume of a prism or cylinder can be calculated by (area of the base) × (height).

① 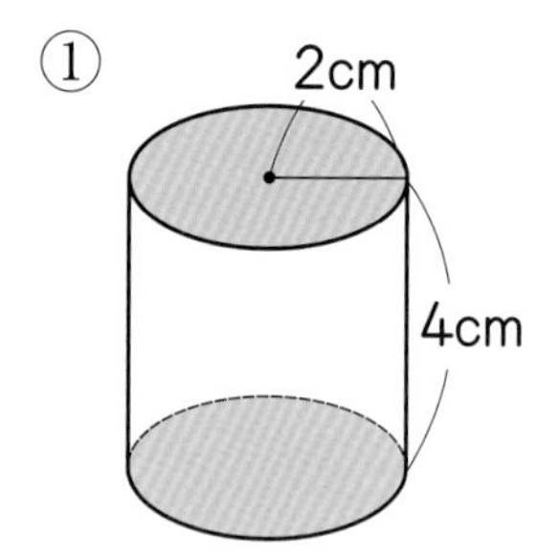

② 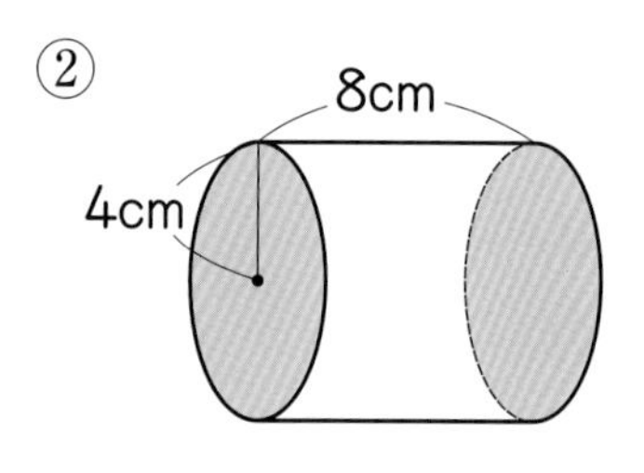

Want to try

Let's find the volume of the following solids.

① 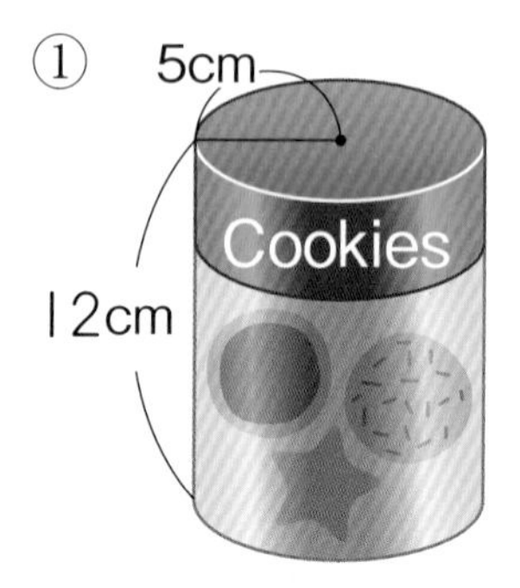

② 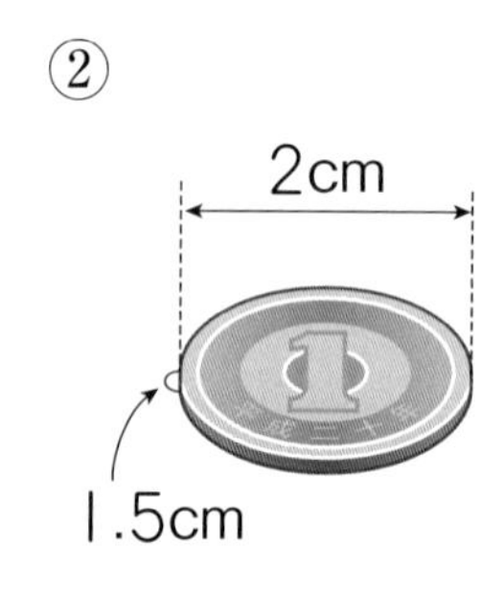

③ 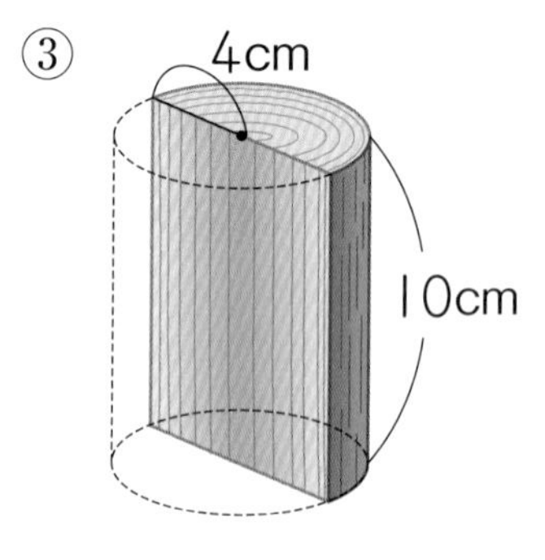

There is a drink mat in the shape of a circle with a diameter of 10 cm and a thickness of 0.8 cm as shown on the right.

Ten of these mats are stacked into a cylindrical shape. Let's find the volume of this cylinder.

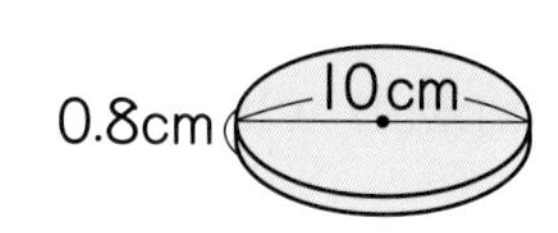

That's it

Compare the volume of various solids

The figures on the right are called pyramids and a cone. There are pyramids with polygonal bases such as a pentagon or so on.

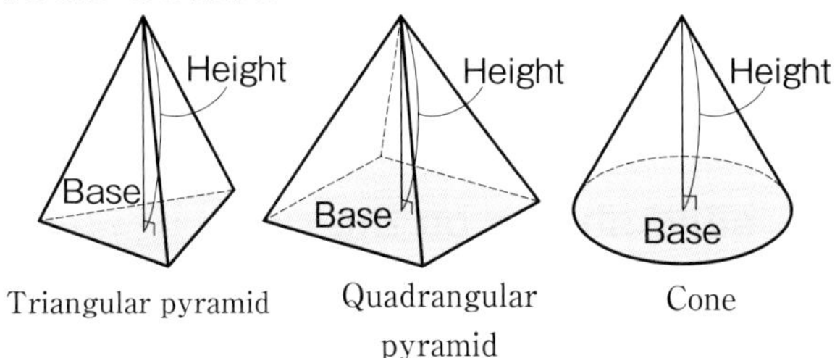

① Let's compare the volume of a quadrangular pyramid and a quadrangular prism when both have the same height and the same squared base.

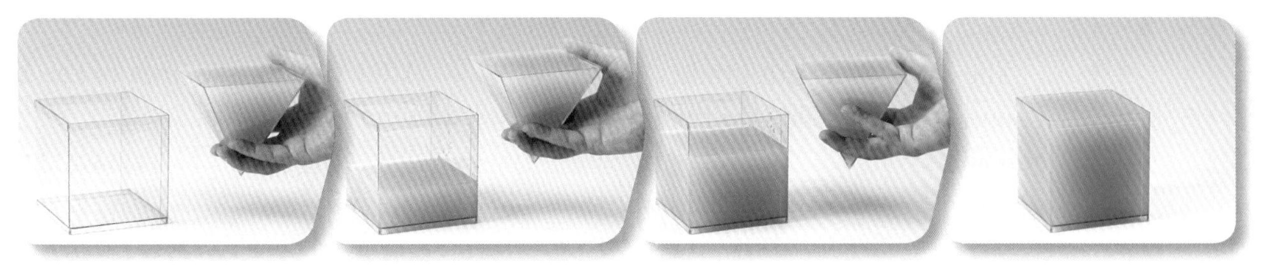

If 1 cup is poured ... If 2 cups are poured ... If 3 cups are poured ...

② Let's compare the volume of a cone and a cylinder when both have the same height and the same base.

If 1 cup is poured ... If 2 cups are poured ... If 3 cups are poured ...

③ What do you understand from the above experiments?

④ Tadashi considered the following formula to calculate the volume of a pyramid or cone.

Let's write the number that applies inside the □. Let's discuss why this math sentence became as it is.

Volume of a pyramid or cone
$= \text{area of the base} \times \text{height} \times \frac{1}{\square}$

Three times the water in the pyramid or cone fits right into a prism or cylinder ...

01604

3 Utilize the formula to find the volume

Want to think

Activity

1 **Let's find the volume of the solid shown on the right.**

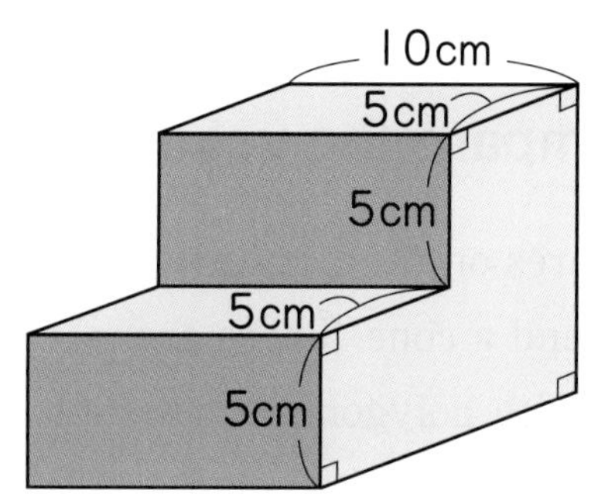

Purpose Can we also find the volume of the solid shown above using the volume formula?

Daiki's idea

The same as in 5th grade, I separated the solid into two cuboids.

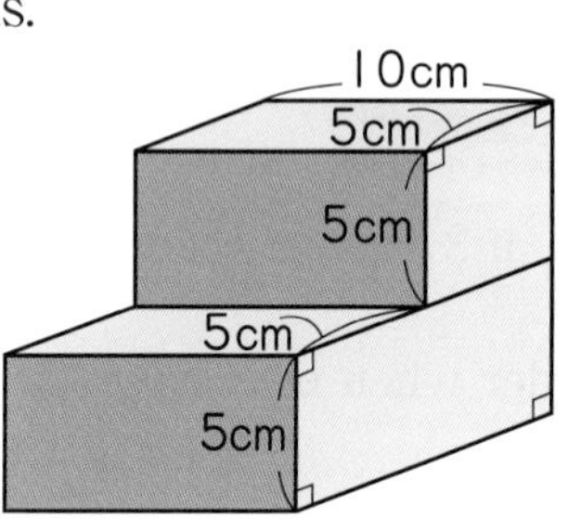

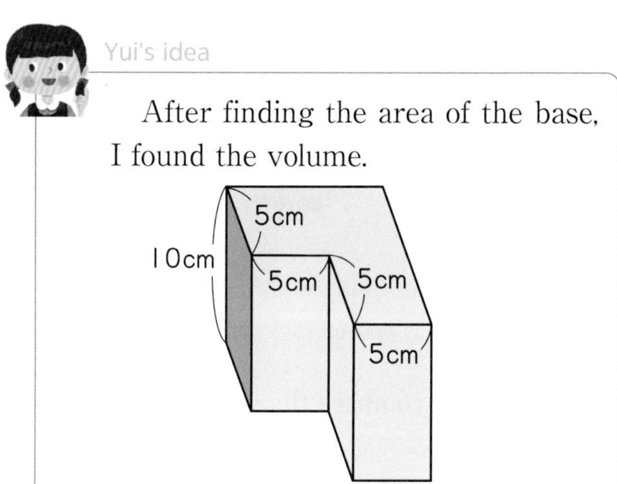

① Let's find the volume using Daiki's idea.

② Let's find the volume using Yui's idea.

Summary

The volume of the above solid, viewed as a prism, can also be found using the formula (area of the base) × (height).

Want to confirm

1 Let's find the volume of the following solids.

①

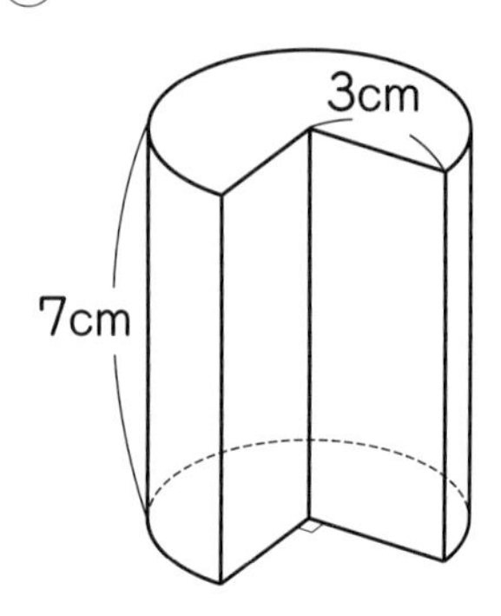

② There is a cylindrical hole in the middle.

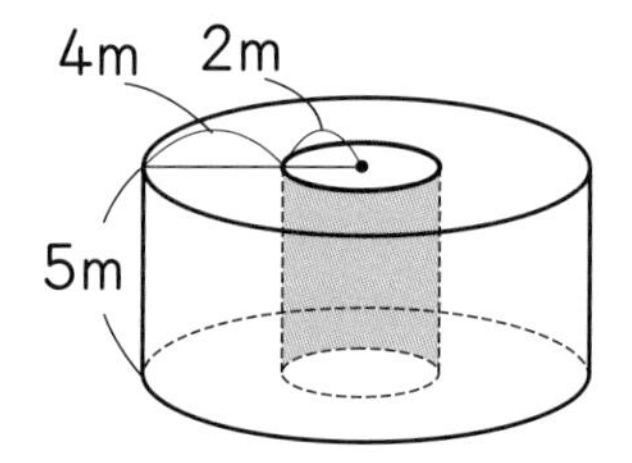

③

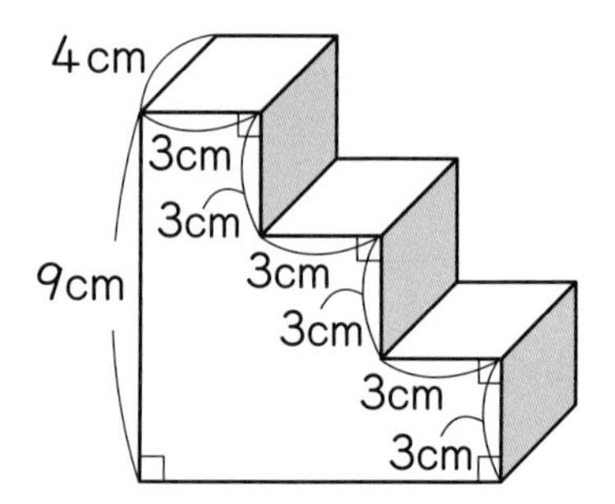

4 Approximate volume

Want to know

1 **Let's find the approximate capacity of the clothes case shown on the right.**

39cm
29cm
30cm
62cm

① What shape does the clothes case look like?

② Let's consider the clothes case as a quadrangular prism with a trapezoidal base, and calculate the approximate volume.

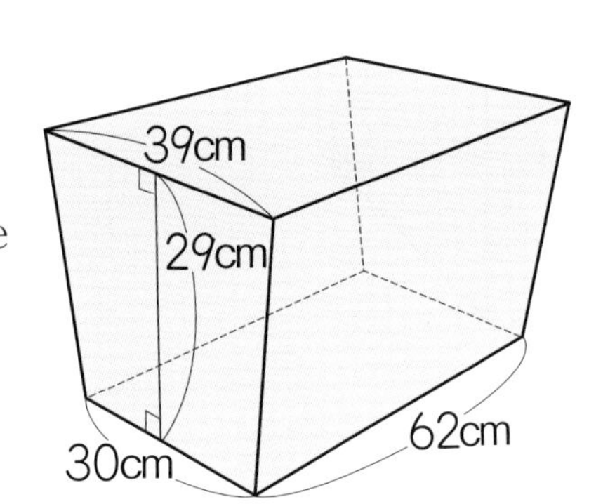

The inner volume of objects like clothes cases is called the capacity.

Want to confirm

There is a cup like the one shown on the right. Let's look at it as a cylinder and find the approximate volume.

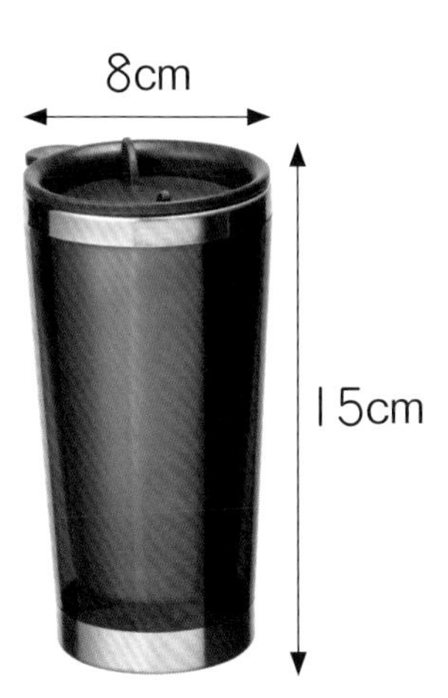

Want to find in our life

Let's find the approximate volume and capacity of various things in your surroundings.

What you can do now

Can find the volume of a prism and a cylinder.

1 Let's find the volume of the following prisms and cylinder.

①

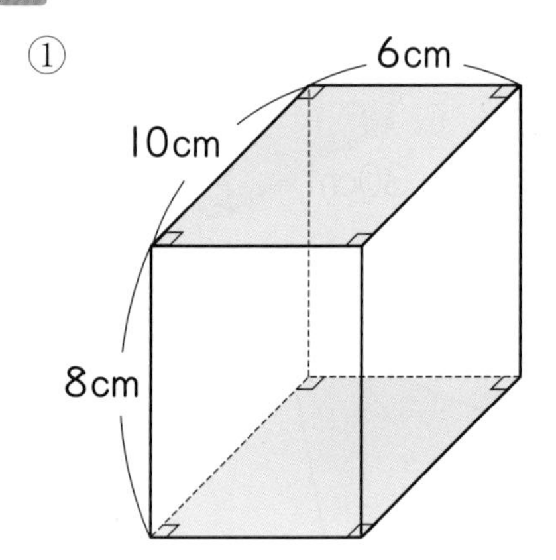

②

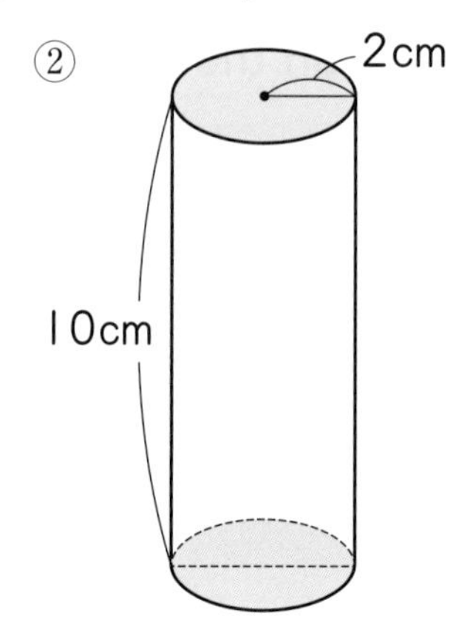

③

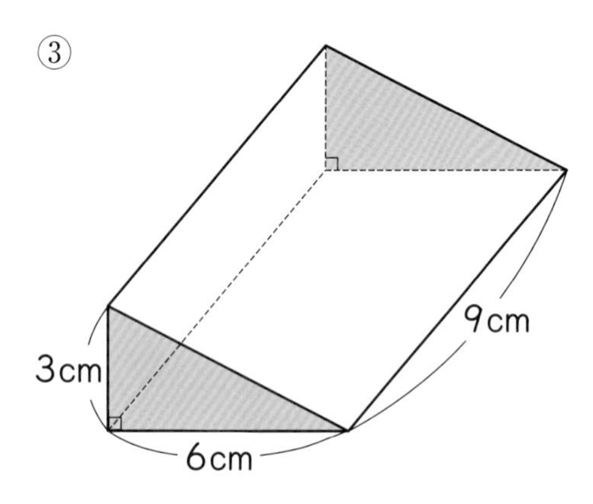

④

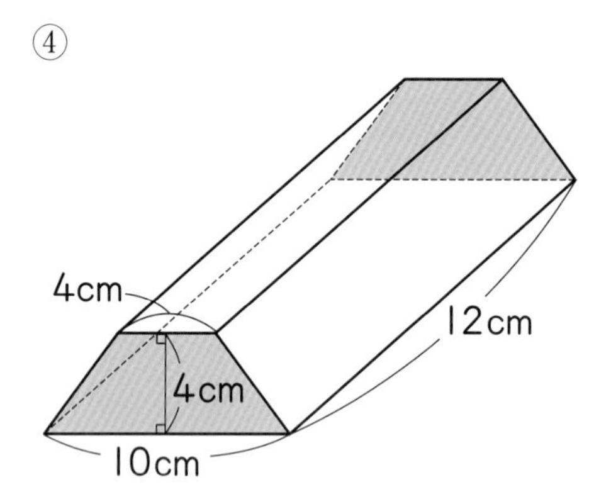

Can utilize the formulas to find the volume.

2 As shown in the figure on the right, there is a solid in which the faces intersect perpendicularly. Let's find the volume of this solid.

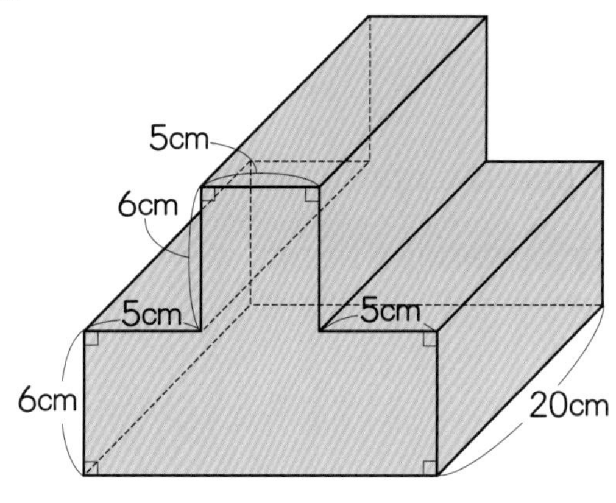

Supplementary problems p.232

Usefulness and efficiency of learning

1 Let's find the volume of the following prisms and cylinder.

☐ Can find the volume of a prism and a cylinder.

① 4cm 8cm 3cm 10cm

② Cylinder

0.5m 1cm

③

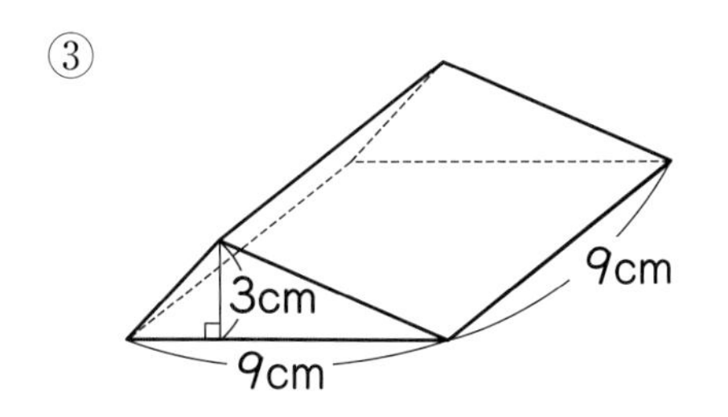

④

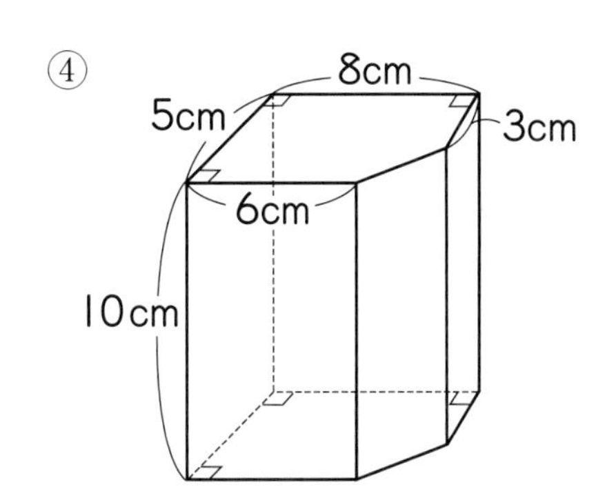

2 As shown in the figure on the right, there is a solid in which the faces intersect perpendicularly. Let's find the volume of this solid.

☐ Can utilize the formulas to find the volume.

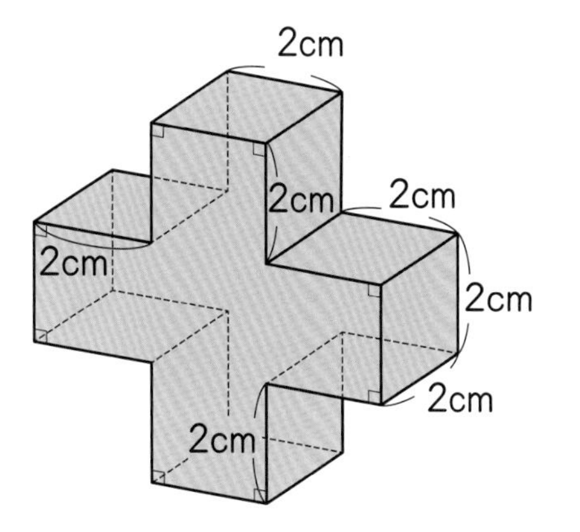

Develop in Junior High School

Problem

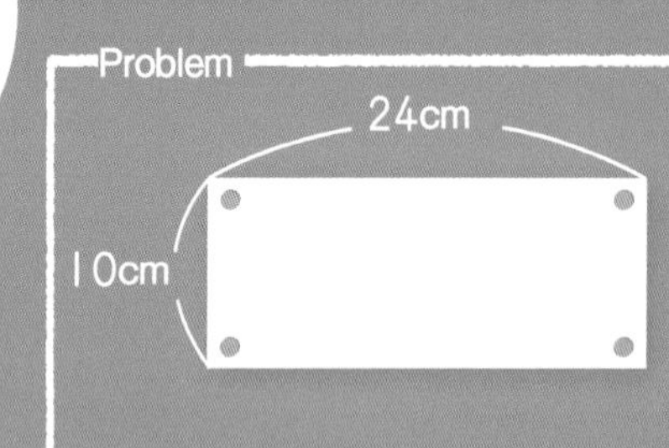

A prism or cylinder can be created using the rectangle shown on the left by folding or rolling the 10cm sides until they overlap. Let's make the largest volume prism or cylinder following this method.

Is the volume the same because the lateral area is the same?

@ What kind of solid can you make?

Triangular prism, quadrangular prism, cylinder....

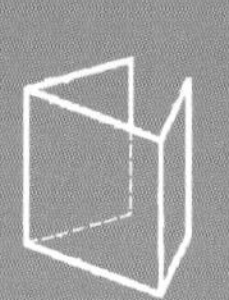

Quadrangular prism

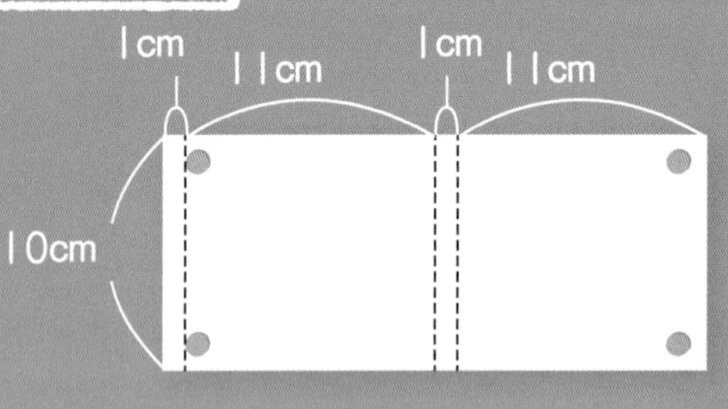

$1 \times 11 \times 10 = 110$

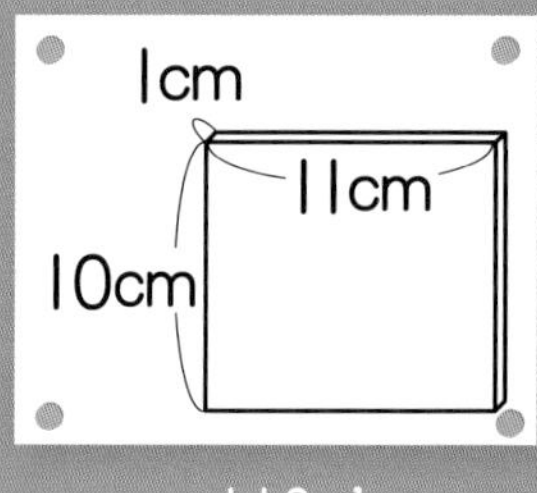

110 cm²

The volume increases as the shape of the base approaches a square.

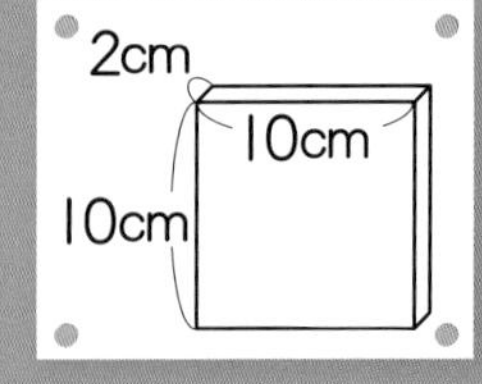

$2 \times 10 \times 10 = 200$ — 200cm³ (Area of the base)

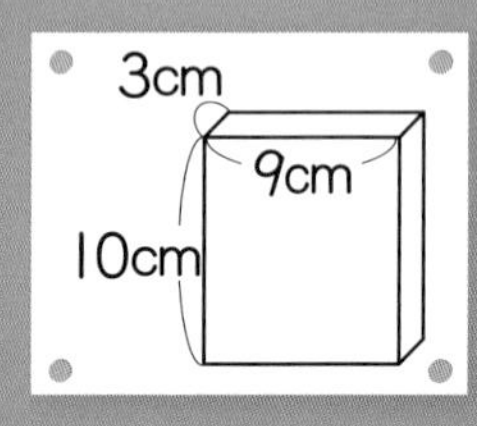

$3 \times 9 \times 10 = 270$ — 270cm³ (Area of the base)

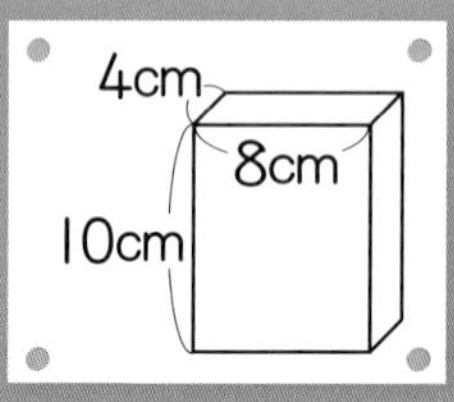

$4 \times 8 \times 10 = 320$ — 320cm³ (Area of the base)

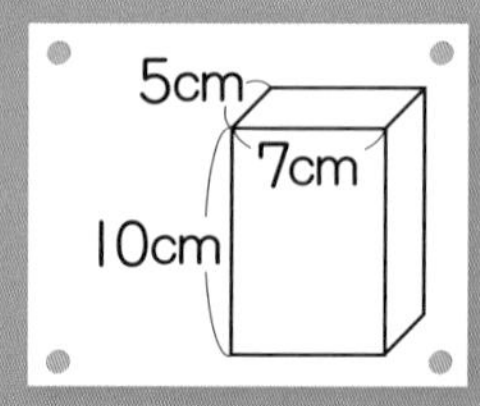

$5 \times 7 \times 10 = 350$ — 350cm³ (Area of the base)

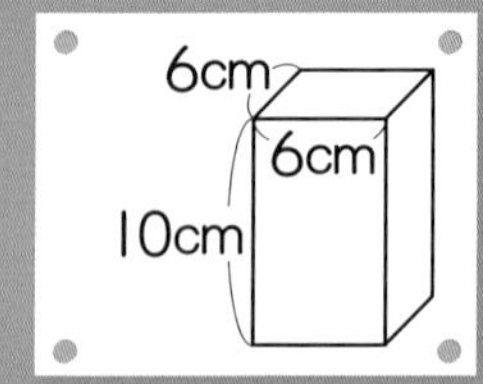

$6 \times 6 \times 10 = 360$ — 360cm³ (Area of the base)

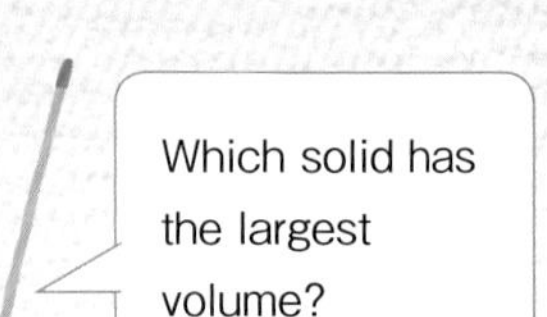

Triangular prism I can do a lot, but ...

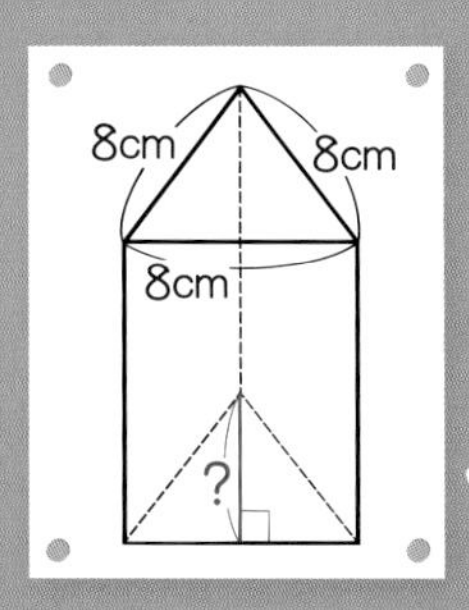

For example, if the shape of the base is an equilateral triangle, the height of the base is unknown, so the area cannot be found.

I can find the area if it's a right triangle.

- Let's make the triangular base a 6cm, 8cm, and 10cm triangle.

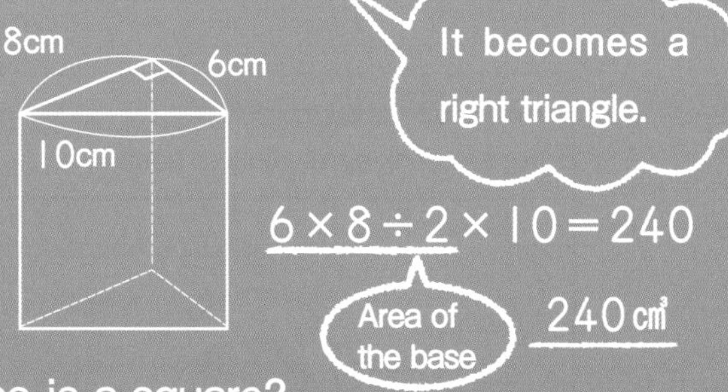

$6 \times 8 \div 2 \times 10 = 240$

Area of the base

240 cm³

Is the volume the largest when the shape of the base is a square?

↓

What happens if the shape of the base is a circle?

Cylinder

The width of the rectangle = Circumference = 24cm

If the radius is x cm

$x \times 2 \times 3.14 = 24$

$x = 24 \div 6.28 = 3.82 \ldots$

Considering a radius of about 3.8cm,

$3.8 \times 3.8 \times 3.14 \times 10 = 453.416$ About 453.416cm³

Area of the base

You may think based on the area of the base.

If the length around the base is the same, I guess the area of the circle will be the largest.

Summary

- Even if the lateral area is the same, the volume is different.
- When the height is the same, the volume of the solid with the largest area of the base is the largest.

Can we make the same taste?

How can we have the same taste even if the cooking quantity changes?

Ratio and its Application

Let's think about how to express by ratio and how to use it.

Making a salad, soup, and rice,

the quantity of ingredients is as follows:

French dressing (for 1 person)

Vinegar ... 2 teaspoons

Salad oil ... 3 teaspoons

Soup (for 1 person)

Water ... 4 cups

Japanese sauce ... 1 cup

Rice (for 1 person)

Rice ... 300 mL

Water ... 360 mL

1 Ratio and value of ratio

Want to explain

1 **Let's try to explain the quantity of ingredients for each of the above dishes using the representing methods for ratios learned so far.**

In soup, we need 4 times more water than Japanese sauce.

Way to see and think

The representing method for a ratio is:

- A is □ times B
- If A is 1, B is □
- If A is 100%, B is □%.

Want to think

2 Let's consider the ratio of vinegar and salad oil when making a French dressing.

French dressing (for 1 person)

Vinegar ... 2 teaspoons

Salad oil ... 3 teaspoons

Teaspoons

Vinegar	Salad oil
(2 spoons)	(3 spoons)

① How can the ratio between the quantity of vinegar and the quantity of salad oil be represented using two numbers?

"When the quantity of vinegar is 2, then the quantity of salad oil is 3" can be represented by using the " : " symbol as follows,

2 : 3 is read as **"two is to three."** This way of representation is called **ratio.**

2 : 3 is also read "ratio of 2 to 3."

② How many times the quantity of the salad oil is that of the vinegar? Let's represent it with a fraction.

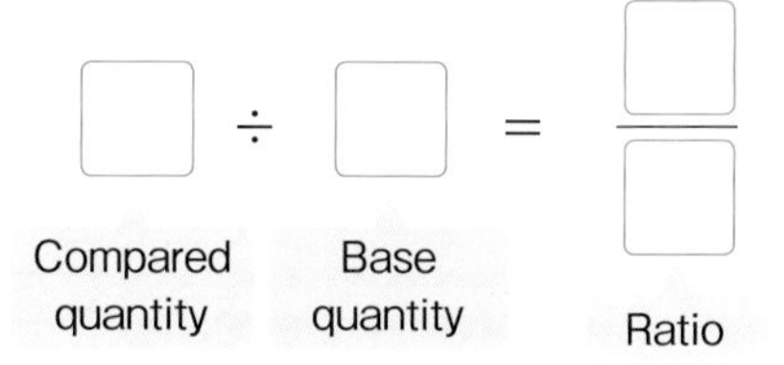

When a ratio is represented as $a : b$, the quotient of a divided by b is called the **value of ratio**. The value of ratio represents how many times b is a.

Value of ratio $a : b$ is the quotient of $a \div b$.

In particular, when a and b are whole numbers, the value of ratio $a : b$ can be represented as $\frac{a}{b}$.

Want to confirm

Let's represent the following ratios as ratio and value of ratio.

① 3 tablespoons vinegar and 5 tablespoons salad oil.

② Tomato ketchup 36 g and mayonnaise 43 g.

2 Equal ratio

Want to think

1 Let's think about the ratio of water to Japanese sauce when making soup.

Soup (for 1 person)
Water ... 4 cups
Japanese sauce... 1 cup

① Hiroto made a 1 person serving using a small cup as shown below. Let's find the value of the ratio.

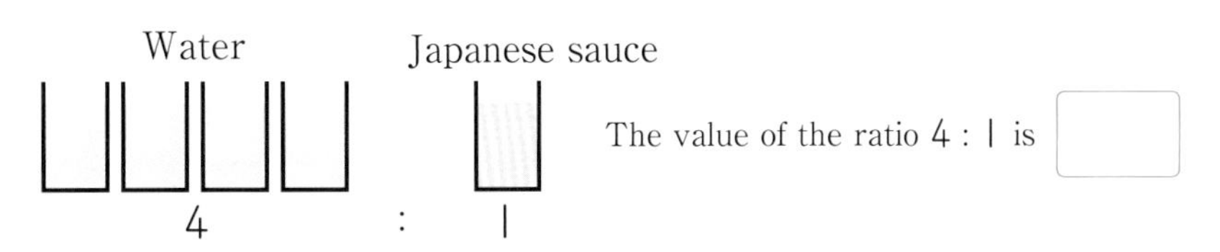

The value of the ratio 4 : 1 is ☐

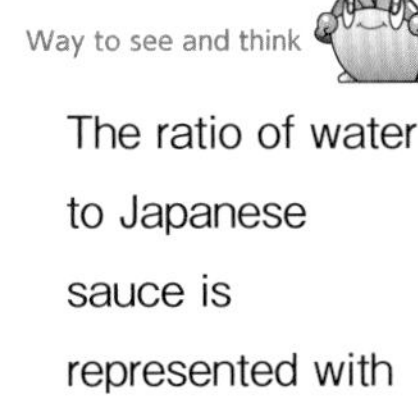

Way to see and think

The ratio of water to Japanese sauce is represented with cup units of the same size.

② Nanami made a 2 people serving with the same cup as Hiroto. Let's find the value of the ratio.

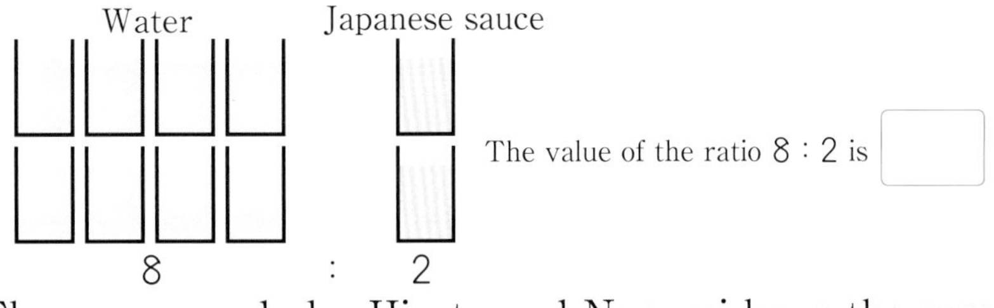

The value of the ratio 8 : 2 is ☐

The value of the ratio is related to the concentration.

Yui

③ The soups made by Hiroto and Nanami have the same concentration?

When the value of the ratio is equal, such as 4 : 1 and 8 : 2, the **two ratios** are said to **be equal**, and are written as follows:

$$4 : 1 = 8 : 2$$

Want to confirm

1 From the following, which have an equal concentration of coffee and milk?

Ⓐ 90 mL of coffee and 60 mL of milk.

Ⓑ 600 mL of coffee and 200 mL of milk.

Ⓒ 3 cups of coffee and 2 cups of milk.

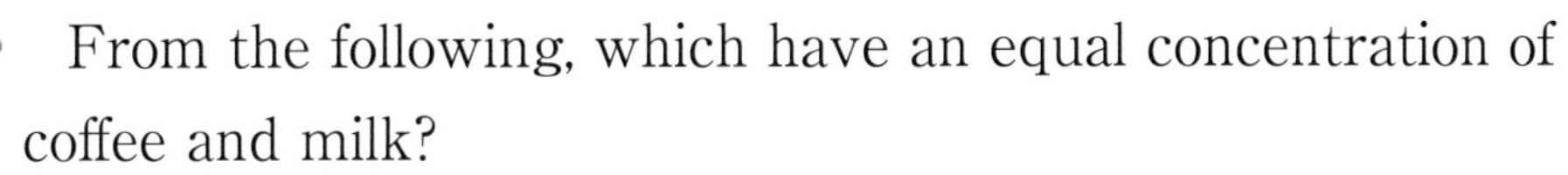

Way to see and think

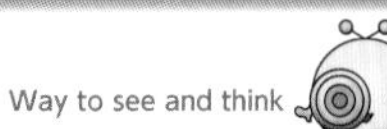

You can use the value of ratio when explaining that two quantities have the same concentration.

2 **There are two combinations of rice and water on the right. Let's examine the relationship between the ratios of rice to water.**

Ⓐ Rice ···60 mL
Water ···72 mL

Ⓑ Rice ···300 mL
Water ··· 360 mL

① From the ratios of rice to water in Ⓐ and Ⓑ, let's find the value of ratio based on the water.

Value of ratio Ⓐ is $\frac{\square}{\square}$ Value of ratio Ⓑ is $\frac{\square}{\square}$.

Purpose As for two equal ratios, what relationship is there?

② Let's explore the relationship between two equal ratios.

$60 : 72 = 300 : 360$

$60 : 72 = (60 \times \square) : (72 \times \square)$

$= 300 : 360$

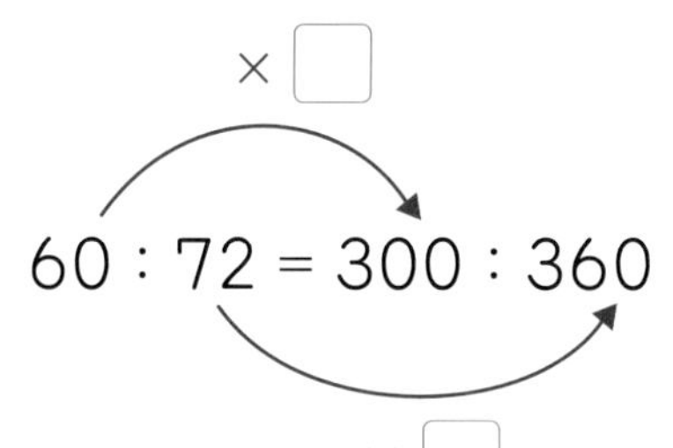

③ On the other hand,

$300 : 360 = 60 : 72$

$300 : 360 = (300 \div \square) : (360 \div \square)$

$= 60 : 72$

÷ □
300 : 360 = 60 : 72
÷ □

Summary

The ratio $a : b$ is equal to other ratio that is created by multiplying or dividing a and b by the same number.

Want to confirm

2 From the following ratios, which is equal to $3 : 1$?

① $6 : 3$ ② $6 : 2$ ③ $1 : 3$ ④ $13 : 10$ ⑤ $9 : 3$

Let's write three ratios that are equal to $6 : 9$.

Want to know How to find an equal ratio

3 **To make a drink, 120 mL of water and 30 mL of pure beverage must be mixed. If you have 60 mL of pure beverage, how many milliliters of water should be added to prepare a drink with an equal concentration?**

If you need x mL of water

$120 : 30 = x : 60$

$x = 120 \times \square$

$= \square$

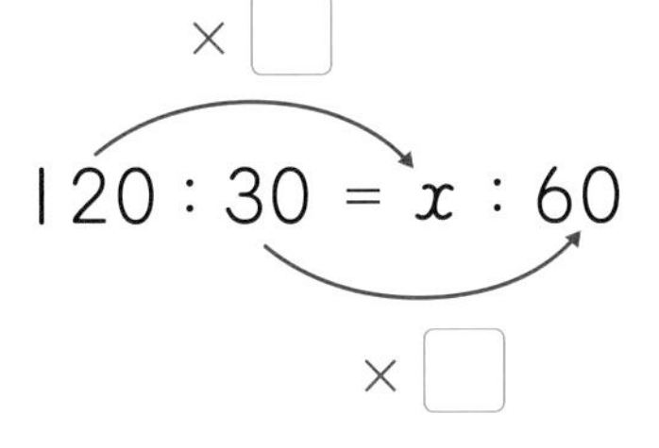

Way to see and think

To find the same concentration, make an equal ratio.

Want to confirm

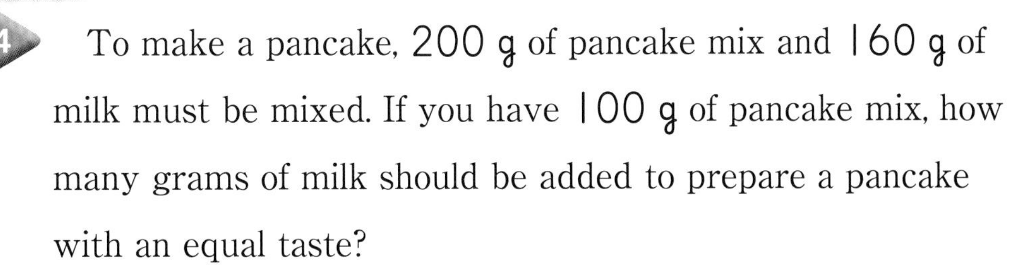

4 To make a pancake, 200 g of pancake mix and 160 g of milk must be mixed. If you have 100 g of pancake mix, how many grams of milk should be added to prepare a pancake with an equal taste?

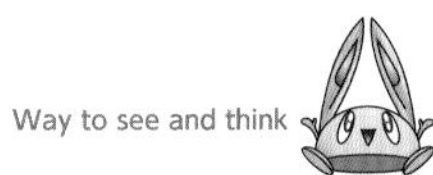

Create an equal ratio of pancake mix to milk.

If you pour in x g of milk,

$200 : 160 = 100 : x$

$x = 160 \div \square$

$= \square$

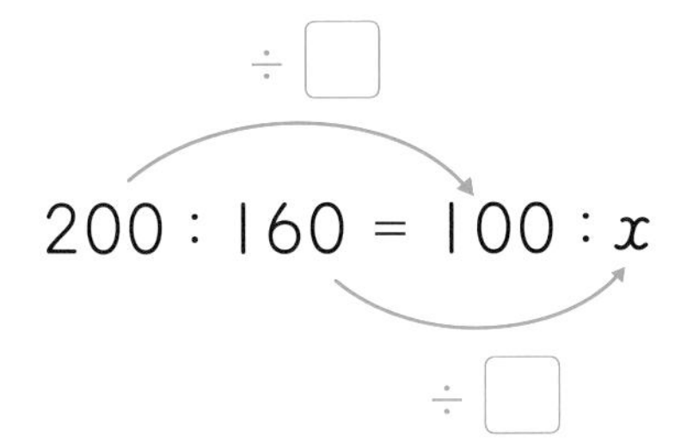

Want to try

Let's find the number that applies for x.

① $2 : 5 = x : 10$

② $4 : 5 = 100 : x$

③ $12 : x = 3 : 5$

④ $x : 20 = 5 : 4$

Want to think How to simplify a ratio

Let's find a ratio that is equal to 12 : 18 and write it with the smallest whole numbers.

Daiki's idea

$$12 : 18 = (12 \div 2) : (18 \div 2)$$
$$= 6 : 9$$
$$= (6 \div 3) : (9 \div 3)$$
$$= 2 : 3$$

Nanami's idea

$$12 : 18 = (12 \div 6) : (18 \div 6)$$
$$= 2 : 3$$

Both use the same equal ratio property.

Reducing a ratio into the smallest whole numbers without changing the value of ratio is called **simplifying a ratio.**

Want to try

Let's simplify the following ratios.

① $1.2 : 3.2 = (1.2 \times 10) : (3.2 \times 10)$

$= \square : \square$

$= \square : \square$

② $\dfrac{2}{5} : \dfrac{3}{8} = \dfrac{16}{40} : \dfrac{15}{40}$

$= \left(\dfrac{16}{40} \times \square\right) : \left(\dfrac{15}{40} \times \square\right)$

$= \square : \square$

Both have been changed to whole number ratios.

Want to confirm

Let's simplify the following ratios.

① $25 : 35$ ② $7 : 28$ ③ $180 : 120$ ④ $0.6 : 2.9$ ⑤ $\dfrac{3}{4} : \dfrac{2}{3}$

Let's measure the length and width of your desk and represent them with a ratio. Also, let's simplify.

Want to think

1 Let's find the height of the tree from the length of its shadow.

① Stick A with a length of 0.8 m and stick B with a length of 2 m were placed vertically in the schoolyard. The lengths of the shadows were 1.2 m and 3 m, respectively. For sticks A and B, let's try to compare the ratio of the length of the stick to that of the shadow.

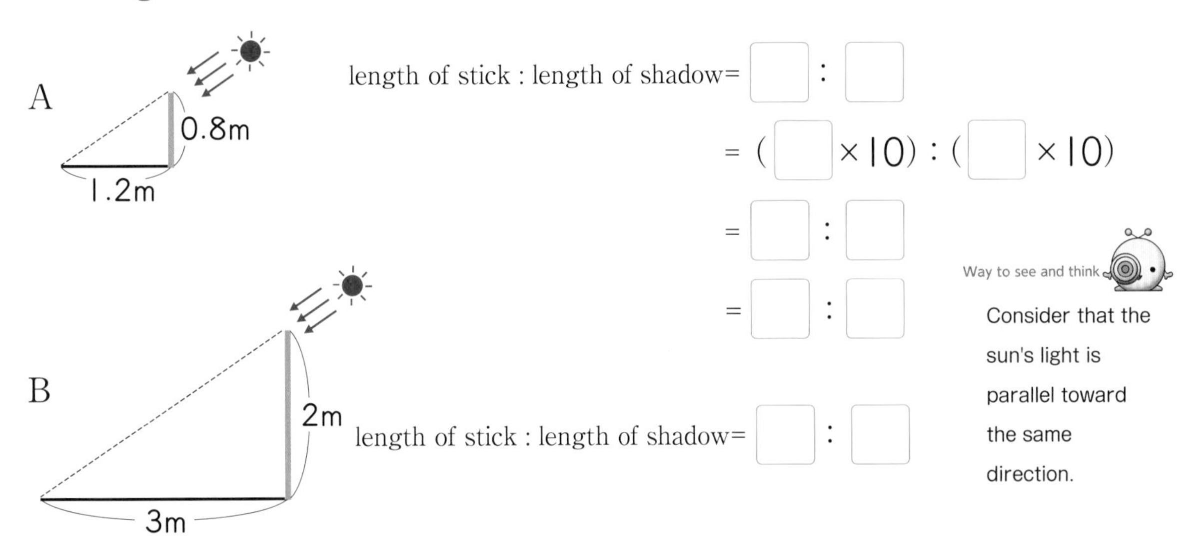

Way to see and think

Consider that the sun's light is parallel toward the same direction.

Want to represent

② In this situation, how many meters is the tree's height when its shadow is 12 m? Let's consider the height of the tree as x m, and write an equal ratio math sentence. Let's find the height of the tree.

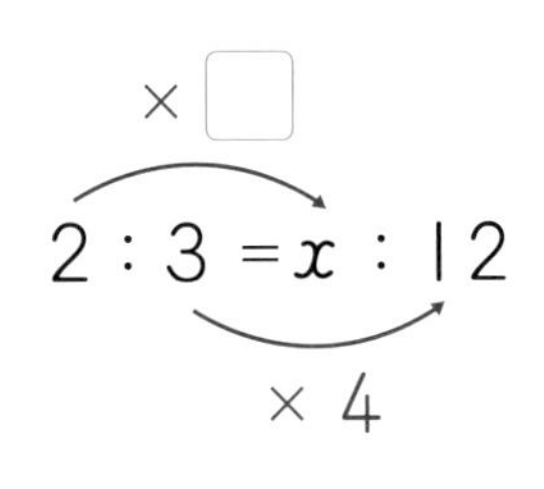

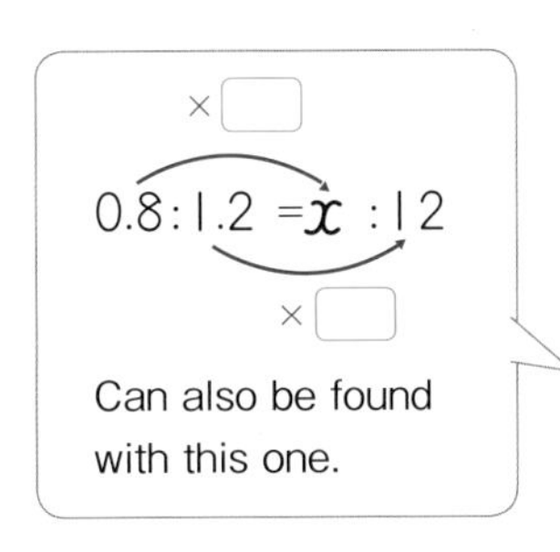

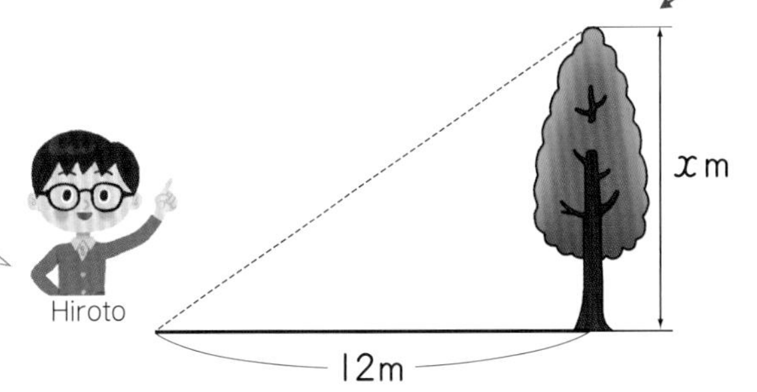

Want to confirm

Based on problem ② from 1, how many meters is the height of the tree if its shadow is 15 m?

Want to know How to divide by a ratio

2 Let's divide a ribbon that is 72 cm long between an elder sister and a younger sister in the ratio 5 : 4. How many centimeters of ribbon will each receive?

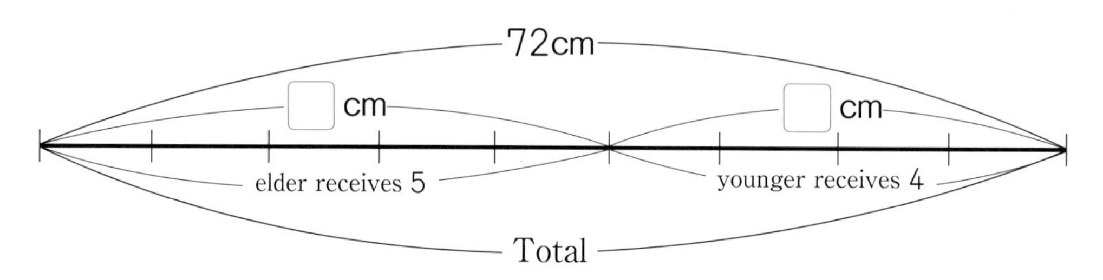

① Let's explain Yui's and Hiroto's idea and find the answer.

Yui's idea

I used the ratio between the elder sister's ribbon and the whole ribbon to find the length of the elder sister's ribbon. If the length of the elder sister's ribbon is x cm, then

$$5 : 9 = x : 72$$

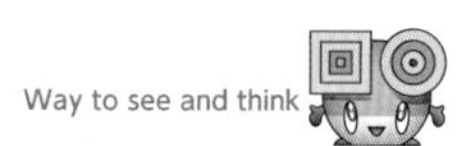

When 5 : 4, the total quantity can be represented as 5 + 4.

Hiroto's idea

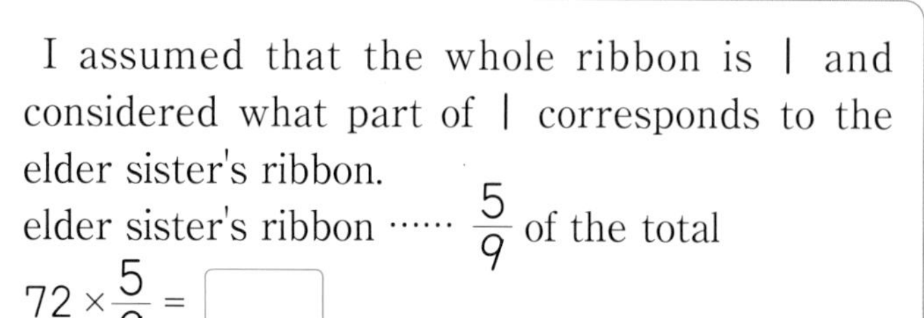

I assumed that the whole ribbon is 1 and considered what part of 1 corresponds to the elder sister's ribbon.

elder sister's ribbon …… $\frac{5}{9}$ of the total

$72 \times \frac{5}{9} = \square$

② Let's find the length of the younger sister's ribbon using a ratio.

Want to confirm

Let's separate 400 mL of milk between Aoi and her father in the ratio 2 : 3. How many milliliters of milk will Aoi receive?

What you can do now

☐ **Can represent the ratio between two quantities as ratio and the value of ratio.**

1 Let's represent the following ratios as ratio and value of ratio.

① The quantity of vinegar and salad oil.

② The length of AB and AC in the triangle ruler.

①
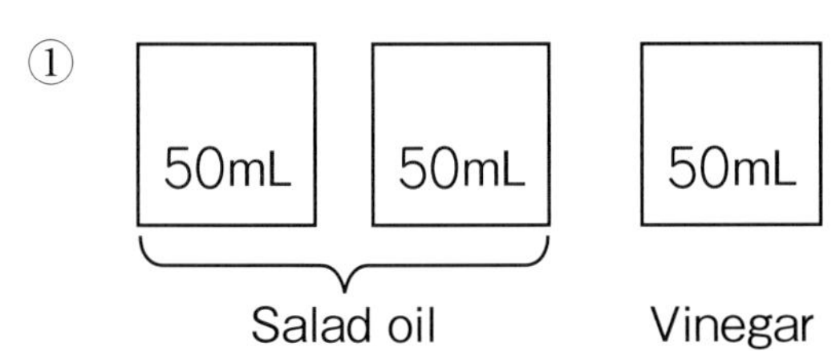

②
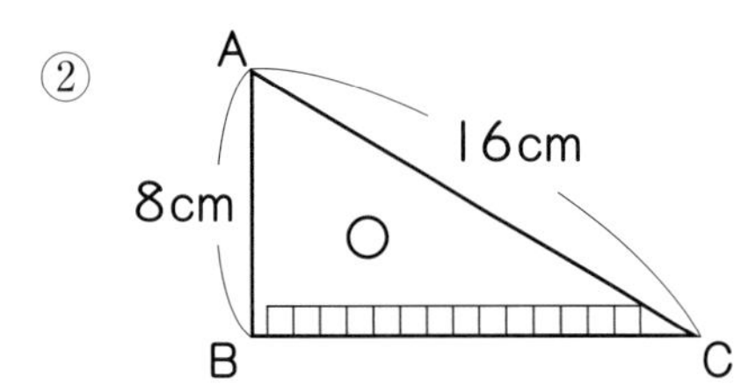

☐ **Understanding the properties of equal ratios.**

2 Let's find the number that applies for x.

① $3 : 5 = x : 10$　　② $7 : 4 = 35 : x$

③ $80 : x = 5 : 8$　　④ $x : 125 = 3 : 5$

☐ **Can use the properties of equal ratios.**

3 The ratio between the length and width of a rectangle is 4 : 7.

① If the length is 28 cm, how many centimeters is the width?

② If the width is 28 cm, how many centimeters is the length?

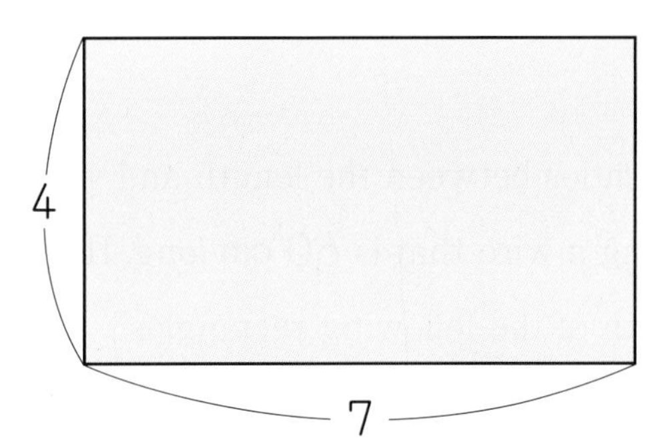

☐ **Can simplify a ratio.**

4 Let's simplify the following ratios.

① $36 : 48$　　② $800 : 1400$

③ $2.4 : 0.8$　　④ $\frac{1}{2} : \frac{2}{3}$

☐ **Can divide by a ratio.**

5 Let's cut a ribbon that is 2 m long by the ratio 2 : 3.

How many centimeters is each section?

Supplementary problems p.234

Usefulness and efficiency of learning

1 400 g of glutinous rice and 40 g of red beans are needed to make red bean rice for 4 people.

① How many grams of glutinous rice and red beans do you need respectively to make red bean rice for one person?

② How many grams of glutinous rice and red beans do you need respectively to make red bean rice for 7 people?

③ There are 600 g of glutinous rice. If you make red bean rice with the same ratio as the one for 4 people, how many grams of red beans do you need?

☐ Can represent the ratio between two quantities as ratio and the value of ratio.

☐ Can simplify a ratio.

2 A lottery box is made so that the ratio of red balls to white balls is 3 : 4. If there are 28 white balls, how many red balls should be needed?

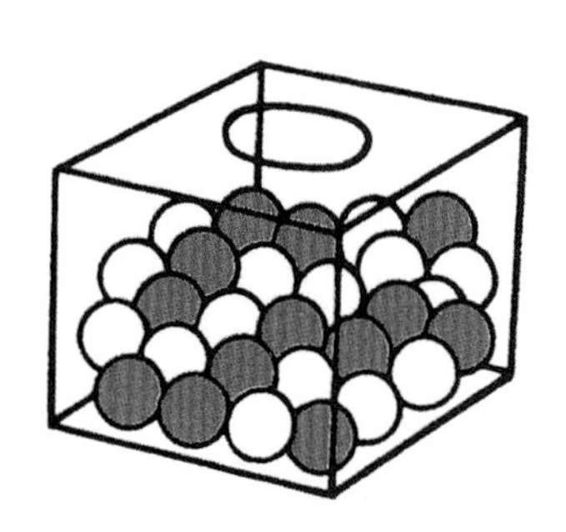

☐ Can use the properties of equal ratios.

3 Considering ratios between the length and width, the following rectangles were made using a wire that is 60 cm long. How many centimeters is the length and width of the following rectangles?

☐ Can divide by a ratio.

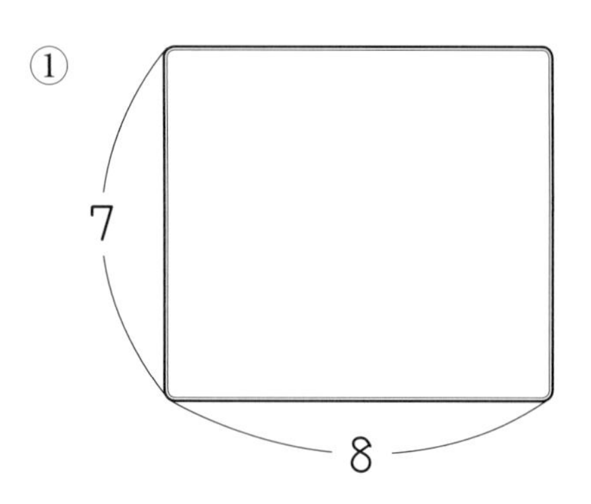

② 4, 11

Let's deepen.

We can make various rectangles with the same length of wire. Which ratio is the best balance?

Hiroto

Utilize in life.

Deepen.

Beautiful ratios

Want to think

There are a lot of things that are said to be beautiful in the world. The ratio 1 : 1.6 is commonly used in some of those beautiful things. This ratio is called the golden ratio. It is used in many structures and art works as a ratio that is harmonic and beautiful. Let's think about this ratio.

Parthenon (Greece)

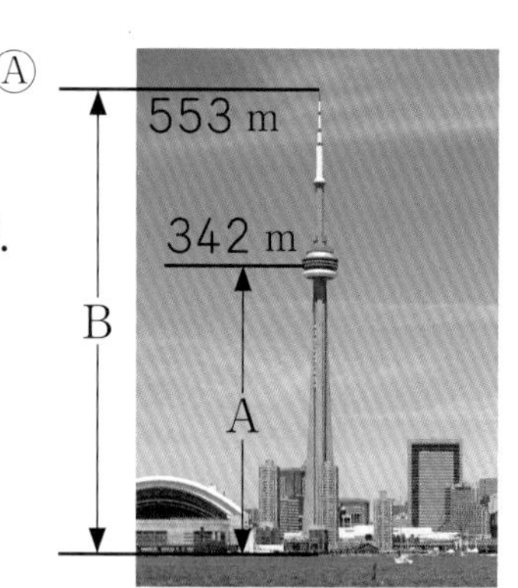

CN Tower (Canada)

① Let's find a ratio between the two lengths in tower Ⓐ. Let's consider A as 1, and find the ratio A : B by rounding off the answer to the tenths place.

A : B = ☐ : 553 = 1 : ☐

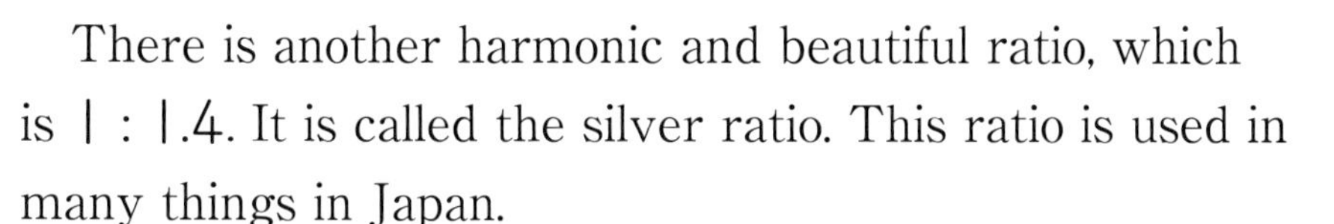

There is another harmonic and beautiful ratio, which is 1 : 1.4. It is called the silver ratio. This ratio is used in many things in Japan.

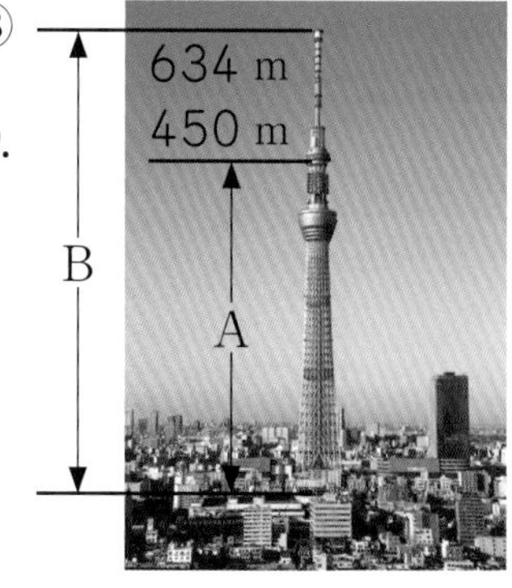

Tokyo Sky Tree

(Sumida-ku, Tokyo)

② Let's find a ratio between the two lengths in tower Ⓑ. Let's consider A as 1, and find the ratio A : B by rounding off the answer to the tenths place.

A : B = ☐ : 634 = 1 : ☐

Want to deepen

Let's measure the length and width of this textbook, and find the ratio by considering the length as 1. Let's round off to the tenths place.

Let's try to explore the things in our surroundings in the same way.

Find the ?

If we try to make the photo bigger or smaller ...

Problem What is the common point between two figures with the same shape but different sizes?

Enlarged and Reduced Drawings

Let's explore the properties of figures with the same shape and how to draw them.

1 Enlarged and reduced drawings of figures

Want to find

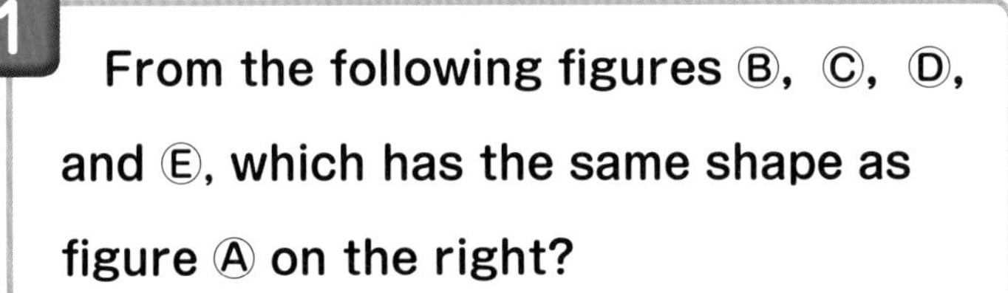

1 From the following figures Ⓑ, Ⓒ, Ⓓ, and Ⓔ, which has the same shape as figure Ⓐ on the right?

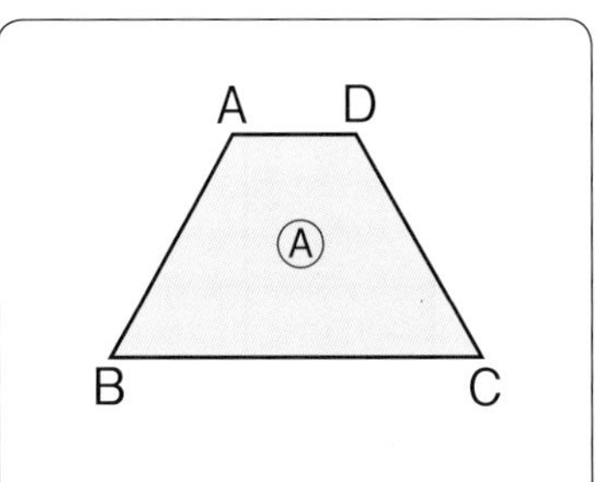

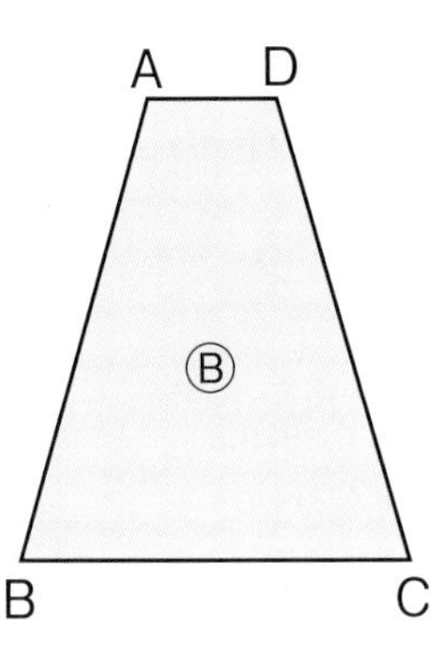

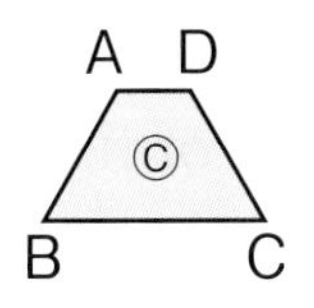

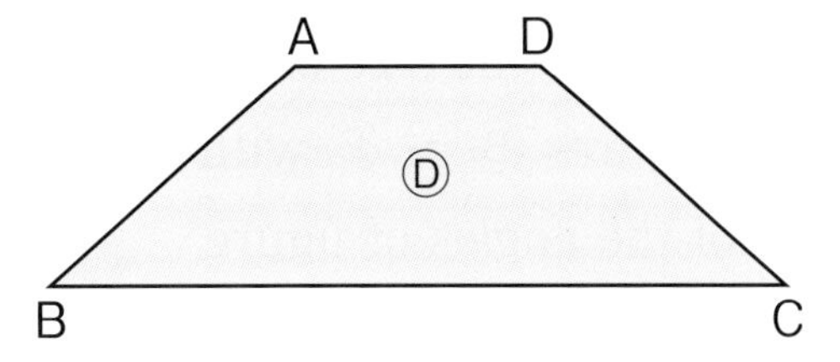

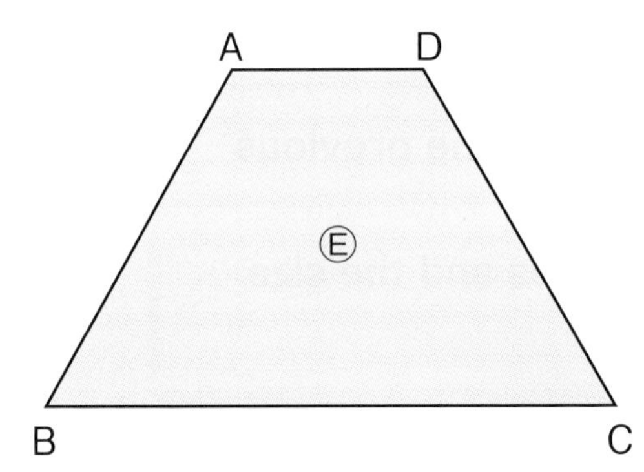

What does the same shape mean?

Ⓐ and Ⓔ look like the same shape although they have different sizes.

All of them are trapezoidal but there are various sizes.

What about the length of the sides and size of the angles?

Want to summarize

① Let's measure the length of the sides and size of the angles of the 4 figures on the previous page and summarize them in the table below.

	Length of sides (cm)			Size of angles (Degrees)	
	Side AB	Side BC	Side DA	Angle A	Angle B
Ⓐ	2	3	1	120	60
Ⓑ					
Ⓒ					
Ⓓ					
Ⓔ					

Way to see and think

The figures are organized in a table by "length of sides" and "size of angles."

Want to compare

② Let's compare the length of the sides of figures Ⓑ ~ Ⓔ with the length of the sides of figure Ⓐ.

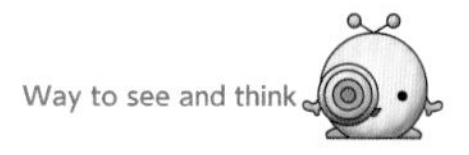

If you compare the length of the sides using a ratio, you can understand the property.

③ Let's compare the size of the angles of figures Ⓑ ~ Ⓔ with the size of the angles of figure Ⓐ.

From ② and ③, it seems that Ⓐ, Ⓒ, and Ⓔ have the same shape.

Daiki

Want to explore

2 The following figures are Ⓐ and Ⓔ from the previous page. Let's investigate the length of the sides and the size of the angles in detail.

Way to see and think

Let's try to think using the "ratio" you learned before.

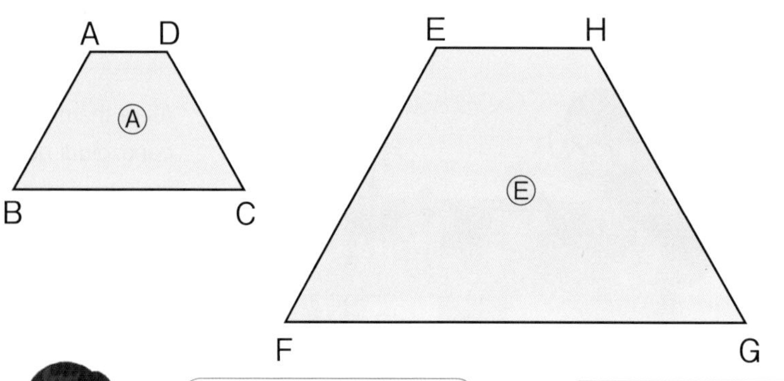

The name of the vertices in Ⓔ were changed to make it easier to compare.

Hiroto

The common property of figures with the same shape but different size is the angle...

The ratio of the length of corresponding sides...

Nanami

Purpose What are the side's and angle's properties in figures with the same shape but different sizes?

① Let's represent the length of the corresponding sides of Ⓐ and Ⓔ with simplified ratios.

Side BC : Side FG = ☐ : ☐

Side AB : Side EF = ☐ : ☐

Side DA : Side HE = ☐ : ☐

② How many times the length of the corresponding side of Ⓔ is the length of the side of Ⓐ?

③ Let's compare the size of the corresponding angles.

When the corresponding angles are respectively equal, and all the lengths of corresponding sides are extended in the same ratio, then it's called an **enlarged drawing.** If shortened in the same ratio, then it's called a **reduced drawing.**

Summary

In enlarged or reduced drawings, the lengths of the corresponding sides share the same ratio and all the corresponding angles are equal.

Figure Ⓔ is a two times enlarged drawing of figure Ⓐ, and figure Ⓐ is a $\frac{1}{2}$ reduced drawing of figure Ⓔ.

Congruent

Reduced

Enlarged

When two figures are congruent, the ratio of the length of the two corresponding sides is 1:1.

Want to confirm

1 Which of the following figures is an enlarged or reduced drawing of Ⓕ? Also, how many times is it enlarged, or how much is it reduced?

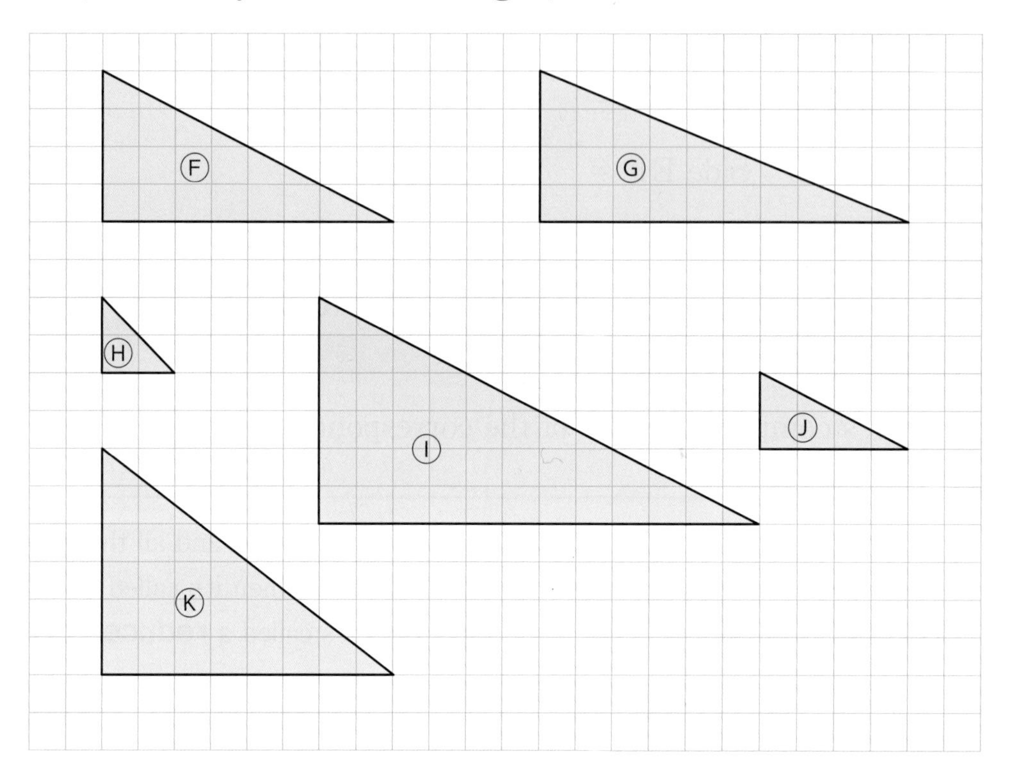

2 Rectangle EFGH was drawn increasing the length and width of rectangle ABCD by 1 cm. Let's answer the following questions about these two figures.

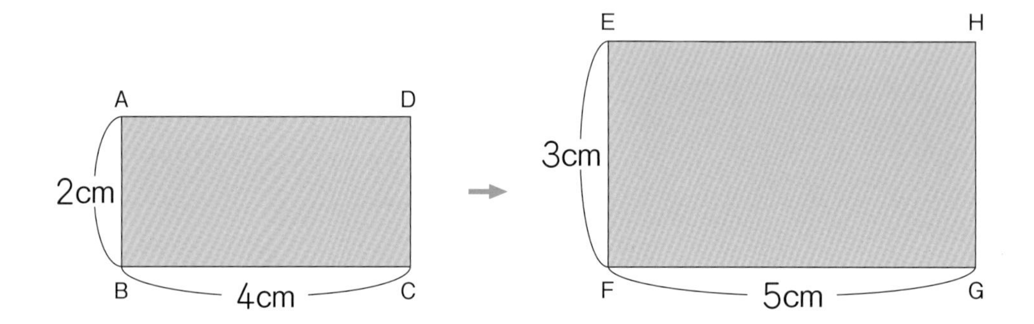

① Can rectangle EFGH be an enlarged drawing of rectangle ABCD?

② In order to enlarge rectangle EFGH as 1.5 times of rectangle ABCD, how many centimeters should the width be?

2 How to draw enlarged and reduced drawings

Want to think How to draw using a grid

1 **Let's think about how to draw enlarged triangle DEF which is 2 times of triangle ABC. Point E, corresponding to point B, is already located on the grid paper below.**

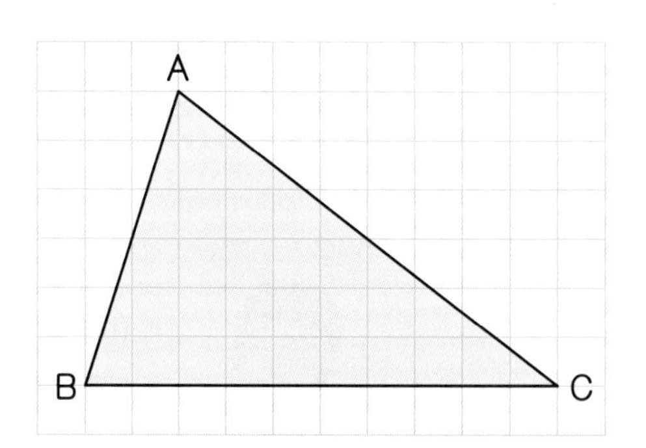

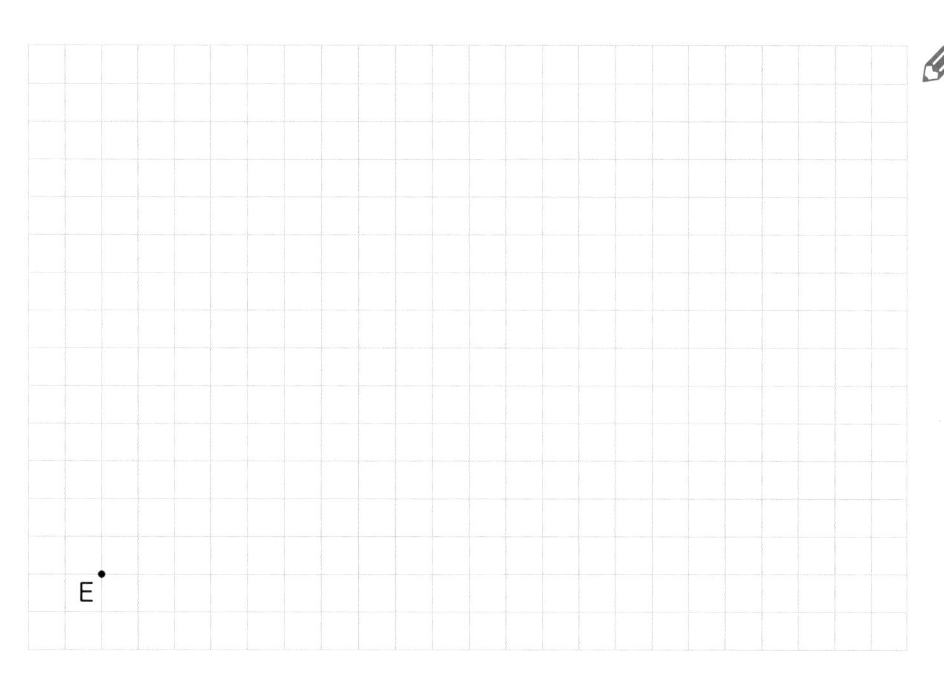

Way to see and think

Point A is two grid points right and six grid points up from point B. If you double enlarged it, these numbers also are doubled.

After drawing an enlarged and reduced drawing, confirm the length of the corresponding sides and the size of the corresponding angles.

Want to confirm

1 Let's draw reduced triangle DEF, which is $\frac{1}{2}$ of triangle ABC on 1. Let's use the grid paper below.

Way to see and think

Reduce the base and height of the triangle by $\frac{1}{2}$.

2 Let's think about how to draw enlarged triangle DEF which is 2 times of triangle ABC.

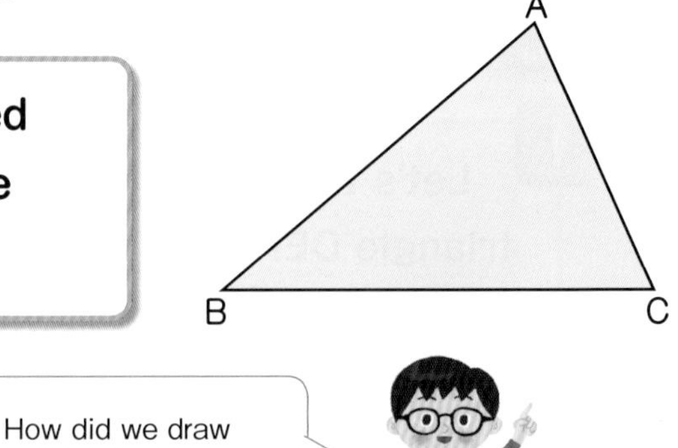

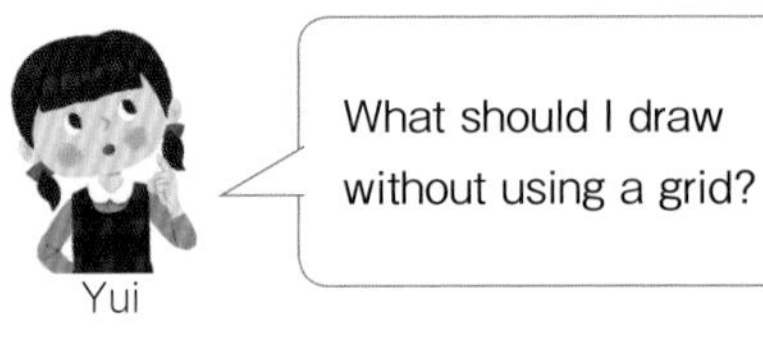

How did we draw congruent triangles?

Purpose Can we draw an enlarged drawing using the length of the sides and the size of the angles?

① Line EF, whose length is twice the length of side BC, is already drawn. Let's think about where point D, which corresponds to point A, should be placed and then continue with the drawing.

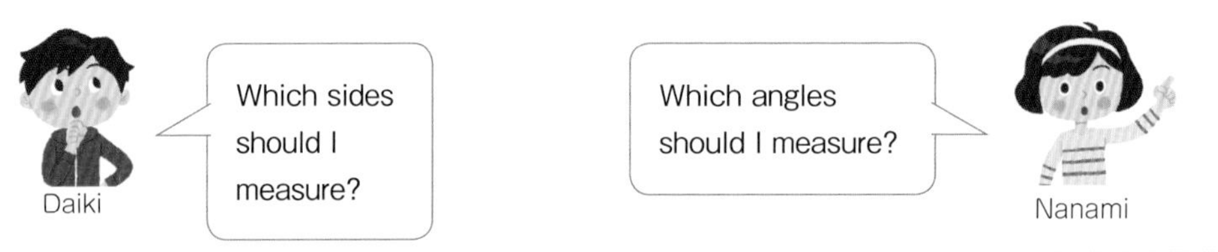

E F

Want to explain

② Let's explain the drawing methods of the following children.

Nanami's drawing method

I drew it by enlarging all three sides as twice as long.

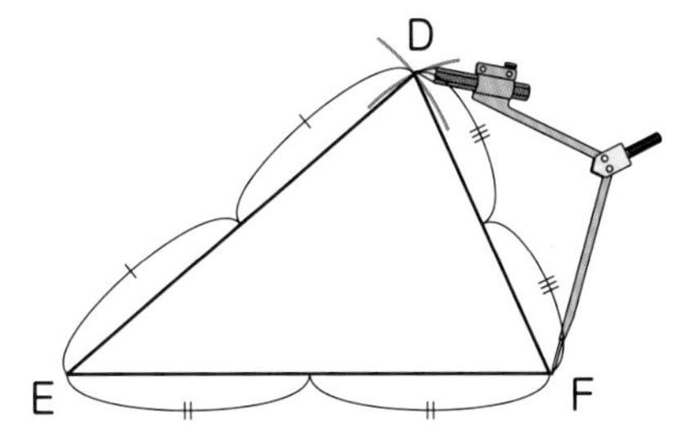

Daiki's drawing method

I drew it by enlarging 2 sides as twice as long and using the angle between those 2 sides.

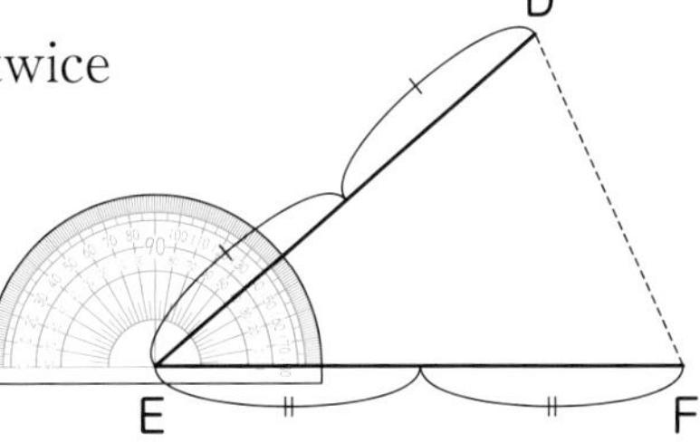

Yui's drawing method

I drew it by enlarging 1 side as twice as long and using both angles at the end of the line.

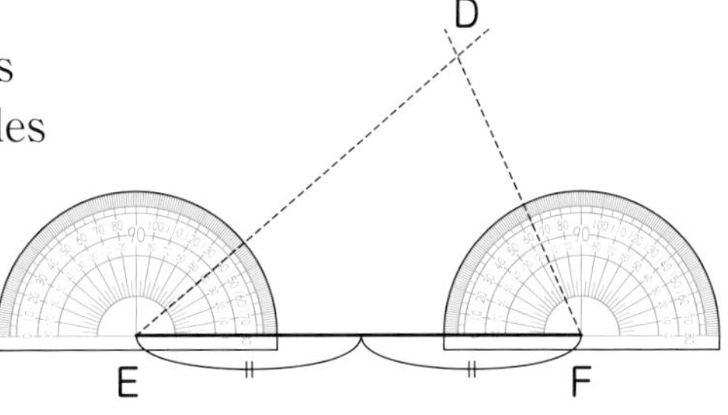

We can use the drawing methods for congruent triangles.

Summary

Enlarged drawings can be drawn using the same angles and enlarging the length of the sides through the methods for drawing congruent figures.

Want to think How to draw a reduced drawing

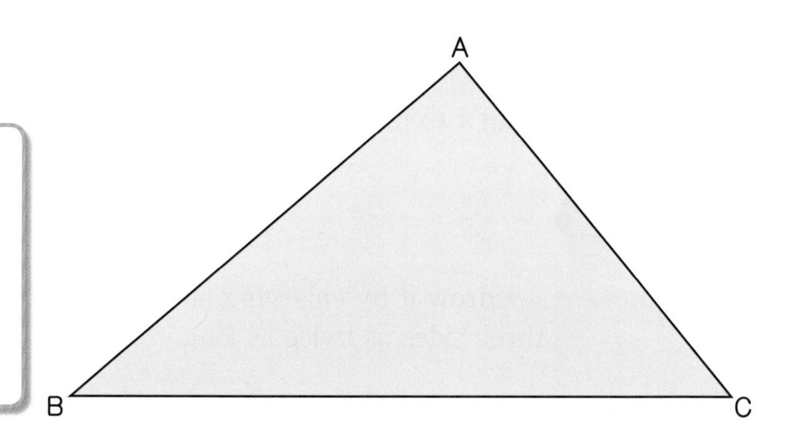

3 Let's think about how to draw reduced triangle DEF, which is $\frac{1}{3}$ of the triangle on the right.

① Let's draw reduced triangle DEF considering your own method. Also, explain how you drew it to a classmate.

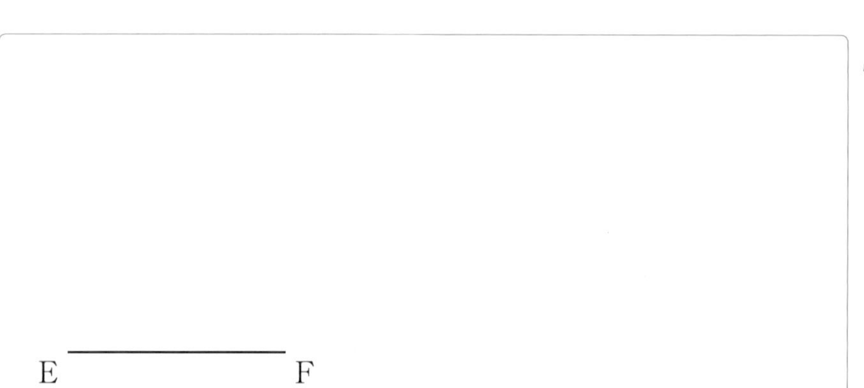

Can I draw it in the same way as the drawing method for enlarged drawings?

② Is your drawing method similar to those on the previous page?

Want to confirm

2 Let's draw a 2 times enlarged drawing and $\frac{1}{2}$ reduced drawing of the quadrilateral shown on the right.

A
D
B
C

Want to think How to draw using a center point

Let's draw enlarged triangle DBE, which is 3 times of triangle ABC.

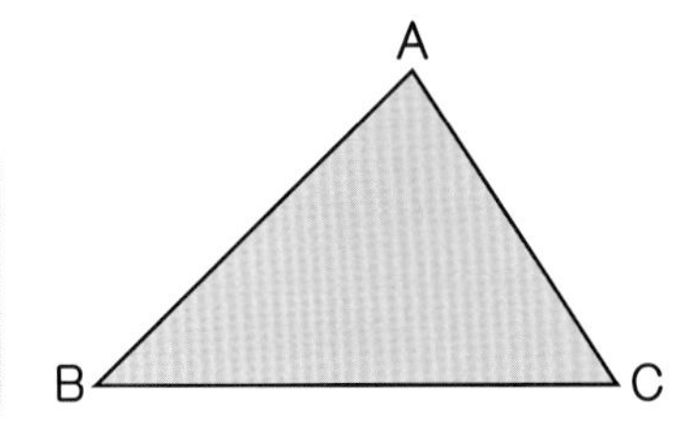

① Let's extend side BA and draw point D, which is the corresponding point to point A. Let's extend side BC and draw point E, which is the corresponding point to point C.

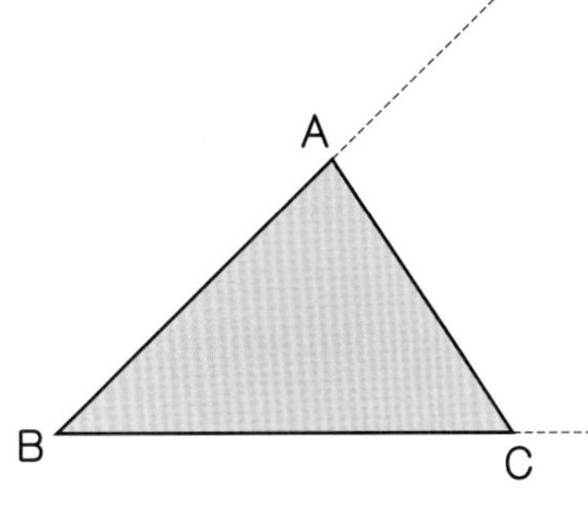

② Let's confirm that enlarged triangle DBE is 3 times of triangle ABC.

Like in the above example, you can enlarge or reduce drawings using 1 point and its connected lines. This reference point is called **center point (center of enlarging or reducing)**.

Want to try

Let's use the diagram in ① exercise 4 to draw a reduced triangle that is $\frac{1}{2}$ of triangle ABC. Use point B as the center point.

Want to confirm

Let's use point B as the center point to draw a 2 times enlarged drawing and a $\frac{1}{2}$ reduced drawing of quadrilateral ABCD.

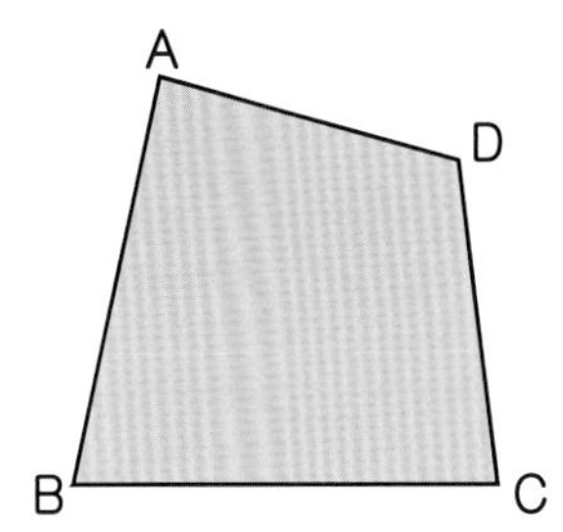

Want to think How to draw when the center point is inside the figure

5 **Let's consider point E as the center point and think about how to draw enlarged quadrilateral FGHJ that is 2 times of quadrilateral ABCD.**

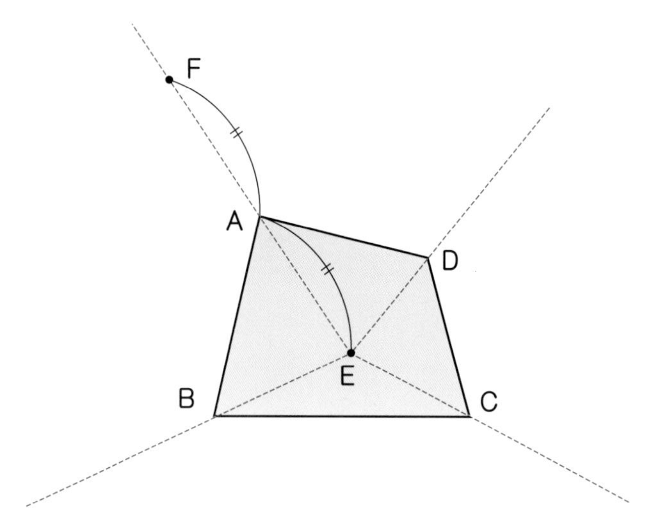

① Straight line EA was extended and point F was drawn, which is the corresponding point to point A. Let's continue and finish the drawing.

② Why can you draw the enlarged figure using this method? Let's explain the reasons.

Want to try

5 In the diagram from **5**, let's use point E as the center point and draw a $\frac{1}{2}$ reduced drawing of quadrilateral ABCD.

Want to confirm

Let's use point D as the center point to draw a 2 times enlarged drawing and a $\frac{1}{2}$ reduced drawing of triangle ABC.

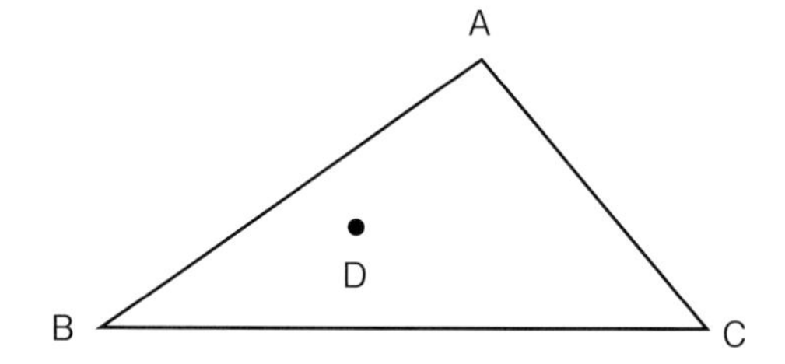

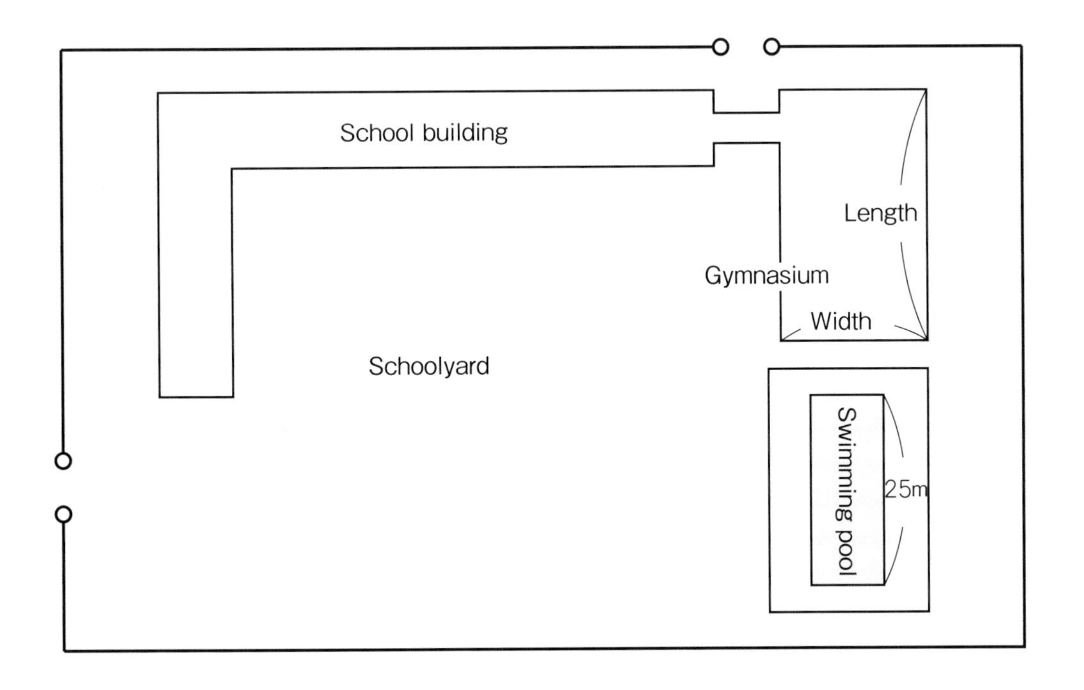

3 Applications of reduced drawings

Want to think

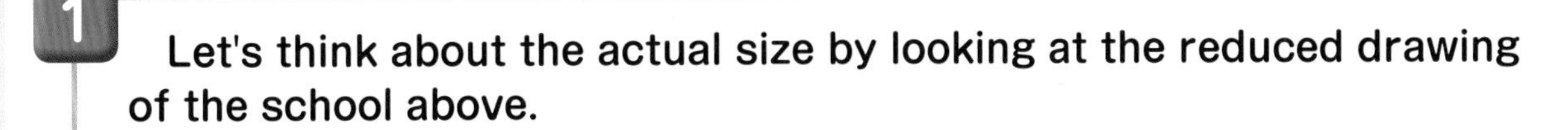

1 Let's think about the actual size by looking at the reduced drawing of the school above.

① The actual length of the swimming pool is 25 m. How many centimeters and millimeters is the length on the reduced drawing? By how much was it reduced?

② How many meters is the actual length of 1 cm on the reduced drawing?

The ratio that represents how much it is reduced by from the actual length is called **reduced scale.** The picture above is a reduced drawing using $\frac{1}{1000}$ reduced scale.

There are 3 ways to show a reduced scale:

Ⓐ $\frac{1}{1000}$　　Ⓑ 1 : 1000　　Ⓒ 0　10　20　30m (scale bar)

③ What is the actual length and width of the gymnasium in meters?

Length $3.3 \times 1000 =$ ☐ (cm) $=$ ☐ (m)

Width $2 \times 1000 =$ ☐ (cm) $=$ ☐ (m)

Want to confirm

1 We came to play at the pond in the park. Sakura walked from point C to point B as shown on the right. What should we do to find the distance from point B to point A? Point A has a cedar in the opposite side of the pond.

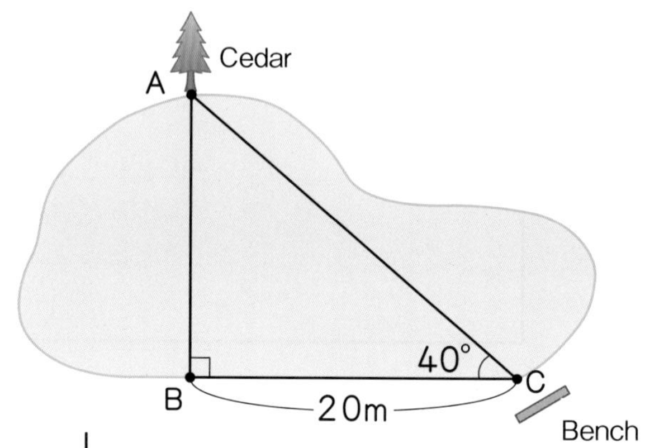

① Let's follow the steps below to draw a $\frac{1}{500}$ reduced drawing of the right triangle ABC.

(1) Find the length of straight line BC and draw it.

(2) Through point B, draw a perpendicular line to straight line BC.

(3) Open angle C by 40 ° and place point A.

(4) Draw right triangle ABC.

② Let's measure straight line AB in the reduced figure from ① and find the actual distance to the cedar.

Want to try

2 In the figure shown on the right, how many meters is the actual height of the tree? Let's explain how to find the answer.

A

50°

B C

7m

What you can do now

☐ **Understanding the properties of enlarged and reduced drawings.**

1 In the following diagram, which figures are an enlarged or a reduced drawing of another figure? Let's also explain the reason.

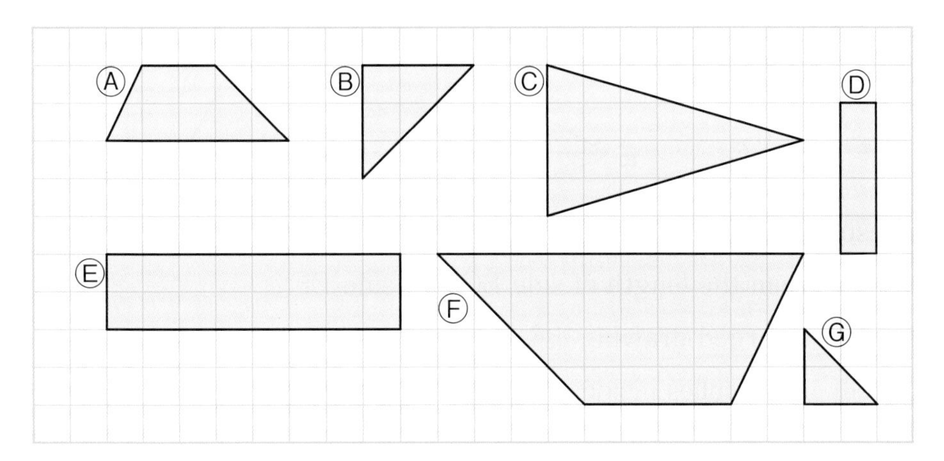

☐ **Can draw enlarged and reduced drawings.**

2 Let's draw a 2 times enlarged drawing and $\frac{1}{2}$ reduced drawing of the triangle ABC shown below.

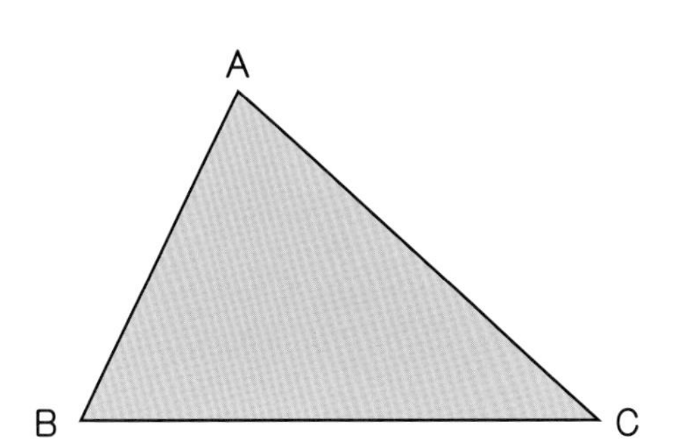

☐ **Can use a reduced drawing.**

3 There is a map of a school that is drawn in $\frac{1}{500}$ reduced scale. In the reduced drawing, the gymnasium has a rectangular shape with a length that is 6 cm long and a width that is 3.2 cm long. How many meters is the actual length and width of the gymnasium?

Supplementary problems p.236

Usefulness and efficiency of learning

1 Triangle Ⓑ is an enlarged drawing of triangle Ⓐ. Let's answer the following questions.

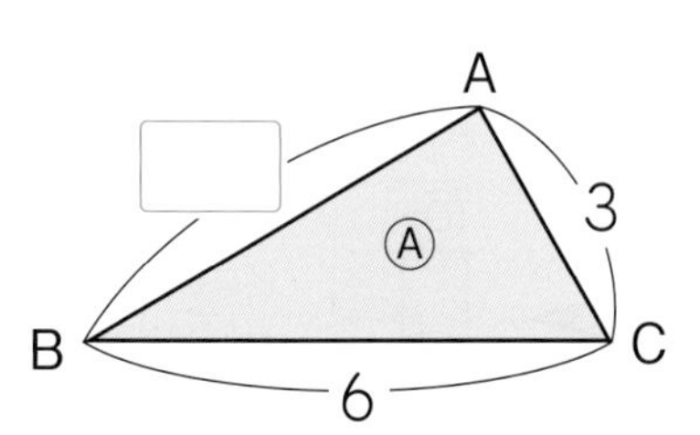

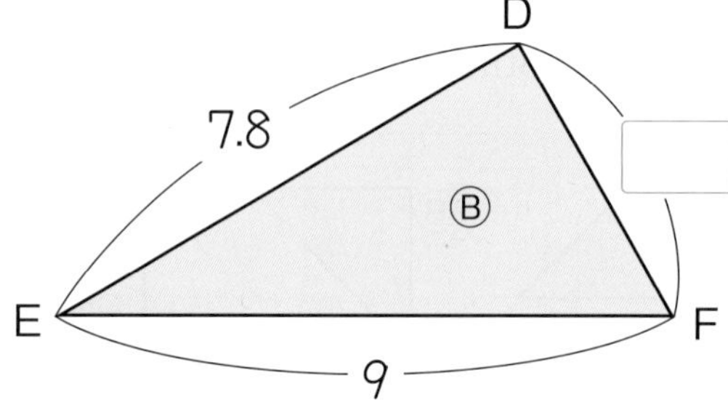

① Which is the corresponding angle to angle B?

② Let's find the ratio between the length of side BC and side EF.

③ How many times Ⓐ is enlarged drawing Ⓑ?

④ Let's find the number that applies inside ☐ in Ⓐ and Ⓑ.

☐ Understanding the properties of enlarged and reduced drawings.

2 Let's draw a 2 times enlarged drawing and $\frac{1}{2}$ reduced drawing of the following quadrilateral.

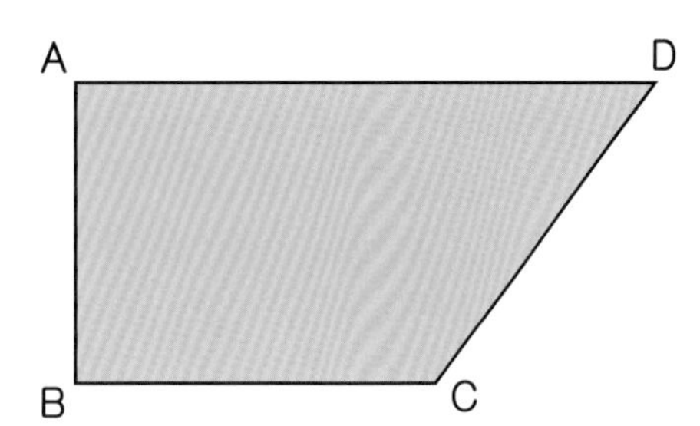

☐ Can draw enlarged and reduced drawings.

3 Let's find the actual height of the building shown in the following diagram.

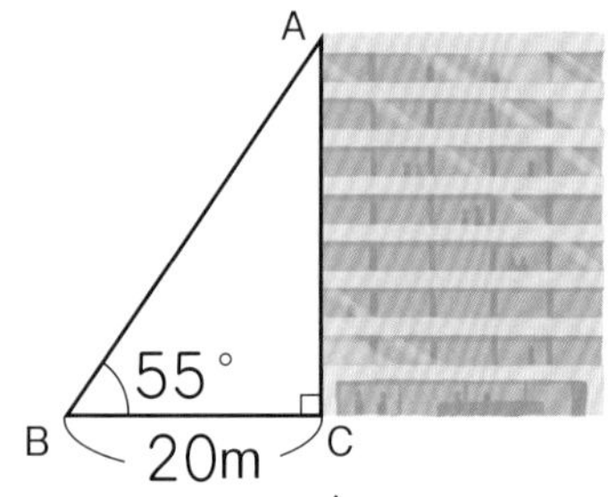

① Let's draw a $\frac{1}{500}$ reduced drawing of the triangle ABC.

② Let's find the actual height of the building.

☐ Can use a reduced drawing.

Let's deepen.

I want to find out the street and distance using an actual map.

Utilize in life.

Deepen.

Let's know actual distance on the map.

—Let's use a scale—

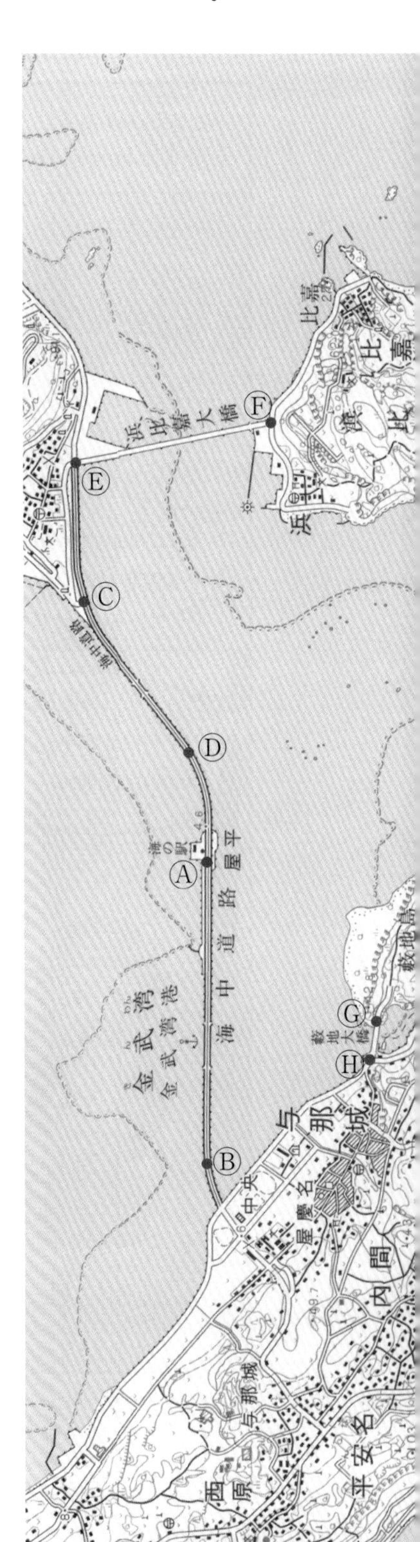

Want to think

The diagram on the right is a map of the road through the sea in Okinawa. The map is in $\frac{1}{50000}$ reduced scale. Let's think about this map.

① How many centimeters represent the actual distance of 5 km on this map?

② What is the actual distance in km from point Ⓐ to point Ⓑ?

③ Let's find the actual distance between the points C and D, E and F, and G and H on the map.

④ Yuta is walking from point Ⓑ to point Ⓐ at 4 km per hour. He left point Ⓑ at 10 : 40 a.m. What time will he arrive to point Ⓐ?

Kaichu-doro (Uruma City, Okinawa Prefecture)

Find the ?

How many sheets of paper are there?

Problem Without counting them all, how can we know the number of sheets in a bundle?

Proportion and Inverse Proportion

Let's explore the properties of two quantities changing and corresponding together.

1 Proportion

Want to think

There is a bundle of sheets of paper. Without counting, let's think about how to explore the number of sheets of paper.

Hiroto: As the number of sheets increases, it becomes heavier ...

Nanami: It becomes thick as the number of sheets of paper increases.

Purpose How can we explore the number of sheets of paper in a bundle?

① Let's try to discuss what changes when the number of sheets of paper increases.

Yui: If the number of sheets is doubled, then the weight will also be doubled.

Daiki: If the number of sheets is doubled, then the thickness will also be doubled.

Want to explore

② Let's think about how to investigate.

Yui's idea

I investigate the relationship between the number of sheets of paper and weight.

(1) I measured the weight when the number of sheets is 10, 20, 30, and so on.

(2) I can consider which relationship exists between the number of sheets of paper and weight using a summary table.

Way to see and think

Let's recall how to learn science.
Find a problem.
↓
Think about how to investigate.
↓
Try to predict the results.
↓
Experiment and get results.
↓
Summarize what you found.

③ Let's weigh each number of sheets of paper, and fill in the table below.

Number of Sheets of Paper and Weight

Number of sheets of paper (sheets)	10	20	30	40	50
Weight (g)					

④ Let's look at the table in ③ and discuss what you notice.

⑤ Yui conducted an experiment and summarized with the following report. What can you say from the results of the experiment?

〈Experiment report〉

- Experiment day : Monday, November 10
- Theme: Examine the relationship between the number of sheets of paper and its weight.
- Preparation: Bundle of paper, scale, calculator.
- Method: Weigh the paper in bundles with ascending order 10, 20, 30, ...
  Record each weight in a table.
- Prediction: The weight of the paper is proportional to the number of sheets.
- Result :

Number of Sheets of Paper and Weight

Number of sheets of paper (sheets)	10	20	30	40	50
Weight (g)	70	140	210	280	350

- Analysis : When the number of sheets of paper is doubled from 10 to 20, the weight is also doubled from 70g to 140g. The relationship between other number of sheets of paper and its weight is shown below.

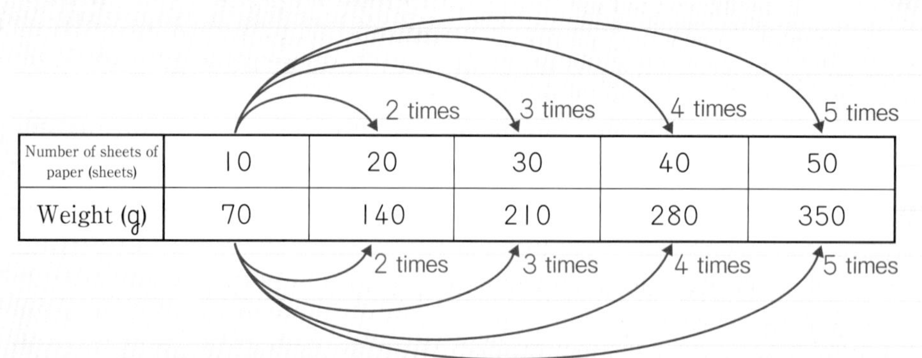

Number of sheets of paper (sheets)	10	20	30	40	50
Weight (g)	70	140	210	280	350

Therefore, the weight of paper is proportional to the number of sheets of paper.

Summary

The weight is proportional to the number of sheets of paper, therefore the total number of sheets can be found by weighing the entire amount of paper.

The thickness is also proportional to the number of sheets of paper.
Let's try to investigate.

Want to try

2

There is a bundle of paper, from the previous page, that weighed 1400 g. How many sheets of paper are there in this bundle? Let's fill in the ☐ with a number and try to explain the ideas of the following children.

Nanami's idea

The weight is 20 times of 70 g. Therefore, the number of sheets of paper is also 20 times more.

☐ × 20 = ☐

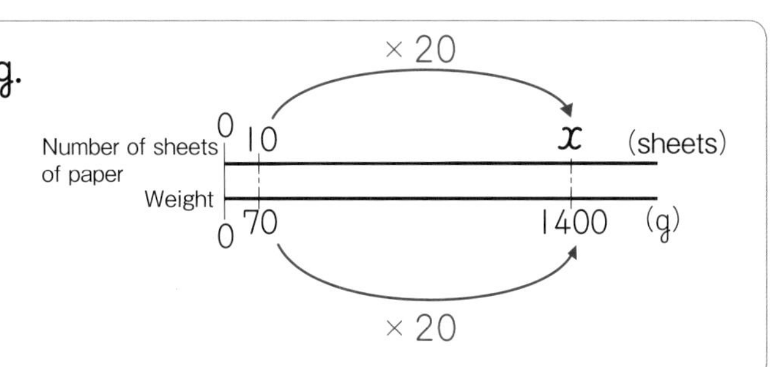

Hiroto's idea

If you find the number of sheets of paper per gram,

$10 \div 70 = \frac{1}{7}$

it is 1400 times of the number of sheets of paper per gram.

☐ × 1400 = ☐

× 1400

Number of sheets of paper (sheets)	$\frac{1}{7}$	x
Weight (g)	1	1400

× 1400

Yui's idea

Consider the number of sheets of paper in 1400 g as x and think the ratio between the number of sheets of paper and the weight.

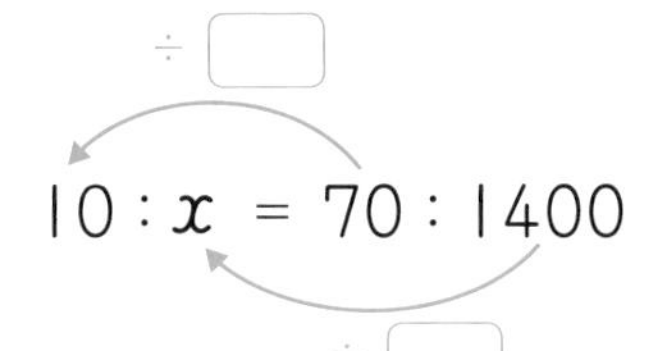

Want to confirm

1 When examining the number of sheets of paper and thickness, the thickness of 420 sheets was 4 cm. How many sheets of paper are there when the bundle of paper is 9 cm thick?

Want to explore

3

After examining the relationship between the length and weight of a wire, the following table was made.

Let's investigate the relationship between x and y, when the length of wire is x m and the weight is y g.

Length and Weight of a Wire

Length x (m)	2	3	4	5	6	9	18
Weight y (g)	40	60	80	100	120	180	360

① When the value of x changes by 2 times, 3 times, ... , how does the corresponding value of y change?

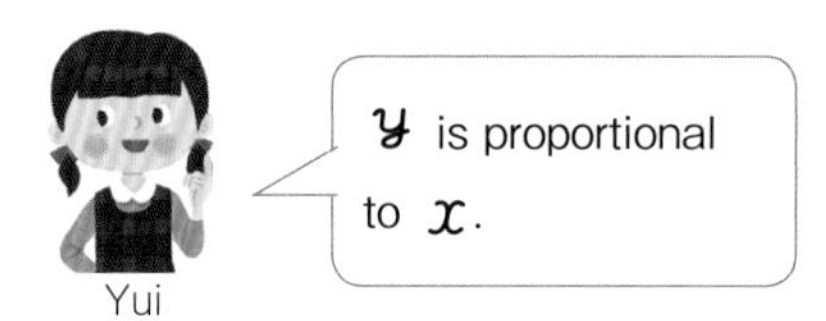

What is the value of y when the value of x changes by 1.5 times or $\frac{1}{2}$ times?

Purpose When y is proportional to x, how does the value of x and y change?

② When y is proportional to x and the value of x changes by 1.5 times and 2.5 times, how does the value of y change?

2.5 times　1.5 times　$\frac{1}{3}$ times　$\frac{1}{2}$ times

Length x (m)	2	3	4	5	6	9	18
Weight y (g)	40	60	80	100	120	180	360

□ times　□ times　□ times　□ times

③ When y is proportional to x and the value of x changes by $\frac{1}{2}$ times $\frac{1}{3}$ andtimes, how does the value of y change?

□ also applies to decimal numbers and fractions.

Summary

When y is proportional to x, if the value of x changes by □ times, then the value of y also changes by □ times.

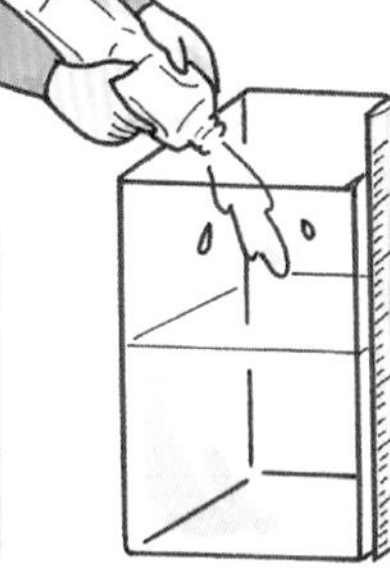

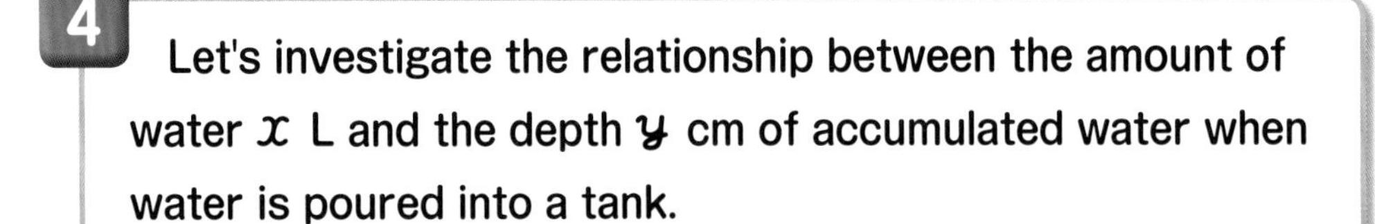

4 **Let's investigate the relationship between the amount of water x L and the depth y cm of accumulated water when water is poured into a tank.**

Amount and Depth of Water inside the Tank

Amount of water x (L)	1	2	3	4	5	6	7	8	9
Depth y (cm)	2	4	6	8	10	12	14	16	18

① Can you say that the depth of water y cm is proportional to the amount of water x L?

② Let's explore how much the value of y increases when the value of x increases by 1.

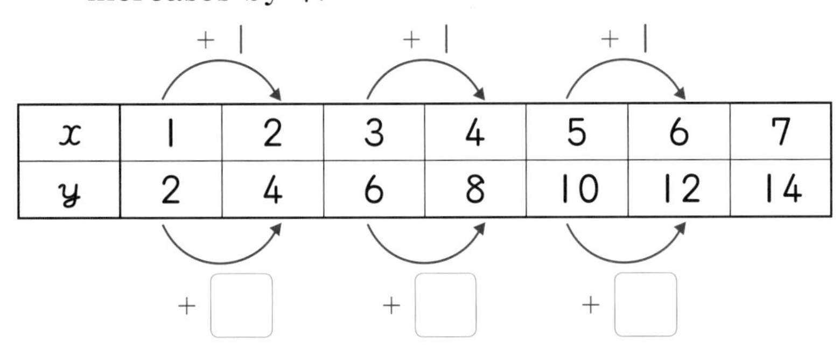

x	1	2	3	4	5	6	7
y	2	4	6	8	10	12	14

③ Let's calculate $y \div x$ using the corresponding values of x and y in the above table. What does the quotient of $y \div x$ represent?

④ Let's investigate the relationship between the amount and depth of water, using the fact that the depth of water per liter is 2 cm. Let's represent the relationship between x and y in a math sentence.

Depth of water y (cm)		Depth of water per liter (cm)		Amount of water x (L)
2	←	2	×	1
4	←	2	×	2
□	←	2	×	3
□	←	2	×	□
□	←	2	×	□
⋮		⋮	⋮	⋮
y	←	2	×	x

$y = \square \times x$

When there are two quantities x and y, and y is proportional to x, their relationship can be represented with the following math sentence:

$$y = \text{constant number} \times x$$

The "constant number" in a proportional relationship represents:

① How much the value of y increases when the value of x increases by 1.

② The quotient of $y \div x$.

③ The value of y when the value of x is 1.

⑤ Using the math sentence in ④, let's find the depth of the water when 10 L and 20 L of water are poured into the tank.

Want to represent

5 **Considering the wire on page 164, let's represent the relationship between length x m and weight y g in a math sentence.**

Length and Weight of a Wire

Length x (m)	1	2	3	4	5	6	
Weight y (g)	20	40	60	80	100	120	

① Let's find the quotient of $y \div x$.

② Let's represent the relationship between x and y in a math sentence.

③ Let's find the weight of a wire that is 12 m long. $y = \square \times \square$

Want to confirm

2 The relationship between time and distance when a car travels at 40 km per hour is shown in the table below. Let's look at this table and answer the following questions.

Time and Distance with a Traveling Speed of 40km per hour

Time x (hours)	1	2	3	4	5	6	
Distance y (km)	40	80	120	160	200	240	

① Is the distance proportional to the traveling time?

② Let's represent, in a math sentence, the relationship between the time x hours and distance y km traveled by the car.

③ How many hours will this car need to travel 560 km?

Want to explore

6 **Let's investigate the relationship between x and y, when the length of one side of an equilateral triangle is x cm and its perimeter is y cm.**

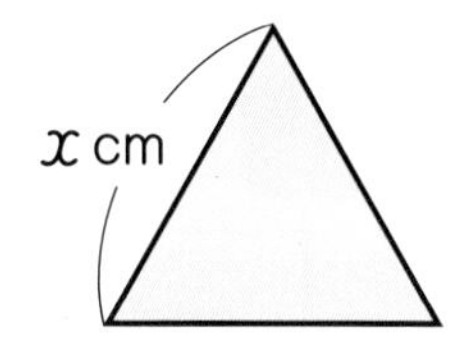

Length of one Side and Perimeter of an Equilateral Triangle

Length of one side x (cm)	1	2	3	4	5	6	
Perimeter y (cm)	3	6					

① Let's fill in the table.

② Is y proportional to x?

③ Let's represent the relationship between x and y in a math sentence. Also, what does the constant number represent?

When y is proportional to x, the relationship can also be represented with the following math sentence:

$$y = x \times \text{constant number}$$

④ Let's find the perimeter when the length of one side is 20 cm and 35 cm.

⑤ Let's find the length of one side when the perimeter is 51 cm.

Want to confirm

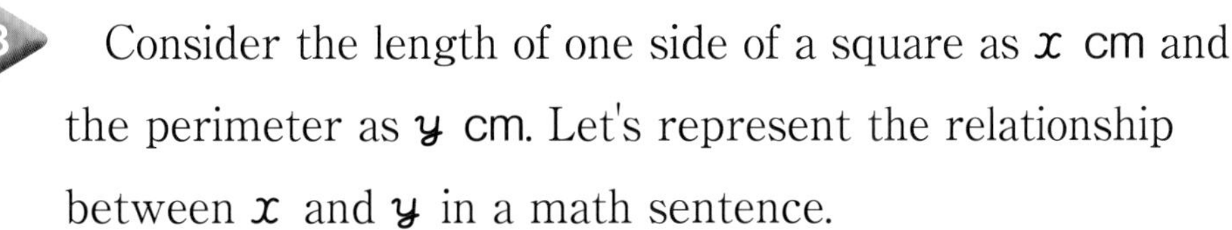

3 Consider the length of one side of a square as x cm and the perimeter as y cm. Let's represent the relationship between x and y in a math sentence.

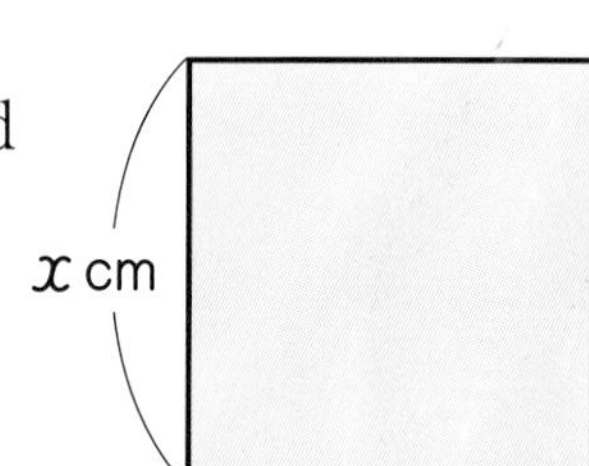

Want to try

4 Write the following relationships between x and y in a table, and represent them in a math sentence. Also, what does the constant number represent?

① Diameter of a circle as x cm and circumference as y cm.

② Number of 50 yen cards as x and total cost as y yen.

③ Length of one side x cm and perimeter y cm of a regular hexagon.

2 Graph of a proportion

Want to know

1 **The table below shows the relationship between the amount of water x L and depth y cm inside a tank. Let's answer the following questions.**

Amount of Water and Depth inside the Tank

Amount of water x (L)	1	2	3	4	5	
Depth y (cm)	2	4	6	8	10	

Way to see and think

The relationship between x and y, represented in the table, can be plotted by a pair of x and y values, just as in a line graph.

① Let's represent the relationship between x and y in a math sentence.

Want to represent

② Let's draw in the following diagram, the points that represent the corresponding pair of x and y values from the above table.

Amount of Water and Depth inside the Tank

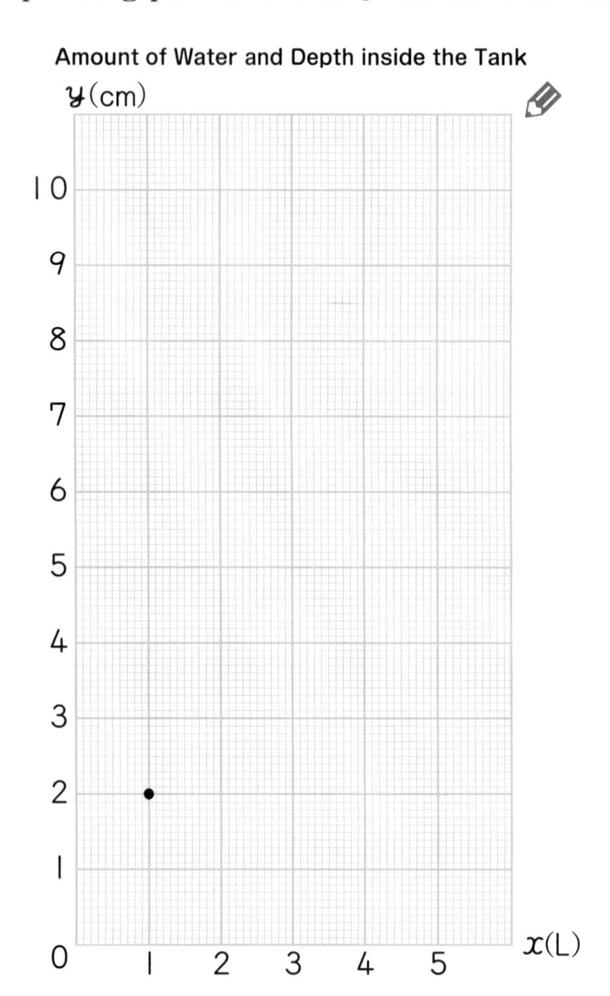

Purpose What kind of graph represents the proportional relationship?

③ How are the graph's points lining up?

④ Let's complete the table below and draw the points that represent the corresponding pair of x and y values on the diagram in the previous page.

Amount of Water and Depth inside the Tank

Amount of water x (L)	0	0.1	0.2	0.5	1	2.4	3.9	
Depth y (cm)	0				2			

When the points that represent the corresponding pair of x and y values are connected, it becomes a straight line as the one shown on the right.

This straight line is the graph of

$y = 2 \times x$.

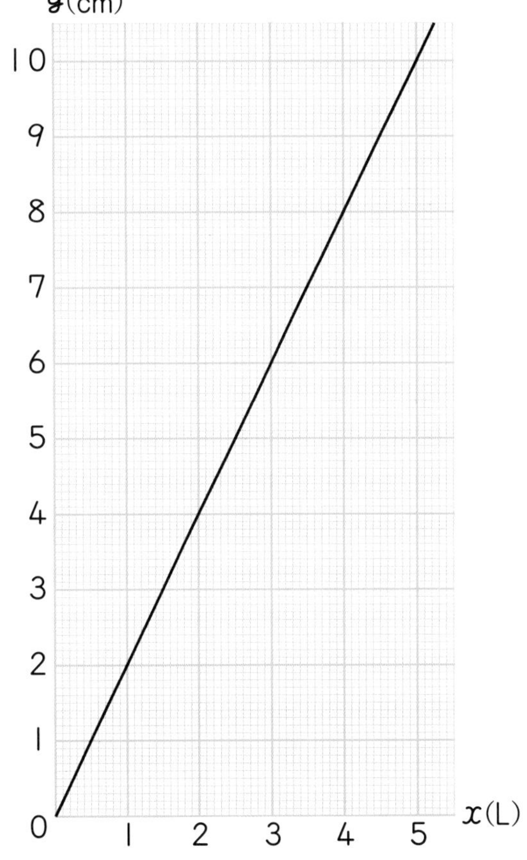

⑤ Let's find, from the graph, the depth when the amount of water is 1.5 L.

Also, let's find, from the graph, the amount of water when the depth is 5 cm.

Summary

When you represent a proportional relationship with a graph, it becomes a straight line that goes through the intersection point of the vertical and horizontal axis at 0.

Want to think Using graphs of a proportion

2

The graph below represents the relationships between the length x m and weight y g of two different wires Ⓐ and Ⓑ. Let's answer the following questions about these relationships.

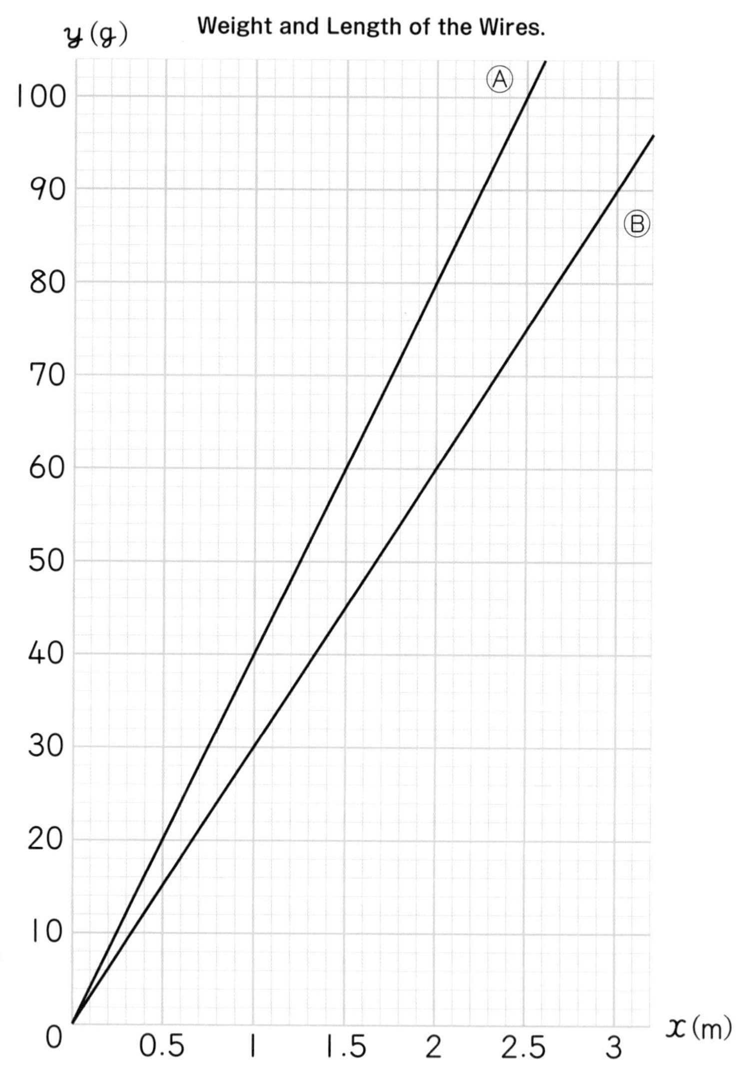

① Can you say which wire weighs more?
Where can you interpret it from the graph?

② Let's read the following length or weight of each wire from the graph.

ⓐ Weight of a 2.4 m wire.

ⓑ Length of a 48 g wire.

③ How many grams is the weight per meter of each wire?

④ Let's represent the relationship between x and y, from Ⓐ and Ⓑ, in a math sentence.

⑤ Which wire, Ⓐ or Ⓑ, is each of the following wires?

ⓐ Wire with a length of 3.8 m that weighs 114 g.

ⓑ Wire with a length of 4.2 m that weighs 168 g.

Want to deepen

⑥ Let's discuss whether the difference between the weight of Ⓐ and Ⓑ is proportional to the length.

3 Using the properties of proportion

Want to think

1 **The table below represents the relationship between the amount of cola and amount of sugar in it. Let's answer the following questions about the relationship between these two amounts.**

Amount of Cola and Amount of Sugar

Amount of cola x (mL)	0	1	50	100	150	180	250
Amount of sugar y (g)	0		6	12	18		

① Is the amount of sugar y g proportional to the amount of cola x mL?

② How many grams of sugar are inside 250 mL of cola?

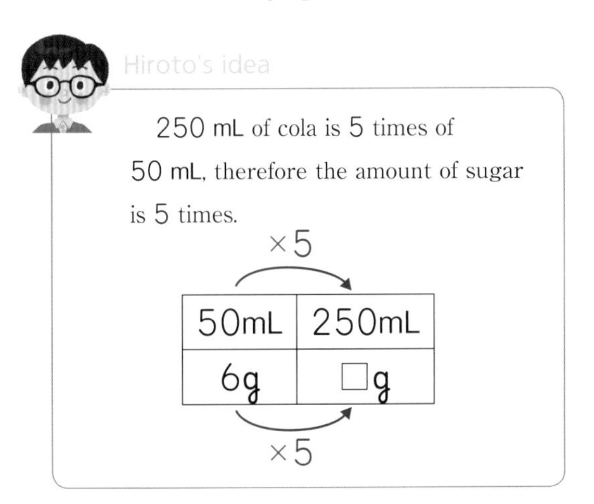

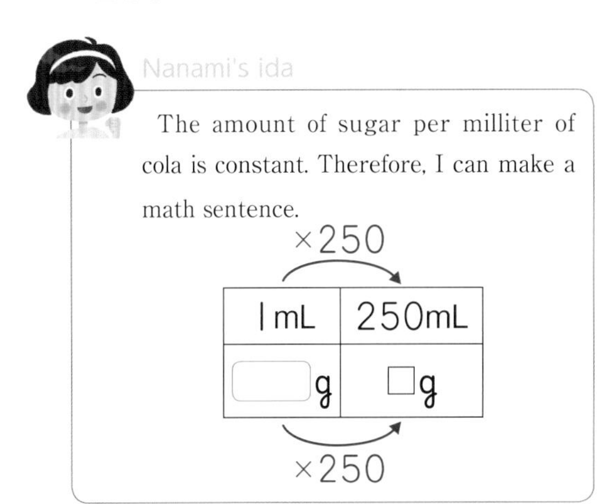

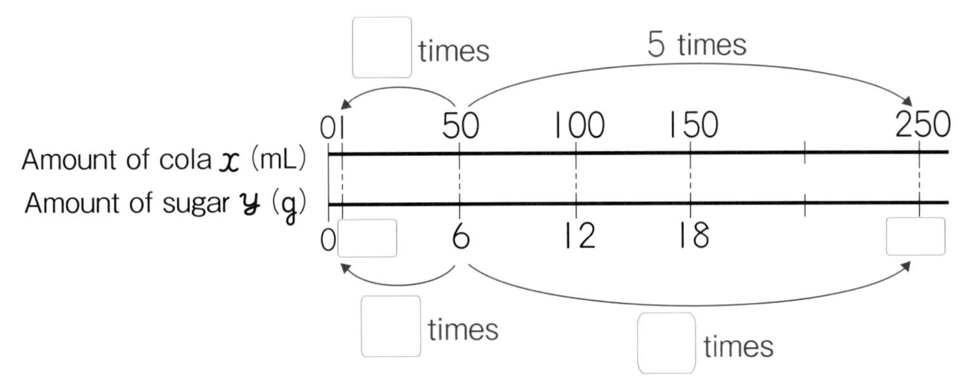

Ⓐ Let's find the answer using Hiroto's idea.

Ⓑ Using Nanami's idea, let's represent the relationship between x and y in a math sentence. y = □ × x

③ How many grams of sugar are there inside 180 mL of cola?

2 **The table below represents the relationship between the number of nails x and its weight y g. Let's answer the following questions about the relationship between these two quantities.**

Number and Weight of Nails

Number of nails x (nails)	0	1	50	100	150	200	250	
Weight of nails y (g)	0	Ⓐ	300	600	900	Ⓑ	Ⓒ	

① Is y proportional to x?

② Let's find the numbers that apply into Ⓐ, Ⓑ, and Ⓒ.

③ Let's represent the relationship between x and y in a math sentence. Also, if the nail's weight is 240 g, can you say how many nails there are?

1 The graph below represents the relationship between the weight x g and length y cm of an extended rubber. Let's answer the following questions about the relationship between these two quantities.

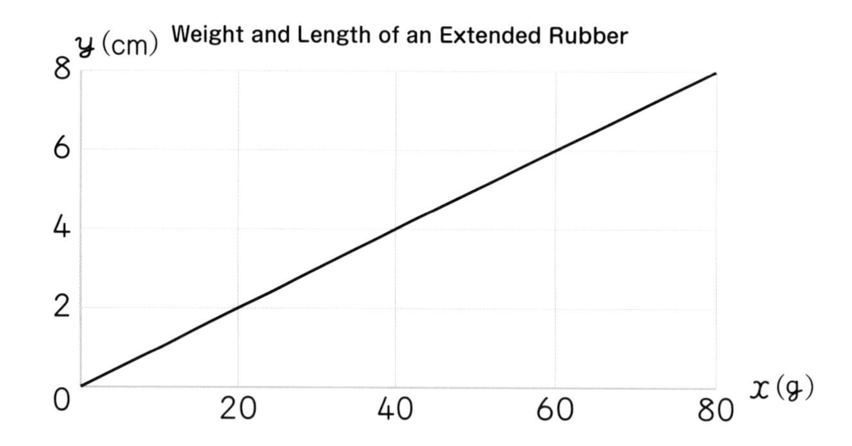

① If the weight is increased by 20 g, how many centimeters will the rubber extend?

② Represent the relationship between x and y in a math sentence.

③ If you put a stone onto the rubber, it extends 13 cm.
Can you say how many grams is the weight of this stone?

Want to explore

1 **Considering a rectangle with an area of 24cm², let's explore how does the width and length change.**

Purpose What is the relationship between the width and length of a rectangle with a fixed area?

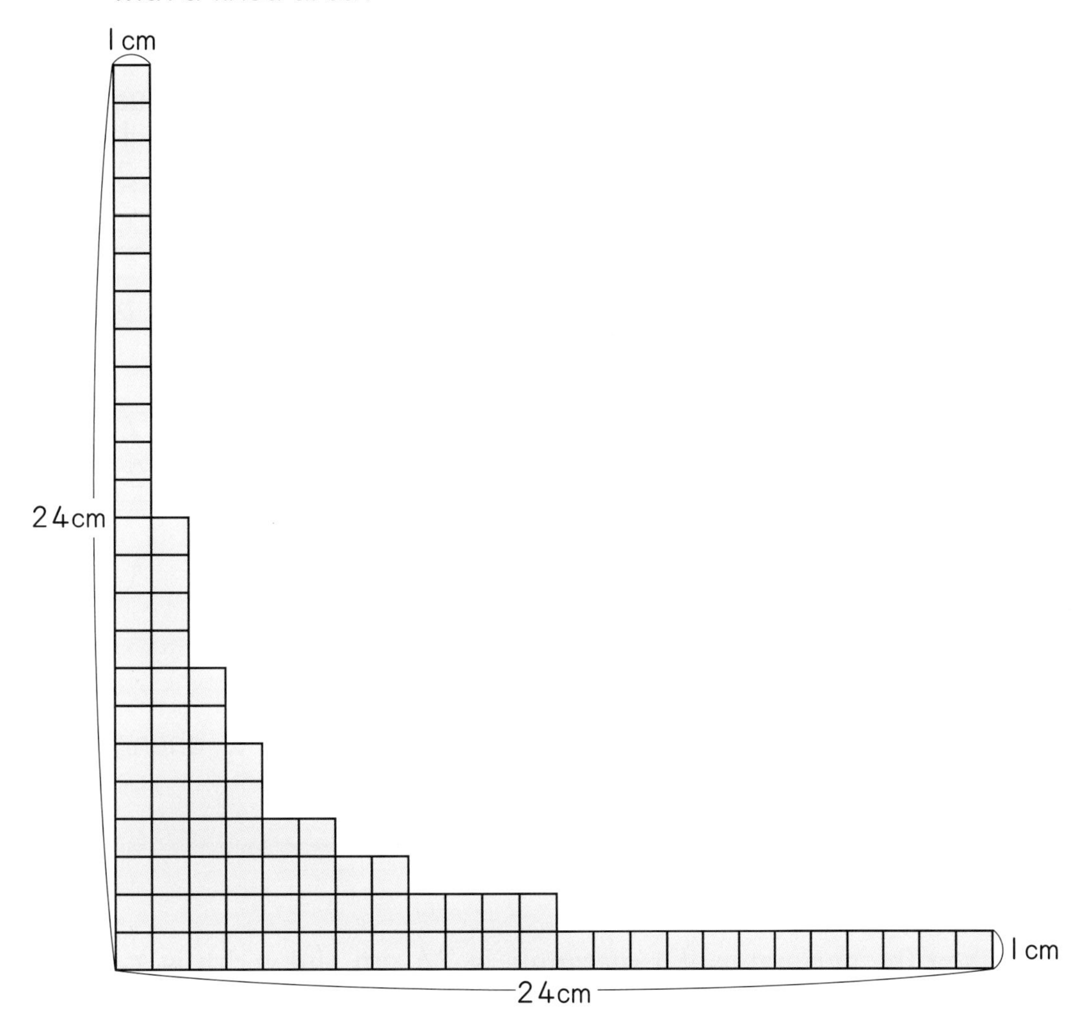

① Let's summarize in the following table, the various rectangles that were created after arranging 24 squares of 1 cm².

Width and Length of a Rectangle with an Area of 24 cm²

Width x (cm)	1	2	3	4	6	8	12	24
Length y (cm)	24							

② If the value of x changes by 2 times, 3 times and so on, how does the corresponding value of y change?

Width x (cm)	1	2	3	4	6	8	12	24
Length y (cm)	24							

2 times, 3 times, 2 times (above); □ times, □ times, □ times (below)

When there are 2 quantities changing together, x and y, if the **value of x changes by 2, 3 times** and the corresponding **value of y changes by $\frac{1}{2}$, $\frac{1}{3}$ times** respectively, we say that **y is inversely proportional to x.**

Proportion can be called **direct proportion** as against inverse proportion.

③ If the value of x changes by $\frac{1}{2}$ times or $\frac{1}{3}$ times, how does the value of y change?

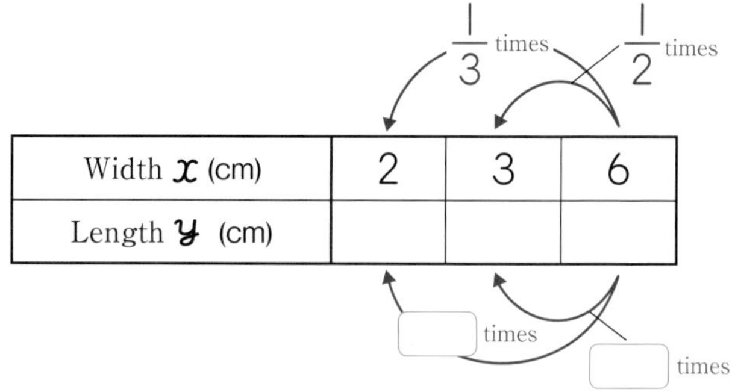

Width x (cm)	2	3	6
Length y (cm)			

Summary

When y is inversely proportional to x, if the value of x changes by □ times, the value of y changes by $\frac{1}{\square}$ times.

Want to confirm

1 When the perimeter of a rectangle is 24 cm, the width is x cm and the length is y cm. Is y inversely proportional to x?

Width and Length of a Rectangle with Perimeter of 24cm

Width x (cm)	1	2	3	4	5	6
Length y (cm)	11	10	9	8	7	6

2

Let's represent, with a math sentence and a graph, the relationship between x and y when the area of a rectangle is 24 cm², the width is x cm, and the length is y cm.

Purpose What is the math sentence and graph of an inverse proportion?

Width and Length of a Rectangle with a Fixed Area of 24 cm²

Width x (cm)	1	2	3	4	6	8	12	24
Length y (cm)	24	12	8	6	4	3	2	1

① What kind of rule is there between x and y?

② Let's find the product of the corresponding x and y values. What does the product of x and y represent?

③ Let's represent the relationship between x and y in a math sentence.

Width (cm)		Length (cm)		Area (cm²)
1	×	24	=	24
2	×	12	=	24
3	×	8	=	☐
4	×	6	=	☐
x	×	y	=	☐

Summary

When there are 2 quantities changing together, x and y, and y is inversely proportional to x, its relationship can be represented with the following math sentence:

$$x \times y = \text{constant number}$$

④ Let's find the value of y when the value of x is 5.

$5 \times y = 24$

$y = 24 \div 5$

$= ☐$

When y is inversely proportional to x, it can also be represented with the following math sentence:

$$y = \text{constant number} \div x$$

⑤ Let's draw in the following diagram, the points that represent the corresponding pair of x and y values from the table in the previous page.

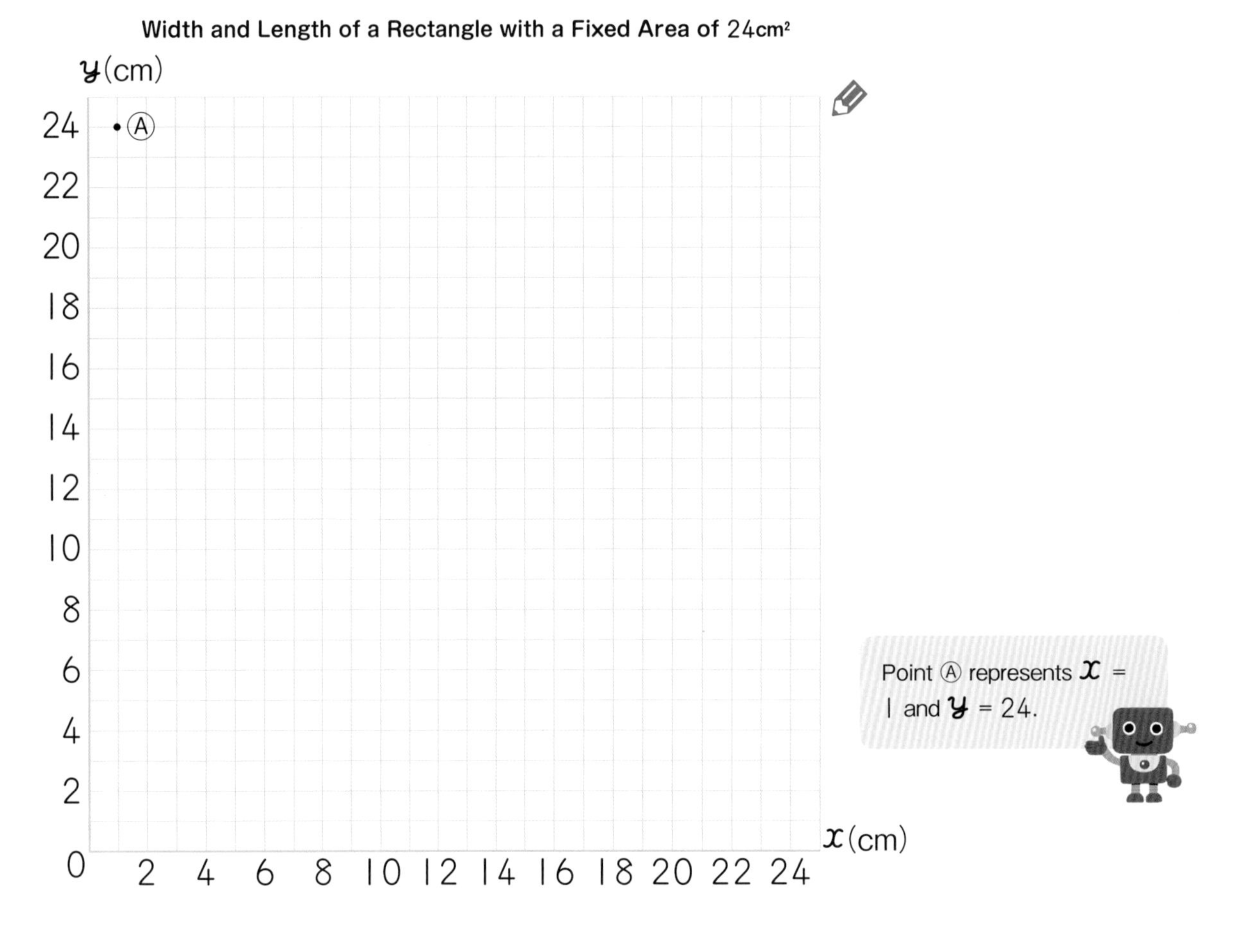

⑥ Let's complete the table below when x is 1.5, 2.5, 7.5, and 12.5, and draw the points that represent the corresponding pair of x and y values on the above diagram.

Width and Length of a Rectangle with a Fixed Area of 24cm²

Width x (cm)	1.5	2.5	7.5	12.5
Length y (cm)				

⑦ Let's compare with the proportional graph on page 169.

When the points from the previous page are drawn finely, it becomes as follows.

Width and Length of a Rectangle with a Fixed Area of 24cm²

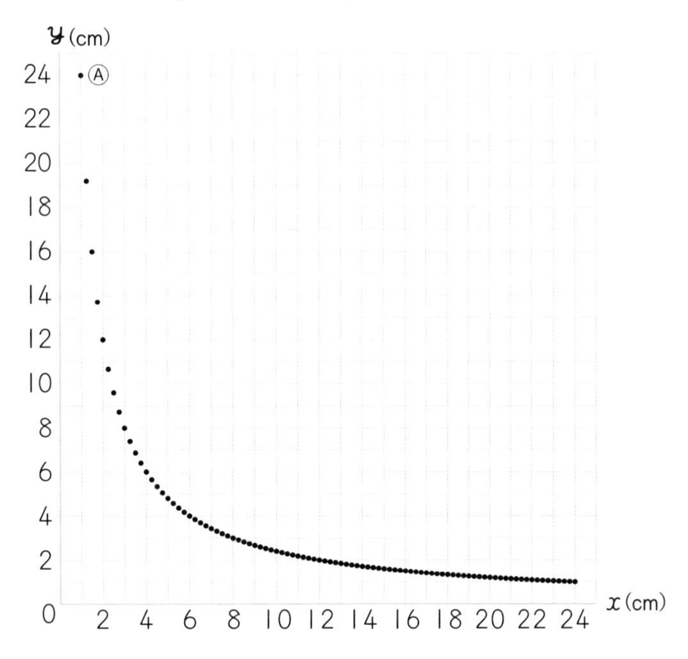

Want to think

3 **There is a project that takes 60 days to be completed by 1 person that makes the same amount of work per day. If the number of people working is x and the number of days is y, let's think about the relationship between x and y.**

① Let's represent the relationship between x and y in a math sentence.

② How many days does it take to complete the work when 5 people are working?

③ How many people are needed to complete the work in 10 days?

Want to confirm

2 Let's investigate the relationship between speed and time when the distance is 36 km.

① Let's summarize the relationship between speed and time in the following table.

Speed and Time

Distance per hour (km)	1	2	3	4	6	9	12	18	36
Time (hours)									

② Is time inversely proportional to the speed?

③ Let's consider speed as x km per hour and time as y hour, and represent the relationship between x and y in a math sentence.

What you can do now

☐ **Understanding the meaning of proportion.**

1 Let's fill in the blanks in the following tables with the number that applies.

① **Number of Pencils and Total Cost**

Number of pencils x (pencils)	0	1	2	3	4	5	
Total cost y (yen)	0	50	100				

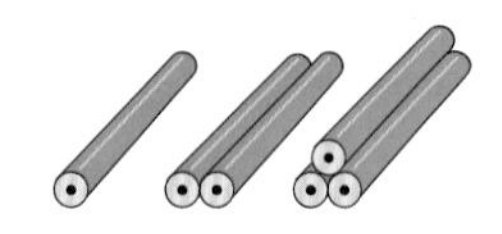

② **Walking Time and Distance**

Time x (hours)	0	1	2	3	4	5	
Distance y (km)	0	4	8				

☐ **Can represent a proportional relationship with a math sentence and graph.**

2 There is a ribbon that costs 80 yen per meter.

① Let's represent the relationship between the length x m and cost y yen of the ribbon in the following table.

Length and Total Cost of the Ribbon

Length x (m)	0	1	2	3	4	5	
Cost y (yen)	0	80					

② Let's represent the relationship between x and y in a math sentence.

③ Let's represent the relationship between the corresponding x and y values in a graph.

Length and Total Cost of the Ribbon

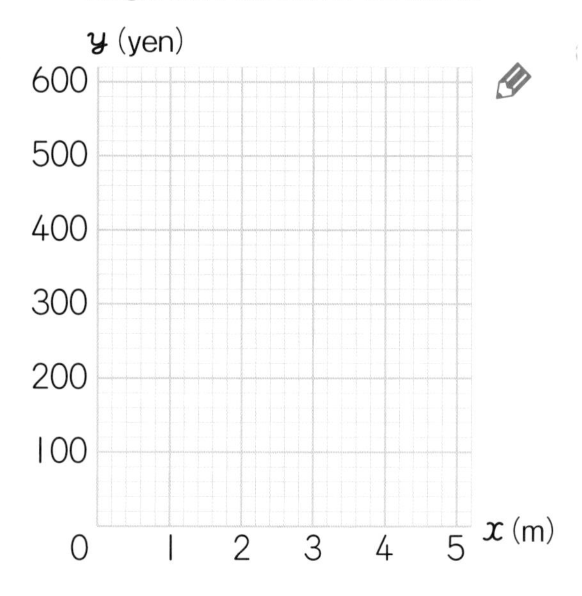

☐ **Can solve problems using proportions.**

3 Let's answer the following questions about the table below.

Length and Weight of a Wire

Length x (cm)	0	1	2	3	4	5	6	
Weight y (g)	0	9	18	27	36	45	54	

① Let's represent the relationship between x and y in a math sentence.

② How many grams is the weight of a wire that is 8 cm long?

③ How many centimeters is the length of a wire that weighs 117 g?

Understanding the meaning of inverse proportion.

4 Let's fill in the blanks in the following tables with the number that applies.

① Base and Height of a Parallelogram with a Fixed Area of 16cm²

Base x (cm)	1		4	5	8	
Height y (cm)		8			2	1

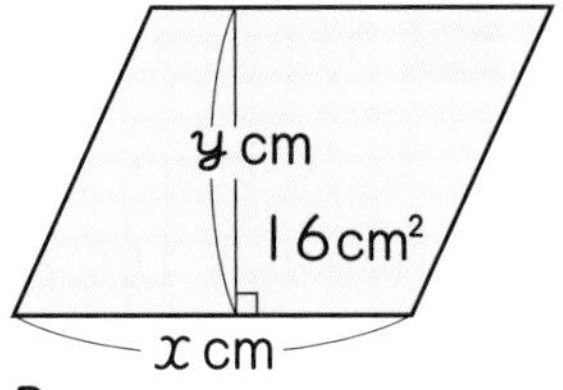

② Number of People and Days to Finish a Work that is Completed by one Person in 45 Days

Number of people x (people)	1	3	5	9	15	45
Days y (days)						

Can represent an inversely proportional relationship with a math sentence and graph.

5 Let's investigate the relationship between speed and time when the distance is 24 km.

① Let's summarize the relationship between speed x and time y in the following table.

Speed and Time

Distance per hour x (km/h)	1	2	3	4	6	8	12	24
Time y (hours)								

② Let's represent the relationship between x and y in a math sentence.

③ Let's represent the relationship between the corresponding x and y values in a graph.

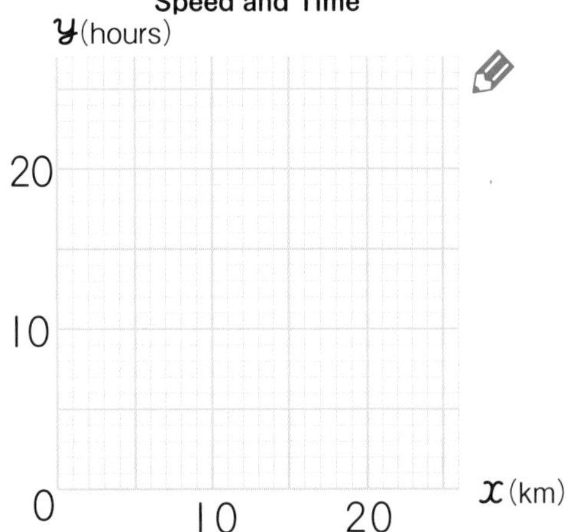

Can solve problems using inverse proportions.

6 Let's answer the following questions about the table below.

Length and Width of a Rectangle with a Fixed Area of 24cm²

Length x (cm)	1	2	3	4	6	9	12	18
Width y (cm)								

① Let's represent the relationship between x and y in a math sentence.

② How many centimeters is the width when the length is 10 cm?

Supplementary problems p.238

Usefulness and efficiency of learning

1 There is a ribbon that costs 150 yen per meter.

① Consider the length of the ribbon as x meters. When the length of the ribbon is 1m, 2m, 3m, and so on, let's find the corresponding cost y yen and summarize it in the table.

Length and Cost of the Ribbon

Length x (m)	0	1	2	3	4	5	6	
Cost y (yen)	0							

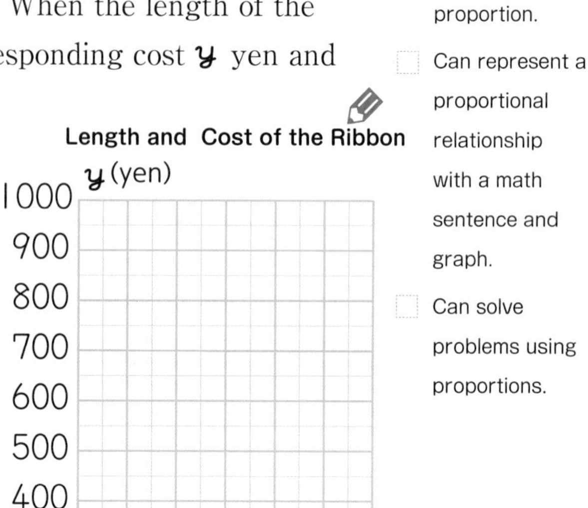

② What is the cost of the ribbon, y yen, proportional to?

③ Let's represent the relationship between x and y in a math sentence.

④ Let's represent the relationship between x and y with a graph.

⑤ How much is the cost when the ribbon's length is 2.5 m?

⑥ How many meters is the ribbon's length when the cost is 1800 yen?

☐ Understanding the meaning of proportion.

☐ Can represent a proportional relationship with a math sentence and graph.

☐ Can solve problems using proportions.

2 The graph on the right shows the relationship between the length x m and weight y g of wire Ⓐ and wire Ⓑ.

Let's answer by looking at the graph.

① Considering the same length, which wire is heavier, Ⓐ or Ⓑ?

② How many grams is the weight per meter of wire Ⓐ and wire Ⓑ?

③ Let's represent, for both wires, the relationship between x m and y g in a math sentence.

④ Which wire, Ⓐ or Ⓑ, weighs 21 g at a length of 3.5 m?

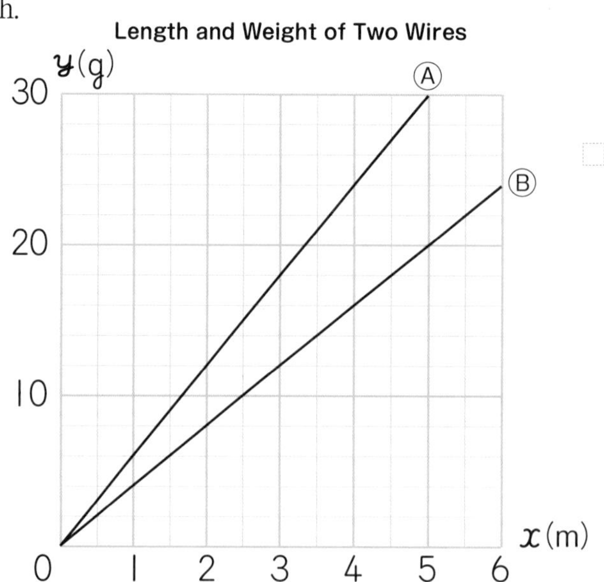

☐ Can represent a proportional relationship with a math sentence and graph.

☐ Can solve problems using proportions.

3 The figure on the right is a parallelogram with an area of 12 cm², base x cm, and height y cm.

Let's answer the following questions.

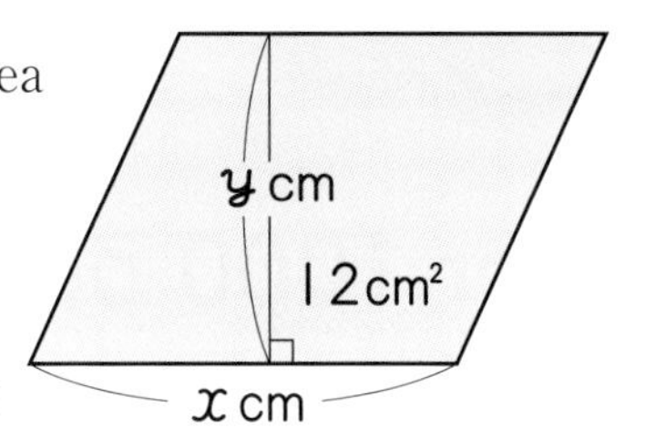

☐ Can understand and represent an inversely proportional relationship with a math sentence and graph.

☐ Can solve problems using inverse proportions.

① The table below shows the relationship between x and y. Let's fill in the blanks with the number that applies.

Base and Height of a Parallelogram with a Fixed Area of 12 cm²

Base x (cm)	1		3	4	6	
Height y (cm)		6			2	1

② Let's represent the relationship between x and y in a math sentence.

③ Let's draw the points that represent the corresponding pair of x and y values from the above table.

④ When the base is 8 cm, how many centimeters is the height?

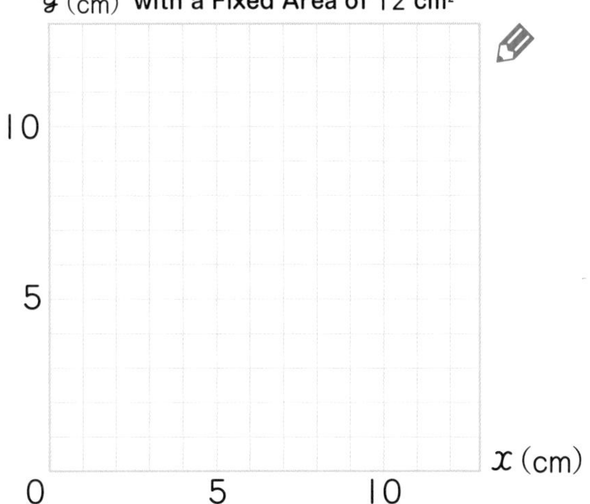

4 You leave Tokyo and go to Shizuoka by the highway. It is about 160 km from Tokyo to Shizuoka. Let's answer the following questions.

☐ Can solve problems using inverse proportions.

① Let's represent the relationship between x and y in a math sentence when the speed is x km per hour and time is y hours.

(Shizuoka City, Shizuoka Prefecture)

② You arrived to Shizuoka in 1 hour and 36 minutes. How many kilometers per hour was your speed?

Curry for 1000 people

Want to explore What size of pot and ingredients are needed?

In home economics class, we made curry for 4 people using the following pot and ingredients:

Ingredients for 4 people

Ingredients	Quantity	
Curry paste	$\frac{1}{2}$ box	115 g
Pig meat (thin slices)		250 g
Onion	2 medium size	400 g
Potato	$1\frac{1}{2}$ medium size	230 g
Salad oil	1 tablespoon	
Water	$4\frac{1}{4}$ cups	850 g

Pot used

Diameter 22 cm

Height 10.5 cm

Capacity 3.3 L

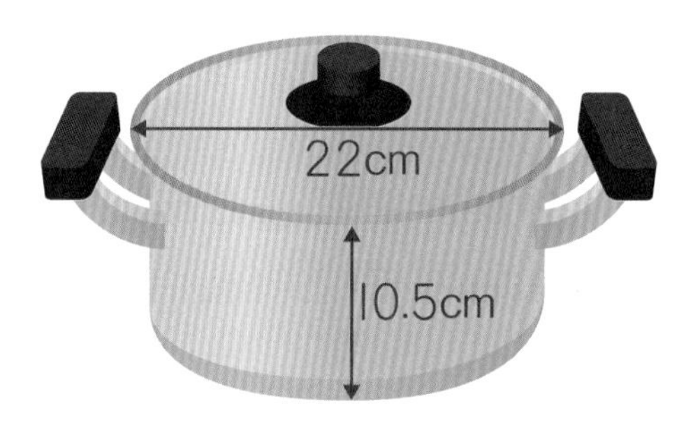

Let's think about the size of the pot and ingredients that are necessary when making curry for 1,000 people in a regional initiative for disaster prevention.

Think by myself

1 Let's think about the amount of materials and size of the pot that is needed to make curry for 1000 people referred to the recipe made in home economics class.

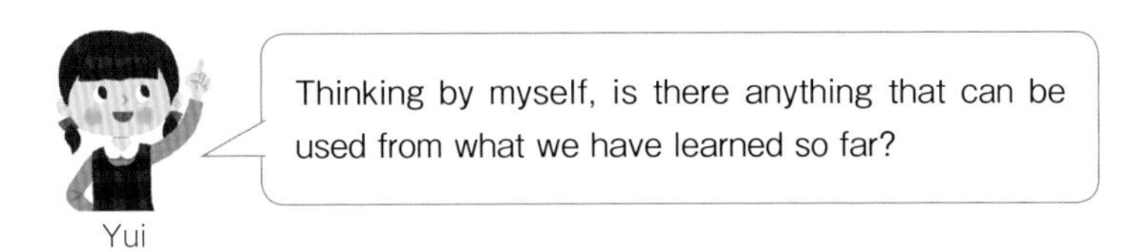

Think in groups

2 Let's separate the class into groups and use what you have learned in mathematics to find the amount of materials and the size of the pot.

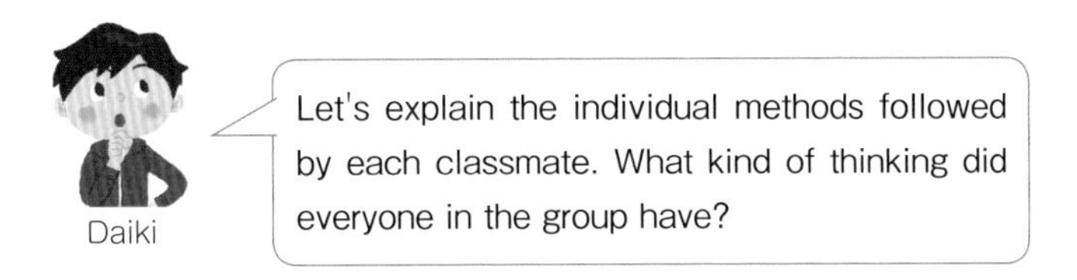

Think as a class

3 Let's present what each group discussed.

Also, discuss with your classmates the difficulty of preparing food for 1,000 people in the case of a disaster.

01605

What is the result of the new physical fitness test?

Is my own record better when compared to everyone?

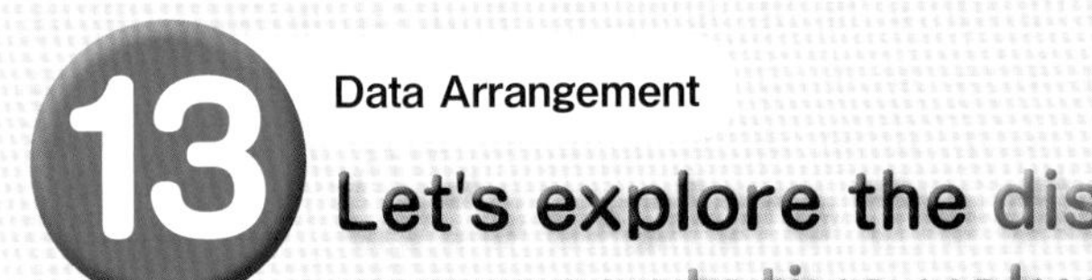

Data Arrangement

13 Let's explore the distribution of data and its representative values.

1 Representative values

Want to explore

1 The following table summarizes the results for the side-step and jump test of two groups in 6th grade. Let's compare which record is better.

Group 1: Side-step and Jump Records

Number	Score	Number	Score	Number	Score	Number	Score
1	56	11	53	21	46	31	31
2	49	12	47	22	57	32	57
3	42	13	44	23	38	33	33
4	46	14	49	24	39	34	37
5	36	15	58	25	50	35	53
6	37	16	53	26	58	36	44
7	49	17	46	27	38	37	43
8	32	18	48	28	46	38	46
9	52	19	38	29	56	39	48
10	39	20	37	30	41		

Group 2: Side-step and Jump Records

Number	Score	Number	Score	Number	Score	Number	Score
1	55	11	38	21	44	31	55
2	44	12	41	22	58	32	49
3	49	13	37	23	49	33	52
4	39	14	50	24	36	34	45
5	37	15	49	25	39	35	38
6	40	16	40	26	40	36	50
7	45	17	38	27	54	37	49
8	54	18	52	28	49	38	41
9	49	19	36	29	38	39	40
10	60	20	49	30	37	40	37

Want to represent

① Let's find the mean for the records of each group.

The mean of the records of Group 1.

The mean of the records of Group 2.

The mean learned so far is called the **mean value.**

Mean value = Total of data values ÷ Number of data

② Which group's record is better?

③ Which is the maximum score for each group? Also, what is the minimum score

Want to discuss

④ From what we have found so far, as for groups 1 and 2, let's discuss which group can be said that has the best records.

2 Considering the tables on page 185, the records of Group 1 were represented as follows. Let's try to think about this graph.

Group 1: Side-step and Jump Records

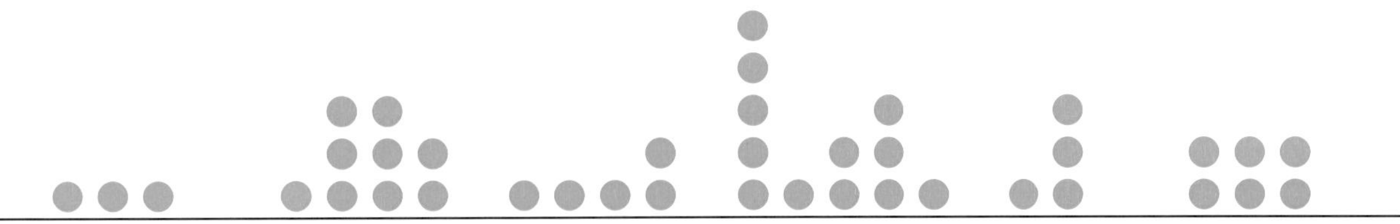

① What does the number written below the line represent?

What does ● represent?

The graph shown above is called a **dot plot.** Vertically, you can see the amount of data. Horizontally, you can understand how the data is distributed.

This ● is called a dot. Drawing a dot is called plotting.

② Let's represent the records of Group 2 using a dot plot.

Group 2: Side-step and Jump Records

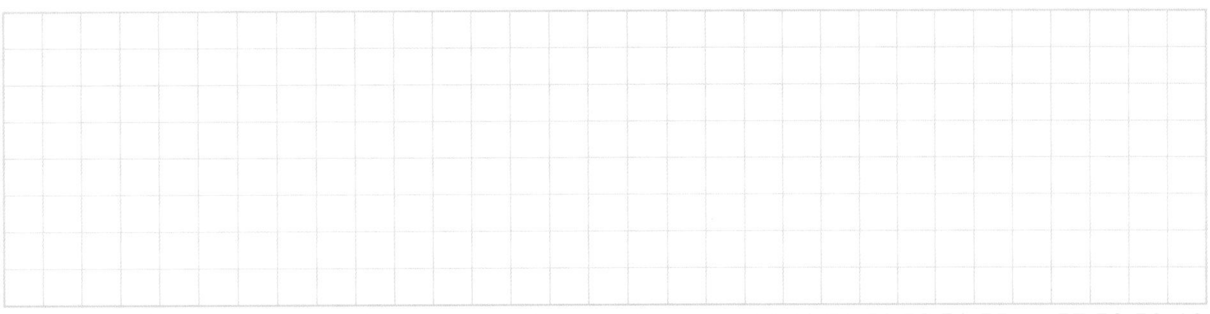

Want to try

Let's look at the dot plots in the previous page and answer the following questions.

① In Group 1, which score has the highest number of children?

Inside the data, the most frequent value is called the **mode value.**

② Let's answer which is the mode value in Group 2.

Want to discuss

Let's discuss whether the children in Group 1 with a score of 49 can be said to have a record that is above the middle of the side-step and jump records on exercise 1 on page 185.

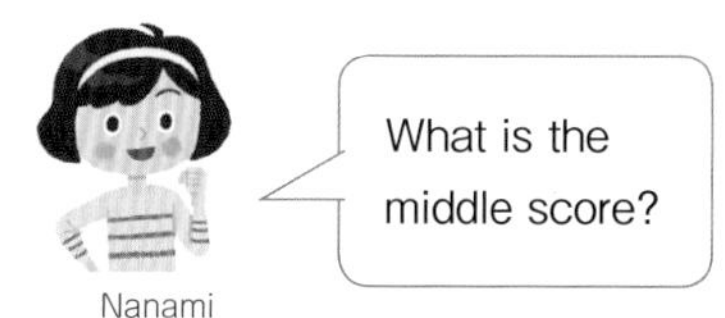

If the data is aligned it seems easy to understand.

When the data is aligned in size order, the value that is located in the middle is called the **median value.**

The median value can be found as follows:

a) when the number of data is odd ... the value in the exact middle.

b) when the number of data is even ... the mean value of the two values in the middle.

① Let's align the data of Group 1 in size order and find the mean value.

② Let's align the data of Group 2 in size order and find the mean value.

Values that represent data, such as the mean value, mode value, and median value, are called **representative value**s.

Want to try

3 The shoe sizes of 20 children of Group 1 in 6th Grade were examined. The shoe sizes were represented in the table below. Let's answer the following questions about this information.

Shoe Sizes in Group 1

(cm)

23.0	21.5	24.0	23.5	23.0	26.0	22.0	22.5	22.5	23.0
21.0	23.5	20.0	24.0	21.5	22.5	23.0	25.0	23.5	22.0

① Let's represent the above table in a dot plot.

② Let's find the mean value, mode value, and median value.

③ Counting from the smallest shoe size, how many centimeters is the shoe in the 12th place?

That's it

Same category as dot plot? — Stem and Leaf Diagram —

Instead of dot plots, the table of Group 1 in 6th grade on page 185 may be organized by tens place to create a stem and leaf diagram as shown below. In the stem and leaf diagram, write the tens place number on the left side of the line and the ones place number on the right instead of the dot.

```
0 |
1 |
2 |
3 | 1 2 3 6 7 7 7 8 8 8 9 9
4 | 1 2 3 4 4 6 6 6 6 6 6 7 8 8 9 9 9
5 | 0 2 3 3 3 6 6 7 7 8 8
6 |
```

Since the ones place number is aligned in size order, the data is also arranged in size order. Therefore the median is easy to find.

2 Frequency distribution table and histogram

Want to represent Frequency distribution table

In Hiroto's school, Group 3 and Group 4 in 6th grade threw a softball in P.E. class. The following table shows the results.

Records at Throwing a Softball of Group 3 (m)

44	33	32	33	22	38	31	32	35	26	14	20	15	30	25	21	33	25	28	26
32	48	40	38	19	15	21	18	16	11	16	20	23	18	18	8	13	6	7	17

Records at Throwing a Softball of Group 4 (m)

14	25	28	20	35	19	22	34	27	24	33	25	34	42	37	19	27	35	11	21
19	16	26	23	15	22	20	20	10	21	17	11	18	16	19	10				

Nanami

The score was the quantity thrown one by one.

The distance is a continuous quantity.

Hiroto

When throwing a softball, 33m 15cm and 33m 98cm are considered 33m.

How can we summarize the distribution of continuous quantities?

Want to explore

1 **The records of Group 3 are summarized in the table on the right so you can see the total distribution. Let's answer the following questions.**

Records at Throwing a Softball of Group 3

Distance (m)	Number of children
greater than or equal to 5 ~ less than 10	3
10 ~ 15	3
15 ~ 20	
20 ~ 25	
25 ~ 30	
30 ~ 35	8
35 ~ 40	3
40 ~ 45	2
45 ~ 50	1
Total	40

① Let's complete the table where there are blank spaces.

② From what number to what number has the largest number of children?

③ How many children threw the softball greater than or equal to 20 m and less than 25 m?

A section (delimiter) such as "greater than or equal to 30 m and less than 35 m" is called a **class**, and the size of the section (delimiter) is called a **class interval.**

Also, the number of data counted for each class is called **frequency.** A table that shows the distribution by class or frequency is called a **frequency distribution table.**

④ Which class has a number of 8 children?

Summary

If you want to summarize distributed data that has continuous quantities, such as a softball throwing record, you can decide the class interval and use a frequency distribution table.

⑤ Let's complete the frequency distribution table for Group 4 shown on the right.

⑥ When comparing the following records of Group 3 and Group 4, which group has a larger number?

Ⓐ Number of children with a distance greater than or equal to 25 m.

Ⓑ Number of children with a distance less than 15 m.

Ⓒ Number of children with a distance greater than or equal to 20 m and less than 30 m.

⑦ From the explored data, let's discuss which group we can say is better, Group 3 or Group 4?

Records at Throwing a Softball of Group 4

Class (m)	Number of children
greater than or equal to 5 ~ less than 10	
10 ~ 15	
15 ~ 20	
20 ~ 25	
25 ~ 30	
30 ~ 35	
35 ~ 40	
40 ~ 45	
45 ~ 50	
Total	

Only with the frequency distribution table, it is difficult.

Should we find the representative values?

Can we represent it with a dot plot?

But, since the distances are continuous quantities...

2

Based on the distribution of throwing records of Group 3 in page 190 and its corresponding frequency distribution table, the graph below was drawn to explore. Let's think about the following questions.

Records at Throwing a Softball of Group 3

Class (m)	Number of children
greater than or equal to 5 ~ less than 10	3
10 ~ 15	3
15 ~ 20	9
20 ~ 25	6
25 ~ 30	5
30 ~ 35	8
35 ~ 40	3
40 ~ 45	2
45 ~ 50	1
Total	40

Records at Throwing a Softball of Group 3

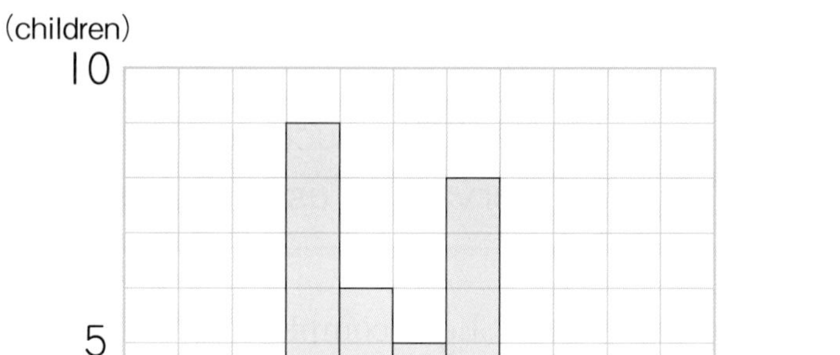

① How many children threw the softball greater than or equal to 30 m and less than 35 m?

② From what number to what number is a class with a number of 1 child?

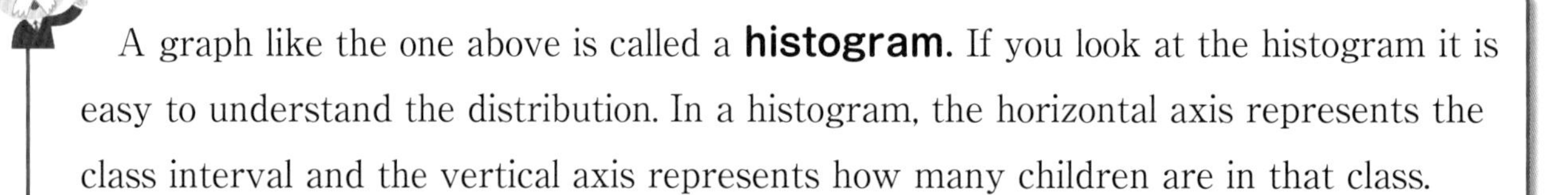

A graph like the one above is called a **histogram.** If you look at the histogram it is easy to understand the distribution. In a histogram, the horizontal axis represents the class interval and the vertical axis represents how many children are in that class.

③ Let's draw the histogram for Group 4.

Records at Throwing a Softball of Group 4

Class (m)	Number of children
greater than or equal to 5 ~ less than 10	0
10 ~ 15	5
15 ~ 20	9
20 ~ 25	9
25 ~ 30	6
30 ~ 35	3
35 ~ 40	3
40 ~ 45	1
45 ~ 50	0
Total	36

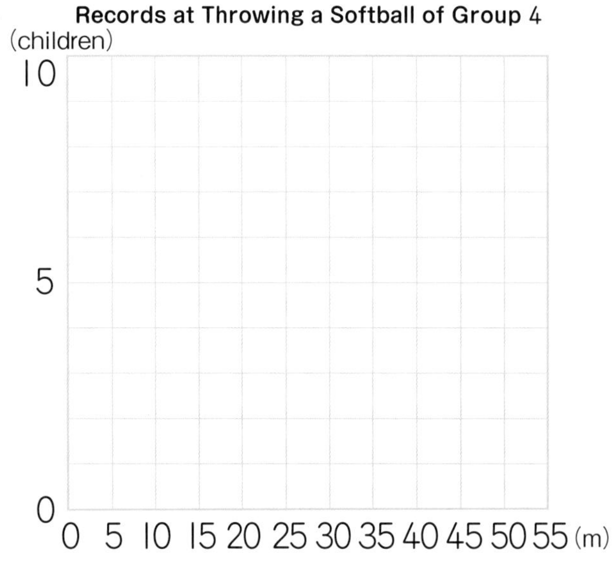

④ Let's compare both histograms and discuss how they are distributed.

⑤ Which class has the largest number of children in each group? Also, what is the percentage of the children in this class, compared to that of the total?

⑥ Based on the child with the best record, which class does the 6th place child belong to in each group?

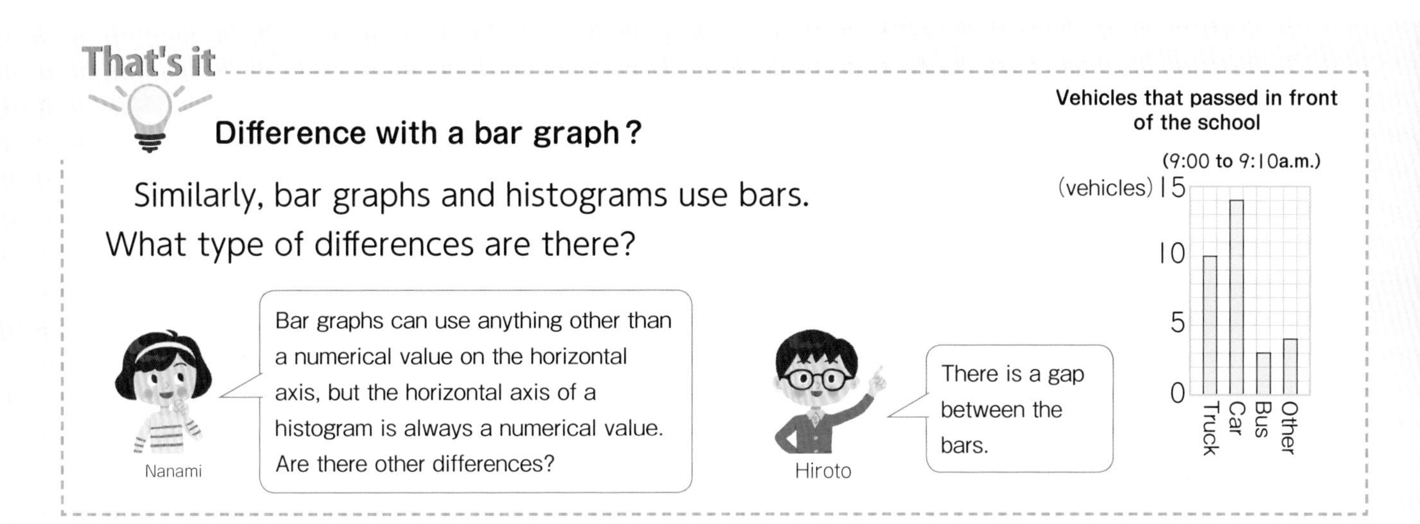

If the class interval changes?

Let's consider what happens when you change the class interval in the frequency distribution table for the throwing records of Group 4.

Records at Throwing a Softball of Group 4 (m)

10	10	11	11	14	15	16	16	17	18	19	19	19	19	20	20	20	21	21	22
22	23	24	25	25	26	27	27	28	33	34	34	35	35	37	42				

Records at Throwing a Softball of Group 4

Class (m)	Number of children
greater than or equal to 10 ~ less than 15	5
15 ~ 20	9
20 ~ 25	9
25 ~ 30	6
30 ~ 35	3
35 ~ 40	3
40 ~ 45	1
Total	36

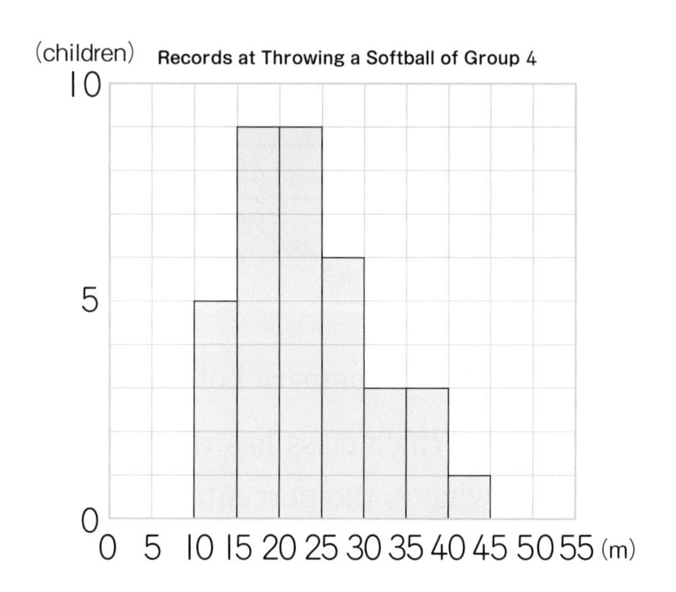

① Create a frequency distribution table with different class intervals as shown in the next page. Let's fill in the blank spaces and draw a histogram.

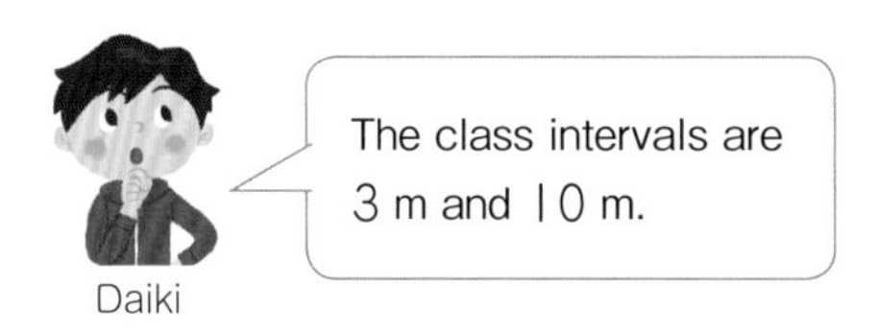

② Let's look at the histogram on the next page and discuss what you notice.

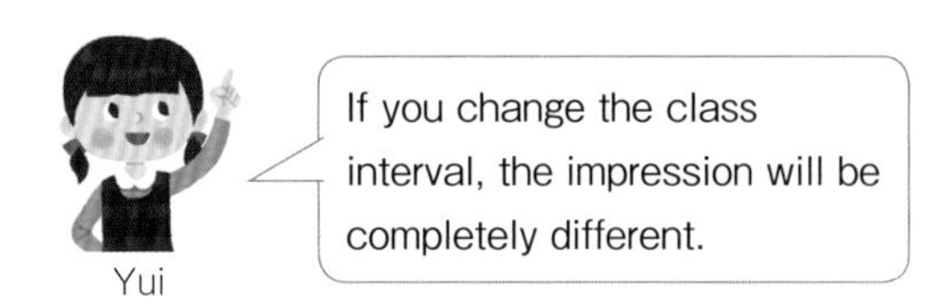

If you don't consider the class interval well, you will not know how the data is distributed.

Records at Throwing a Softball of Group 4

Class (m)	Number of children
greater than or equal to 9 ~ less than 12	4
12 ~ 15	1
15 ~ 18	
18 ~ 21	
21 ~ 24	
24 ~ 27	
27 ~ 30	
30 ~ 33	0
33 ~ 36	5
36 ~ 39	1
39 ~ 42	0
42 ~ 45	1
Total	36

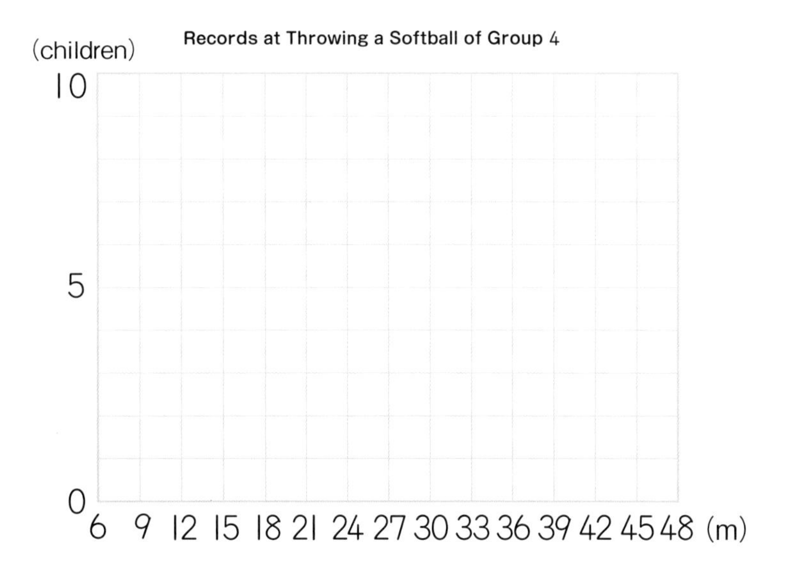

Records at Throwing a Softball of Group 4

Class (m)	Number of children
greater than or equal to 10 ~ less than 20	14
20 ~ 30	
30 ~ 40	
40 ~ 50	1
Total	36

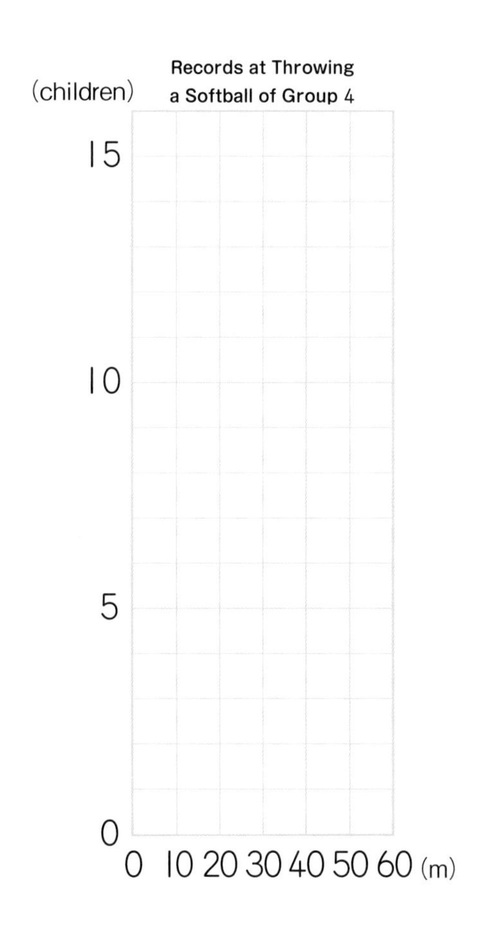

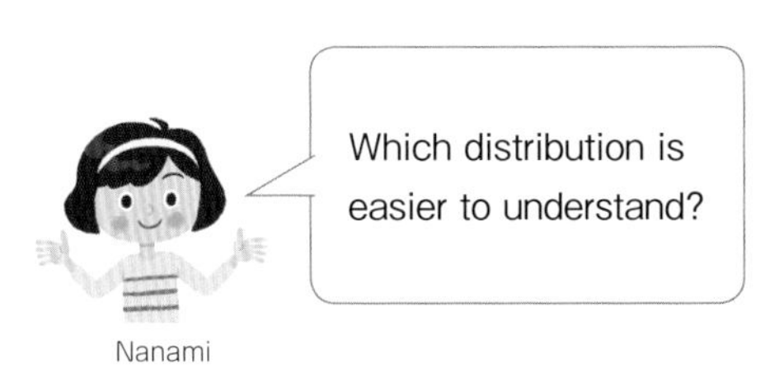

What you can do now

☐ **Understanding the mean value, mode value, and median value. Can draw a dot plot.**

1 The following table shows the records for raising the body in the new physical test of Yumi's group.

Let's draw a dot plot and find the mean value, mode value, and median value.

Records for Raising the Body

Number	Times	Number	Times	Number	Times	Number	Times
1	22	6	13	11	22	16	24
2	11	7	24	12	27	17	26
3	10	8	15	13	20	18	26
4	15	9	28	14	22	19	23
5	21	10	17	15	29	20	22

10 11 12 13 14 15 16 17 18 19 20 21 22 23 24 25 26 27 28 29 30 (times)

☐ **Can make a frequency distribution table and a histogram.**

2 Using the data from **1**, let's answer the following questions.

① Let's complete the frequency distribution table. Also, let's draw a histogram.

Records for Raising the Body

Class (times)	Number of children
greater than or equal to 5 ~ less than 10	
10 ~ 15	
15 ~ 20	
20 ~ 25	
25 ~ 30	
Total	

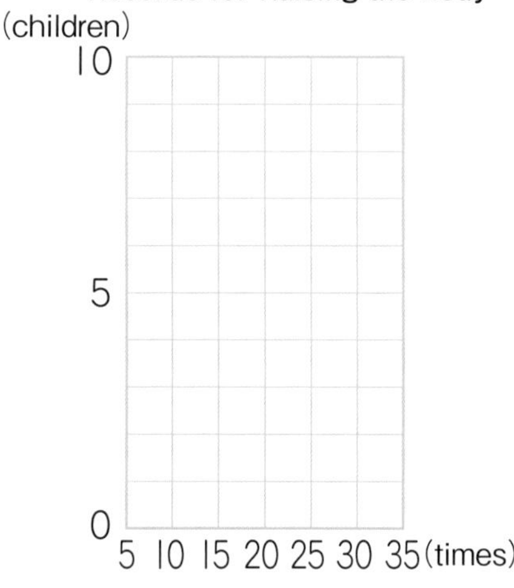

② Let's see the frequency distribution table in ① and answer.

ⓐ How many children raised their body greater than or equal to 15 times and less than 20 times?

ⓑ From the best record, which class does the 7th place child belong to?

Supplementary problems p.239

Usefulness and efficiency of learning

- Understanding the mean value, mode value, and median value. Can draw a dot plot.
- Can make a frequency distribution table and a histogram.

1 The following table shows the results of examining the school's commuting time for Group 1 and Group 2.

Commuting Time for Group 1

Number	Time (min)	Number	Time (min)
1	20	11	28
2	22	12	20
3	19	13	20
4	21	14	22
5	15	15	26
6	13	16	11
7	25	17	24
8	18	18	23
9	20	19	27
10	16	20	10

Commuting Time for Group 2

Number	Time (min)	Number	Time (min)
1	28	11	16
2	15	12	25
3	30	13	23
4	23	14	22
5	13	15	13
6	15	16	29
7	21	17	18
8	11	18	23
9	10	19	23
10	25		

① Let's find the mean value, mode value, and median value for each group. As for the mean value, let's round off to the hundredths place.

② The diagrams on the right is represented as the histograms based on the above tables. Let's answer the following questions about Group 1 and Group 2.

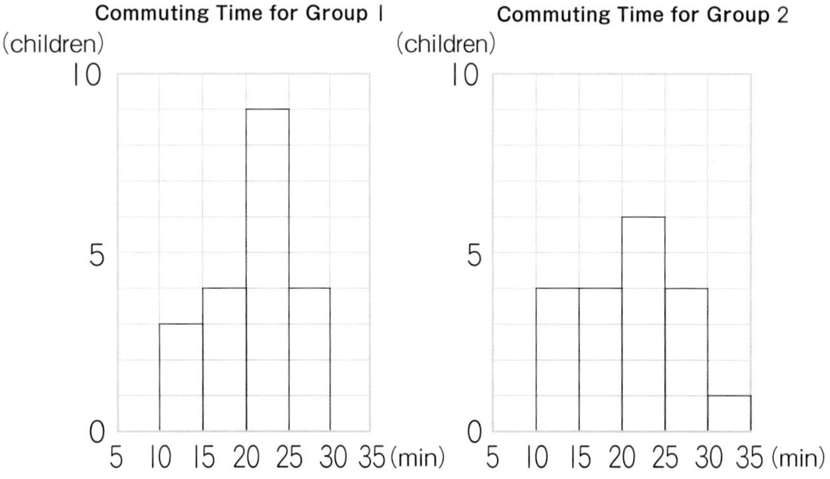

ⓐ How many children does the class greater than or equal to 20 min and less than 25 min have in each group? Also, what is the percentage of the children in this class, compared to that of the total? If the answer is not divisible, round off to the tenths place.

ⓑ From the shortest commuting time, which class does the 8th place child belong to?

③ Let's compare the answers from ① and histograms from ②. Let's write what you noticed.

Let's deepen.

The only thing that can be understood from a histogram is the distribution?

Let's separate and think.

Want to know

Based on the distribution of throwing records of Group 3 on page 192, Yui considered the following.

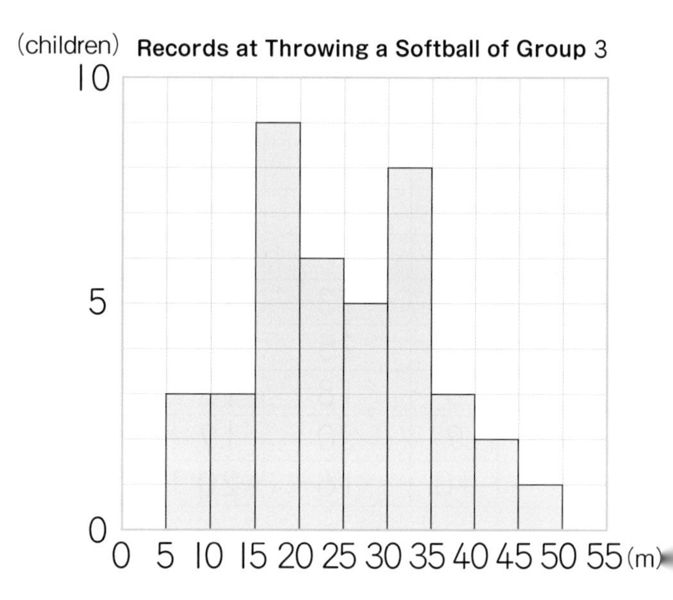

① Yui considered to separate the results from boys and girls. She knew the results for boys, so the histogram was drawn as shown below. Using the result from this histogram, let's draw the histogram for girls.

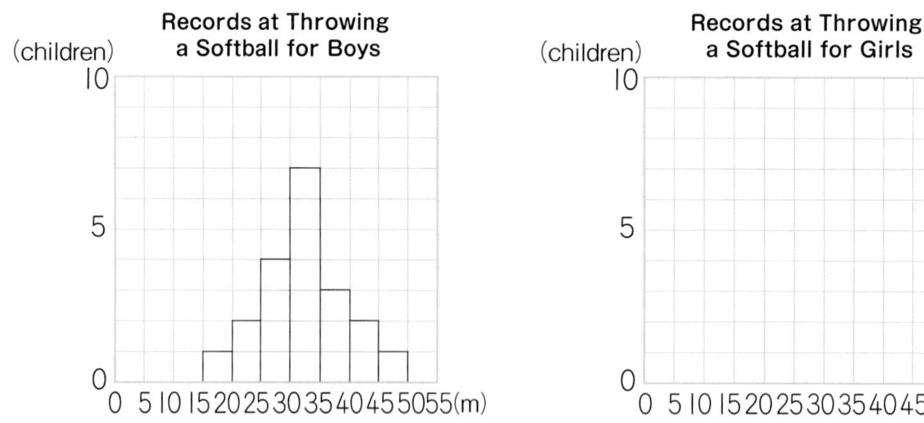

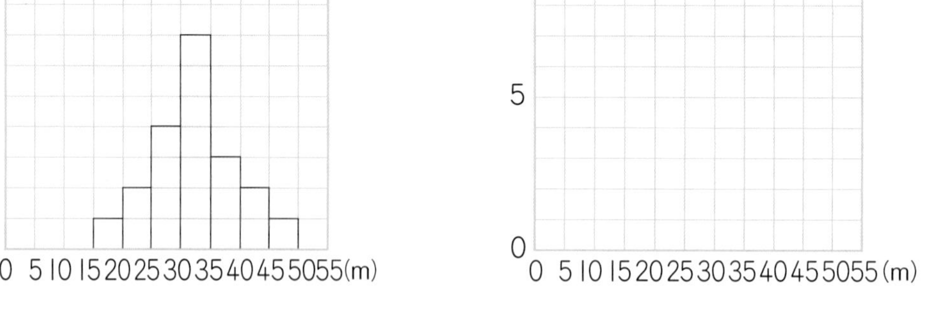

② Let's discuss what you noticed by looking at the graphs divided into boys and girls and the graph in which they are together.

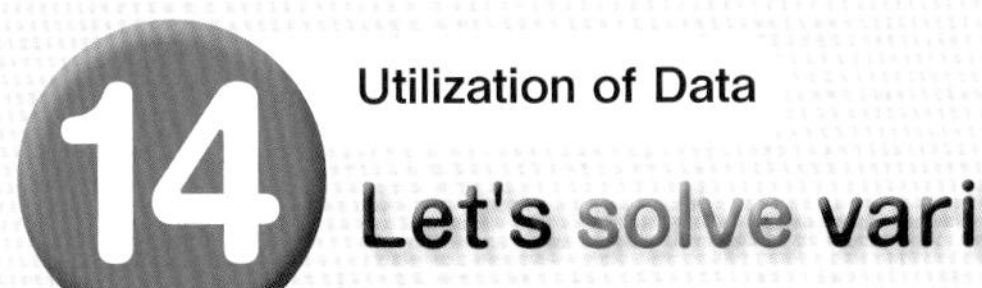

Utilization of Data

14 Let's solve various problems.

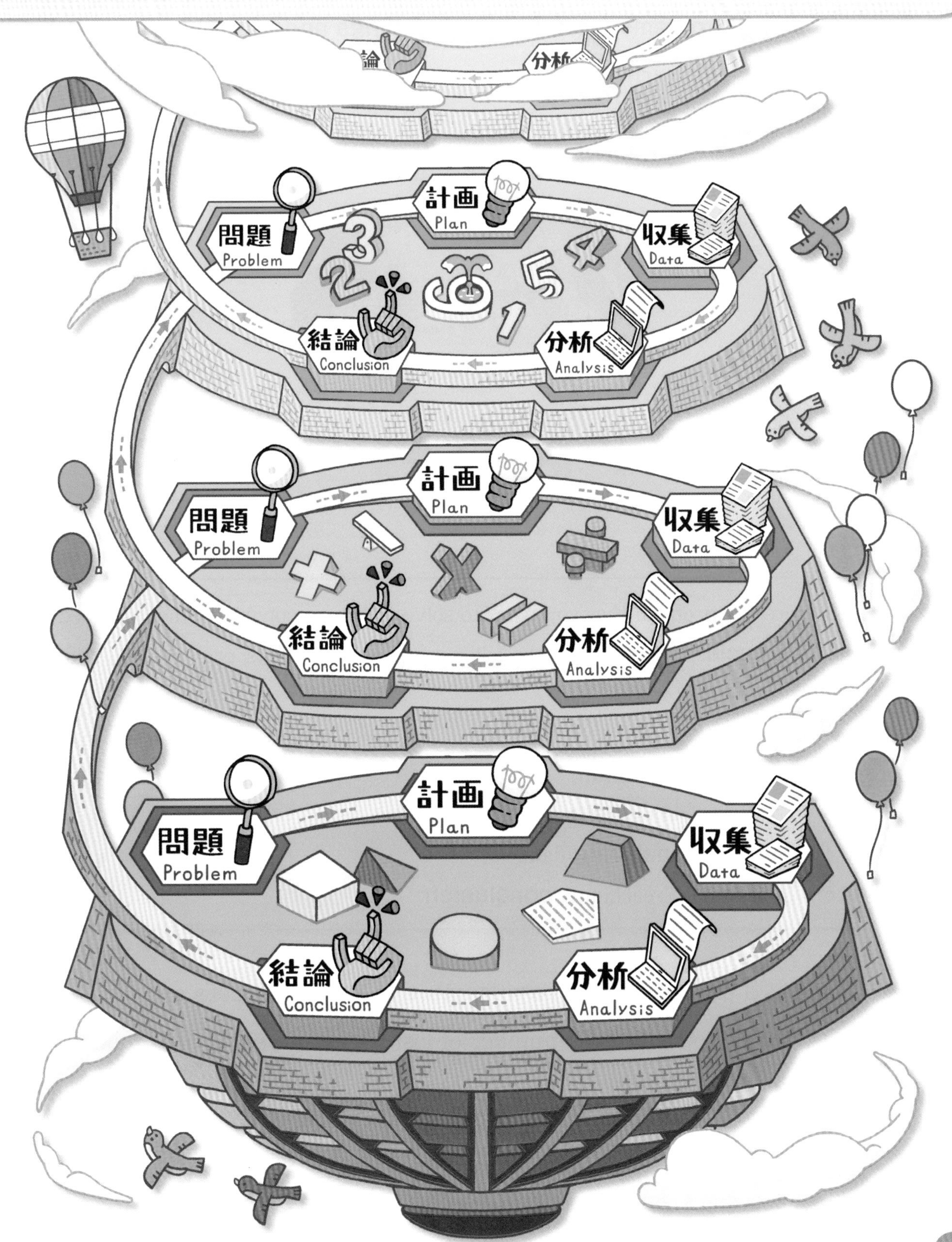

1 PPDAC cycle

Want to explore

1 **Let's find problems from our surroundings and use tables or graphs that have been learned so far to solve them.**

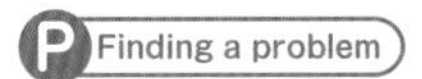

P Finding a problem

If there is a problem that you want to solve, then one solving method is called the **PPDAC cycle**. This cycle has the following 5 ordered steps, each represented by the first letter of a word:

(1) Problem ... finding a **problem.**

(2) Plan ... making a **plan.**

(3) Data ... collecting **data.**

(4) Analysis ... **analyzing** data.

(5) Conclusion ... reaching a **conclusion.**

The illustration in the previous page represents the PPDAC cycle.

Yui has a problem with a large number of lost items. Let's use the PPDAC cycle to reduce the number of lost items and solve the problem.

P Making a plan

① Let's think about which type of steps you will follow to explore lost items.

First, let's explore what kind of lost items there are.

If you know the type of lost item, you can find a method to reduce them.

Let's collect data from last year's lost item book and represent it in tables and graphs.

Is it enough to explore only the type of lost items?

D Collecting data

② The following table shows the data collected from last year's lost items. Let's draw a bar graph in the diagram shown on the right.

Last Year's Lost Items

Type	Number of items
Writing tools	36
Handkerchief	23
Clothing	21
Towels	15
Keys	9
Other	18
Total	122

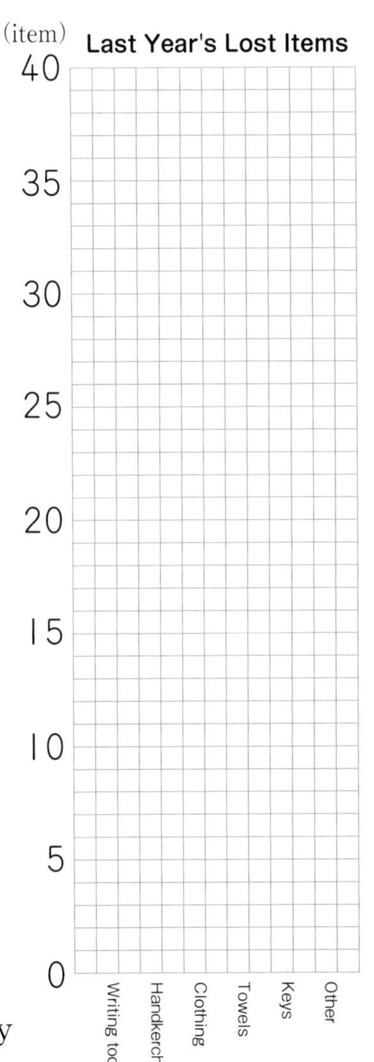

A Analyzing data

③ Let's present what you understand from the table and bar graph in ②.

C Reaching a conclusion

④ From what you have explored until now, let's try to discuss how the lost items can be reduced.

Where is it effective?

Should we investigate places with a large number of lost items?

Can we separate the writing tools by type?

2 **I heard that physical strength has dropped compared to the strength from 15 years ago, so I decided to explore it. Let's think about how to investigate this problem.**

① I want to explore if the physical strength has dropped by comparing the results of the side-step and jump test from current 6th graders Group 1 and past 6th graders Group 1 (15 years ago). Let's discuss how to compare, using the records from current 6th graders Group 1 on page 185 and the records from past 6th graders Group 1 (15 years ago).

Side-step and Jump Records from Past 6th Graders Group 1 (15 years ago)

Number	Score	Number	Score	Number	Score	Number	Score
1	28	11	49	21	54	31	47
2	53	12	43	22	32	32	46
3	29	13	40	23	49	33	28
4	47	14	48	24	35	34	57
5	32	15	56	25	41	35	47
6	30	16	44	26	29	36	47
7	57	17	30	27	40	37	44
8	44	18	29	28	37	38	43
9	43	19	31	29	41	39	31
10	58	20	49	30	53	40	43

② From what you noticed on ①, let's try to discuss what you understand.

It seems that it can be said that the current 6th graders Group 1 has more strength.

I want to try to explore other strength events.

But, is it correct to come to a conclusion only with the side-step and jump test?

Let's try to explore the records at throwing a softball.

③ In the same way, the records at throwing a softball were explored. Can you make a conclusion by looking at this table? Let's discuss if there are other exploring methods.

Records at Throwing a Softball from Current 6th Graders Group 1

Number	Distance (m)	Number	Distance (m)	Number	Distance (m)	Number	Distance (m)
1	11	11	19	21	17	31	20
2	17	12	13	22	21	32	15
3	23	13	29	23	14	33	13
4	23	14	26	24	19	34	17
5	14	15	27	25	20	35	13
6	19	16	19	26	20	36	18
7	32	17	29	27	15	37	20
8	29	18	25	28	14	38	29
9	32	19	22	29	31	39	18
10	31	20	16	30	29		

Records at Throwing a Softball from Past 6th Graders Group 1 (15 years ago)

Number	Distance (m)	Number	Distance (m)	Number	Distance (m)	Number	Distance (m)
1	28	11	18	21	30	31	15
2	28	12	29	22	26	32	33
3	27	13	20	23	22	33	24
4	17	14	18	24	29	34	30
5	15	15	19	25	19	35	18
6	29	16	14	26	23	36	24
7	13	17	24	27	32	37	30
8	20	18	29	28	31	38	27
9	19	19	31	29	26	39	30
10	17	20	13	30	25	40	24

Want to try

2 From your surroundings, let's find things you want to explore, apply the PPDAC cycle, and actually try to investigate.

There are many other events in the new physical fitness test.

2 Utilization of data

Want to explore

1 **The table on the right shows the results from a survey conducted in 52 cities about the amount of money spent on gyozas per household per year (mean from 2015 to 2017). Let's answer the following questions.**

① Let's find the mean for the 52 cities. Which city is closest to the mean value?

② In order to explore the distribution, let's write the number of cities per class in the following frequency distribution table.

Amount of Money Spent on Gyozas

Class (yen)	Cities
greater than or equal to 1000 ~ less than 1500	
1500 ~ 2000	
2000 ~ 2500	
2500 ~ 3000	
3000 ~ 3500	
3500 ~ 4000	
4000 ~ 4500	
Total	

01606

Amount of Money Spent on Gyozas (yen)

City	Amount of money	City	Amount of money
Hamamatsu	4348	Nagano	2082
Utsunomiya	4297	Tottori	2055
Miyazaki	2820	Toyama	2051
Kyoto	2818	Matsuyama	2045
Otsu	2611	Matsue	2023
Sakai	2593	Sendai	2019
Shizuoka	2457	Yokohama	1983
Nara	2387	Yamagata	1943
Maebashi	2351	Nagasaki	1936
Kagoshima	2336	Fukuoka	1917
Tokyo	2329	Tsu	1913
Mito	2326	Hiroshima	1903
Saitama	2312	Takamatsu	1900
Saga	2285	Morioka	1884
Fukui	2274	Wakayama	1882
Niigata	2225	Kumamoto	1881
Osaka	2183	Aomori	1874
Kawasaki	2181	Naha	1871
Kobe	2172	Kanazawa	1869
Chiba	2158	Sagamihara	1862
Nagoya	2143	Kitakyushu	1794
Fukushima	2140	Kochi	1778
Okayama	2138	Yamaguchi	1766
Gifu	2116	Oita	1747
Tokushima	2094	Sapporo	1712
Kofu	2085	Akita	1428

③ Based on the table made in ②, let's draw a histogram.

④ Which class has the largest number of cities?

⑤ In which class is the median value included?

⑥ In which class is the mean value found in ① included?

⑦ Based on the investigation done so far, let's discuss which is the representative value that best describes this data.

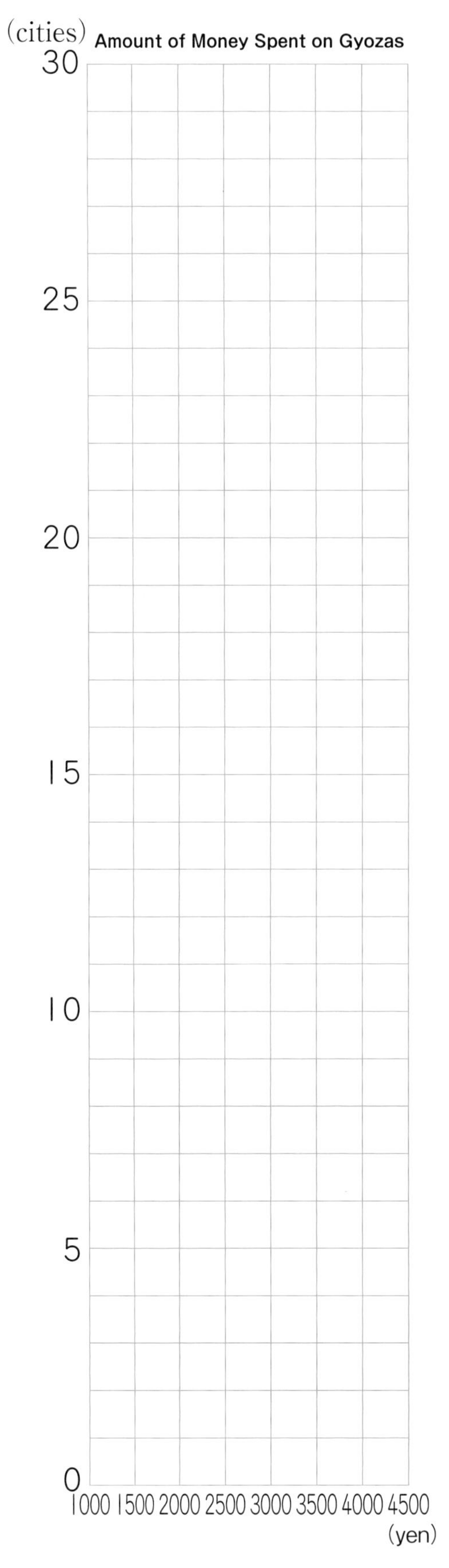

Want to try

In the following cases ①~③, which is the representative value that best describes the data?

① At the indoor shoe store, a decision is taken based on the shoe sizes sold last year. For the shoe sizes that will be prepared this year, they want to decide which size should have the highest amount.

② Based on the new physical fitness test records of all your classmates, you want to find out if your score is higher than the middle scores.

③ As for a relay competition between Group 1 and Group 2, based on the records of each student per group, you want to predict the outcome.

Problem

Let's try to explore the data gathered at the library.

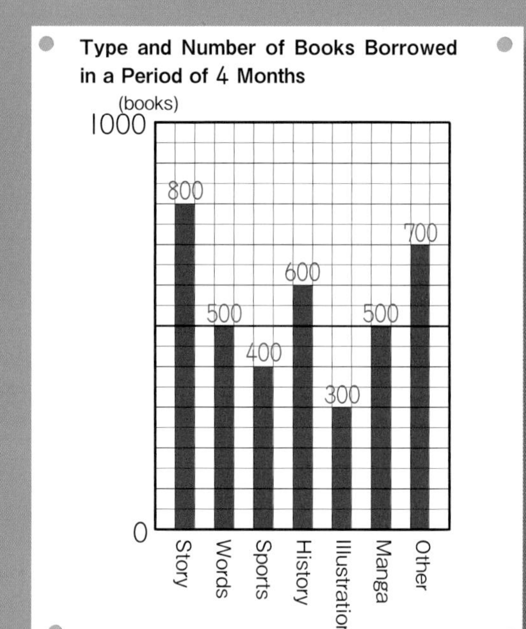

(Bar Graph)

- Story was the most borrowed book.
- The second most borrowed was History.
- The third were Words and Manga.

With a bar graph, the amount can be understood in a glance.

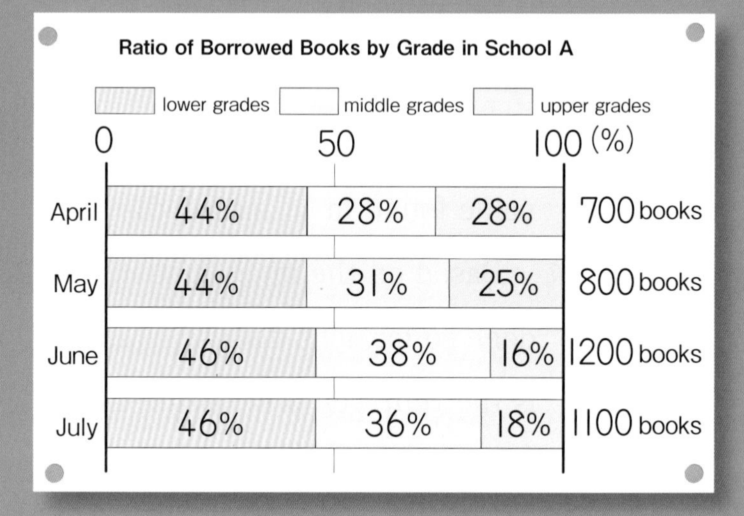

(Strip Graph)

- The percentage for lower grades has hardly changed.
- The percentage for middle grades is increasing.
- Although the percentage for upper grades has decreased, the number of books has hardly changed.

April $700 \times 0.28 = 196$

July $1100 \times 0.18 = 198$

With a strip graph, the percentages can be understood immediately.

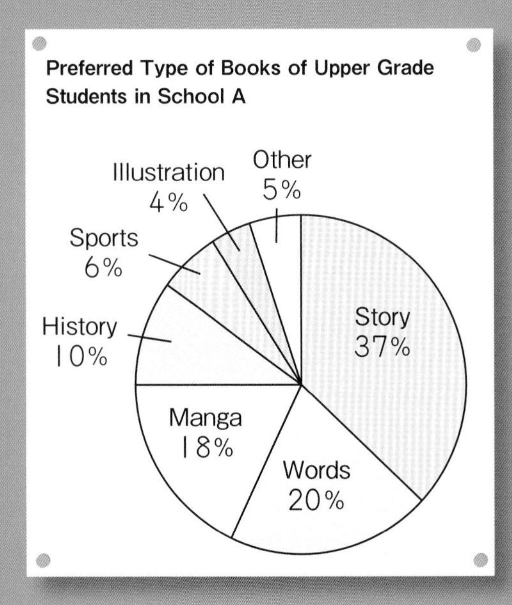

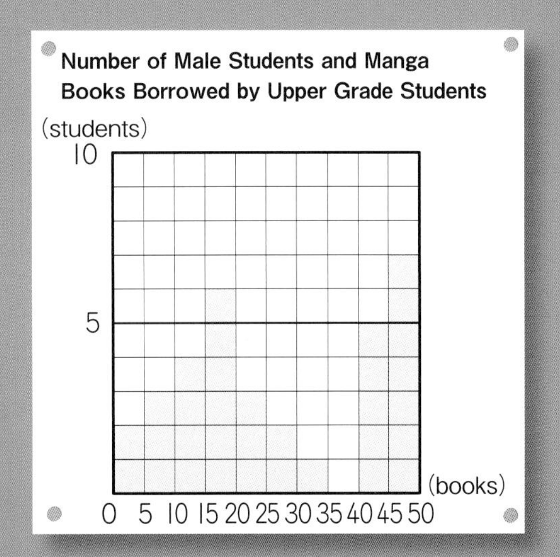

(Circle Graph)

- The number of students who wanted more books of Story is highest.
- Words and Manga are popular among upper grade students.

(Histogram)

- Most boys borrowed between 45～50 books.
- There were 12 boys who borrowed more than 40 manga books.
- The second highest number of boys borrowing books was between 15～20 books.

With a circle graph, the percentages can also be understood immediately. With a circle graph, percentages can be easily compared.

With a histogram, the data distribution can be understood.

Summary

Your understanding changes depending on the graph you see.

↓

Your understanding changes depending on the graph.

Want to connect

The histogram lets you understand how each child borrowed books.

I want to try to explore with a histogram, the type of books borrowed by upper grade girls.

15 Summary

Let's review what you have learned in mathematics.

Whole numbers and decimal numbers

1 Let's summarize about whole numbers and decimal numbers.

① Let's write the number that is 10 times, 100 times, and 1000 times of 1.07.

② Let's write the number that is $\frac{1}{10}$ and $\frac{1}{100}$ of 521.

③ As for the following numbers, how many units are gathered using the number inside [ ] as the reference.

ⓐ 23000 [100]　　ⓑ 23000 [1000]

ⓒ 2.3 [0.1]　　ⓓ 2.3 [0.01]

Fractions

2 Let's summarize about fractions.

① Let's write the equality or inequality sign inside the following □.

ⓐ $\frac{2}{5}$ □ $\frac{3}{5}$　　ⓑ $\frac{2}{5}$ □ $\frac{2}{7}$　　ⓒ $\frac{2}{5}$ □ $\frac{8}{20}$

② Let's write the number that applies inside the following □.

ⓐ $\frac{3}{5}$ has □ sets of $\frac{1}{5}$.　　ⓑ $\frac{9}{7}$ has nine sets of □.

③ Let's change the following mixed fractions into improper fractions and the improper fractions into mixed fractions.

ⓐ $1\frac{2}{3}$　ⓑ $4\frac{3}{5}$　ⓒ $\frac{7}{4}$　ⓓ $\frac{8}{3}$

Whole numbers

3 Let's summarize about the properties of whole numbers.

① From the whole numbers up to 50, let's find numbers that have three divisors.

② Let's find the least common multiple and greatest common divisor from the following pair of numbers.

ⓐ (12, 18)　ⓑ (8, 16)

Whole numbers, decimal numbers, and fractions

4 Let's summarize the relationship between whole numbers, decimal numbers, and fractions.

① Let's change the following whole number and decimal numbers into fractions and the fractions into decimal numbers.

ⓐ 4 ⓑ 0.7 ⓒ 3.08 ⓓ $\frac{13}{25}$ ⓔ $1\frac{3}{4}$

② Let's arrange the following five numbers in ascending order.

$\frac{2}{5}$ $\frac{1}{3}$ $\frac{7}{15}$ 0.3 0.41

Various operations

5 Let's summarize how to calculate.

① Let's solve the following calculations.

ⓐ $4+2\times6-3$ $(4+2)\times6-3$ $4+2\times(6-3)$

ⓑ $4.2+1.5$ $4.2-1.5$ 4.2×1.5 $4.2\div1.5$

ⓒ $64.8+1.8$ $64.8-1.8$ 64.8×1.8 $64.8\div1.8$

ⓓ $\frac{2}{5}+\frac{1}{3}$ $\frac{2}{5}-\frac{1}{3}$ $\frac{2}{5}\times\frac{1}{3}$ $\frac{2}{5}\div\frac{1}{3}$

② Let's find the number that applies for x.

ⓐ $8+x=15$ ⓑ $x\times7=56$

Speed

6 Let's summarize about speed.

① How many meters per second is 240 m per minute? Also, how many kilometers per hour?

② I started walking from the station toward the library that is 1.5 km away. After 15 minutes, I arrived at the park that is 900 m far from the station. If I continue walking with the same speed, how long will it take me to arrive at the library from the park?

Unit of a quantity

7 Let's summarize about the unit of quantities in your surroundings.

① Let's write the unit that applies inside the following □.

ⓐ The cover area of the mathematics textbook is about 540 □.

ⓑ The amount of milk inside a milk pack is 200 □.

ⓒ The weight of one egg is about 50 □.

ⓓ Shinano River is Japan's longest river with a length of about 367 □.

② Let's answer the following questions.

ⓐ Koharu walked 1.6 km. How many meters remains to complete a full walk with a length of 2 km?

ⓑ There is a flowerbed with a length of 1 m and a width of 3 m. How many square meters is the area of this flowerbed? Also, how many square centimeters?

ⓒ There are 4 bottles with a capacity of 500 mL. In total, how many liters can be poured in? Also, how many deciliters is equivalent to it?

Area

8 Let's summarize how to find the area.

① Let's write the area formula for the following figures.

Area of a rectangle = □ × □

Area of a square = □ × □

Area of a parallelogram = □ × □

Area of a triangl = □ × □ ÷ □

Area of a circle = □ × □ × □

② Let's draw two figures with an area of 20 cm^2.

③ Let's find the area of the following colored sections.

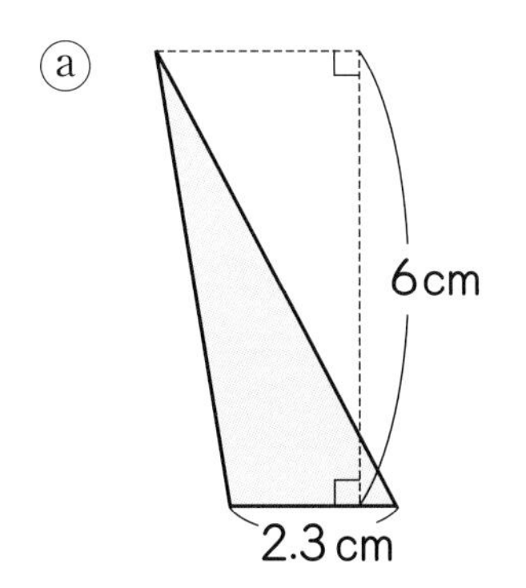

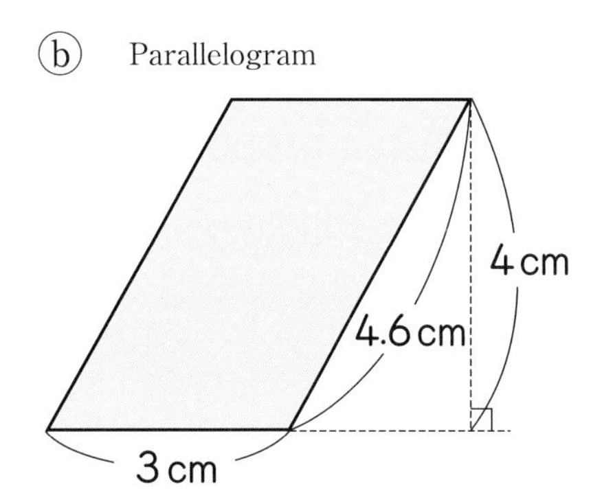

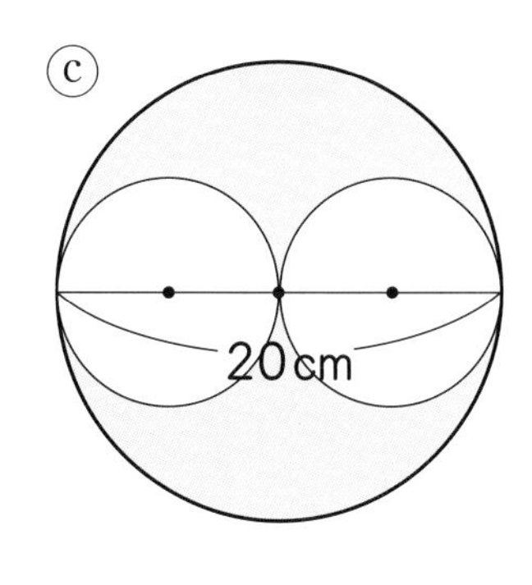

Volume

9 Let's summarize how to find the volume.

① Let's write the formula to find the volume of a cuboid and a cube.

② Let's find the volume of the following solids.

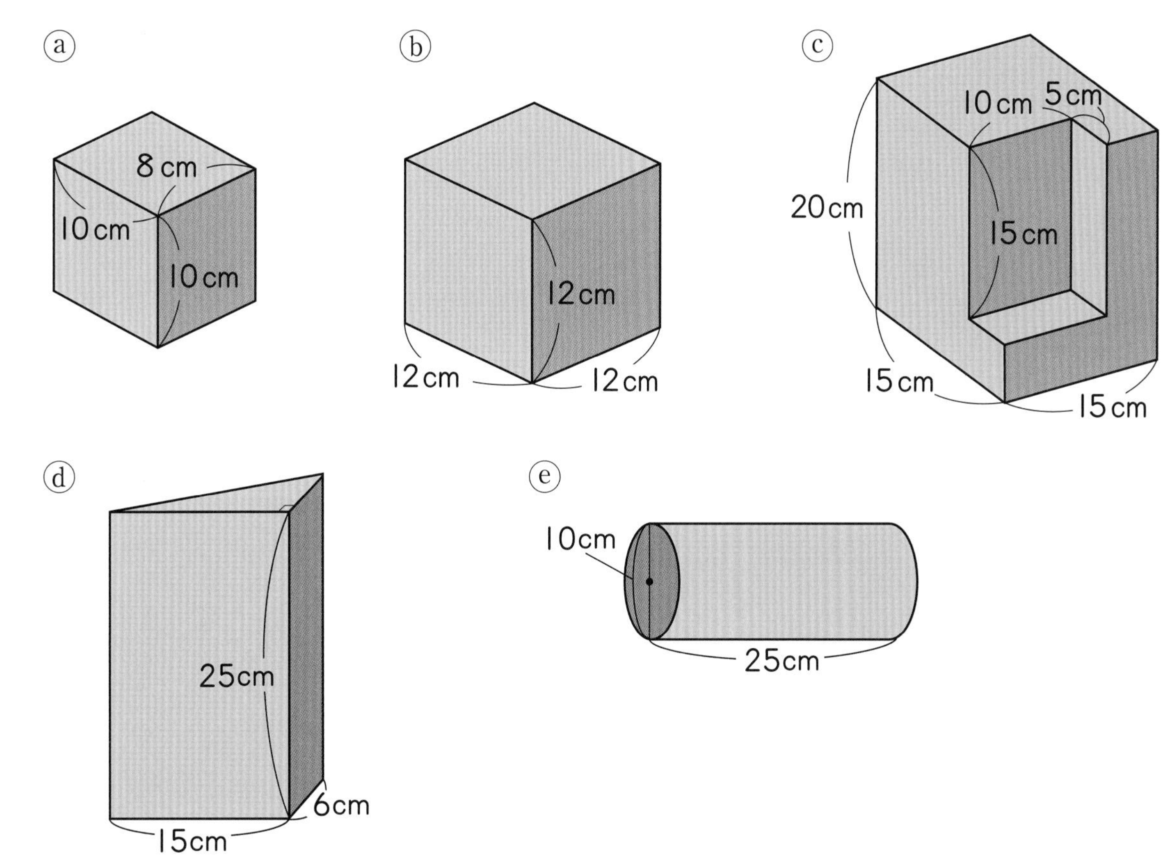

10 Let's summarize the properties of figures.

① Let's write ○ on the properties that apply or × on the properties that do not apply about parallelograms, rhombuses, rectangles, and squares.

	Parallelogram	Rhombus	Rectangle	Square
2 pair of sides are parallel.				
All 4 angles are right angles.				
All 4 sides have the same length.				
2 diagonals intersect perpendicularly.				
The sum of adjacent angles is 180°.				

② Let's write the number that applies inside each ☐.

ⓐ

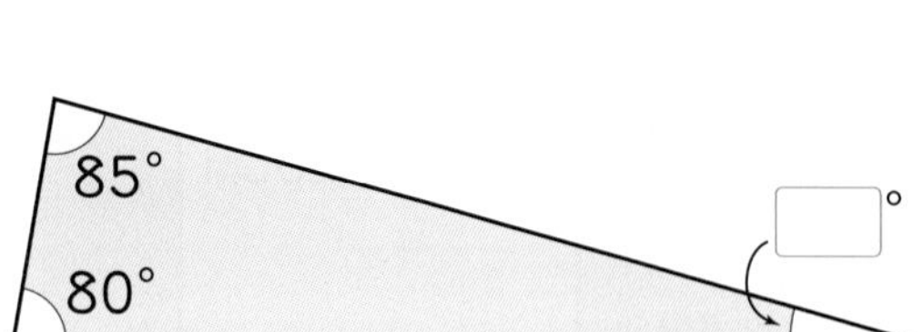

ⓑ

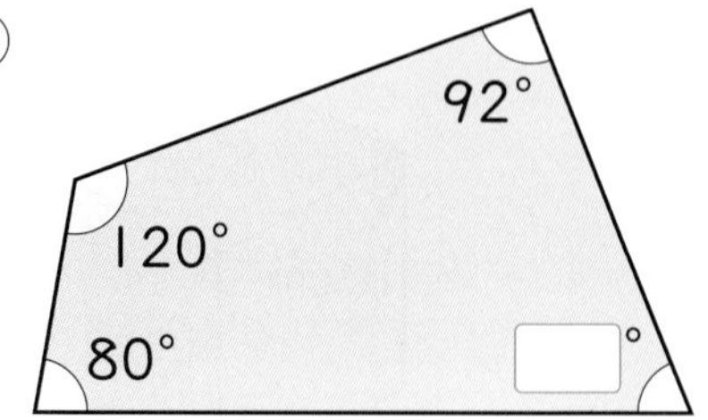

ⓒ Parallelogram

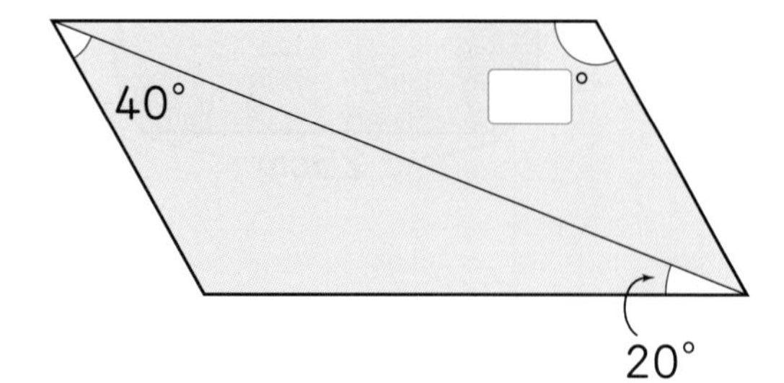

ⓓ Regular hexagon

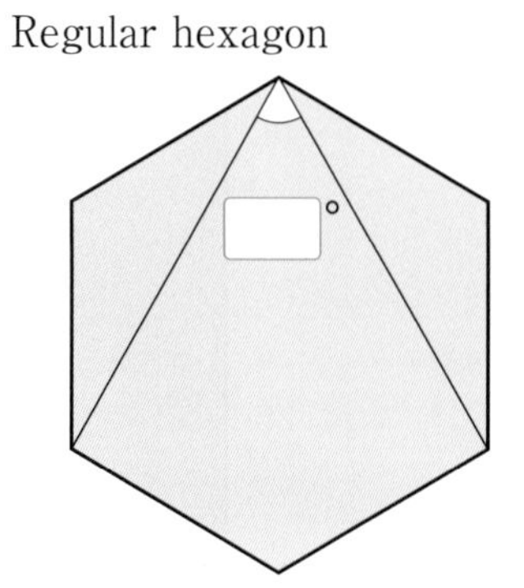

③ Let's explore about the cuboid shown on the right.

ⓐ Which face is parallel to face ABCD?

ⓑ Which side is parallel to side AB?

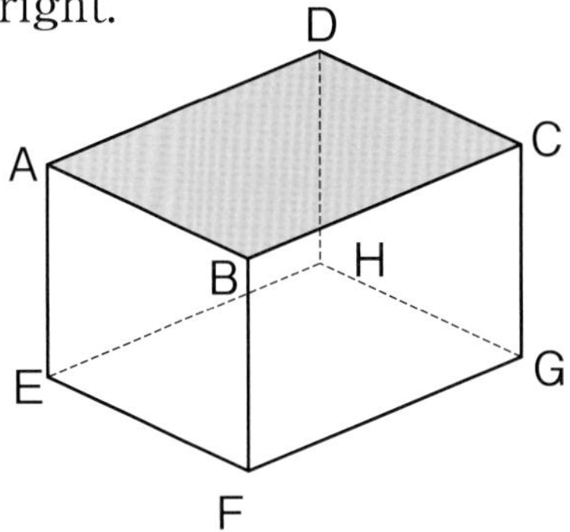

Symmetric figures, Enlarged and reduced drawings

11 Let's draw the following figures.

① A line symmetric figure and straight line XY as the line of symmetry

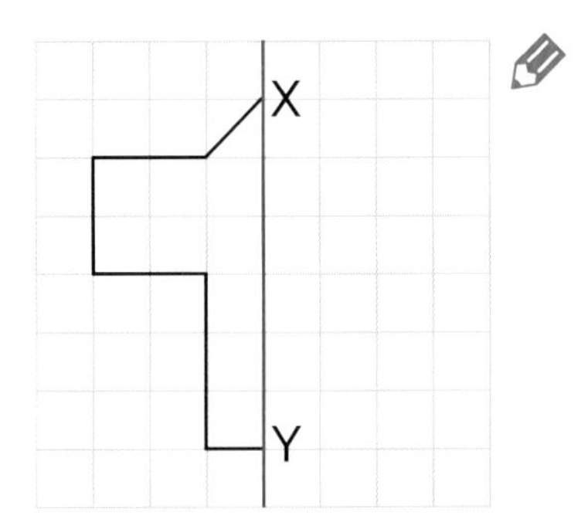

② A point symmetric figure and point O as the point of symmetry

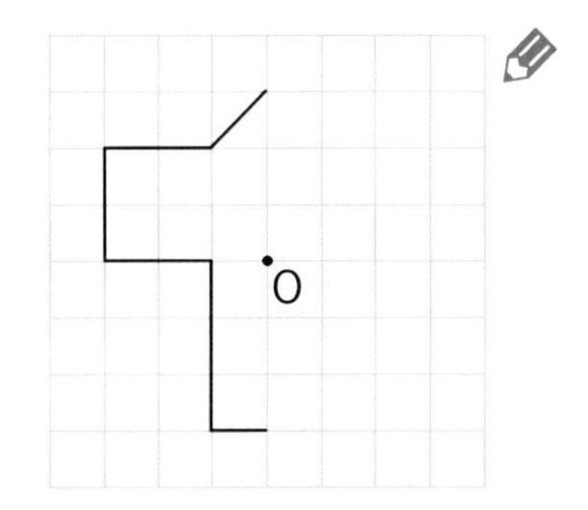

③ 2 times enlarged drawing of Ⓐ

④ $\frac{1}{2}$ reduced drawing of Ⓑ

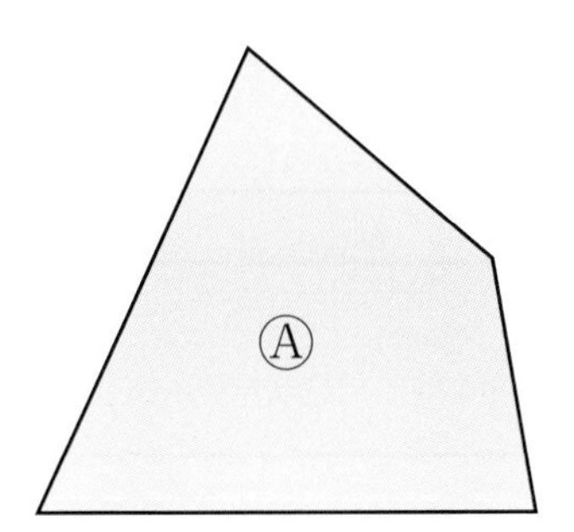

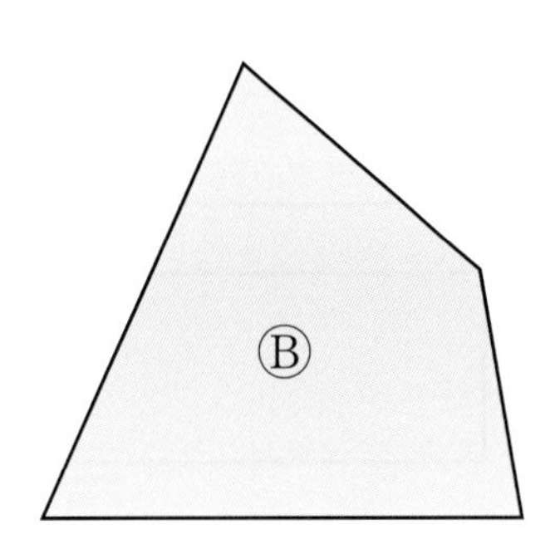

12 Let's summarize how to represent the relationship between quantities.

① What kind of graph should be used to represent the following?

ⓐ Ratio of import value for each type of imported goods.

ⓑ Changes in export values.

ⓒ Rice production in each country.

② The table shown on the right summarizes the number of books and magazines published in one year.

ⓐ As for the ratio of monthly magazines, what is the percentage to all publications in each year? Let's round off to the nearest whole number.

ⓑ Let's represent the ratio of books and magazines published each year using a strip graph. Let's discuss what you noticed.

Number of Books and Magazines

(Hundred million books)

	1995	2015
Books	14.6	10.5
Weekly magazine	19.4	6.7
Monthly magazine	31.2	17.9
Total	65.2	35.1

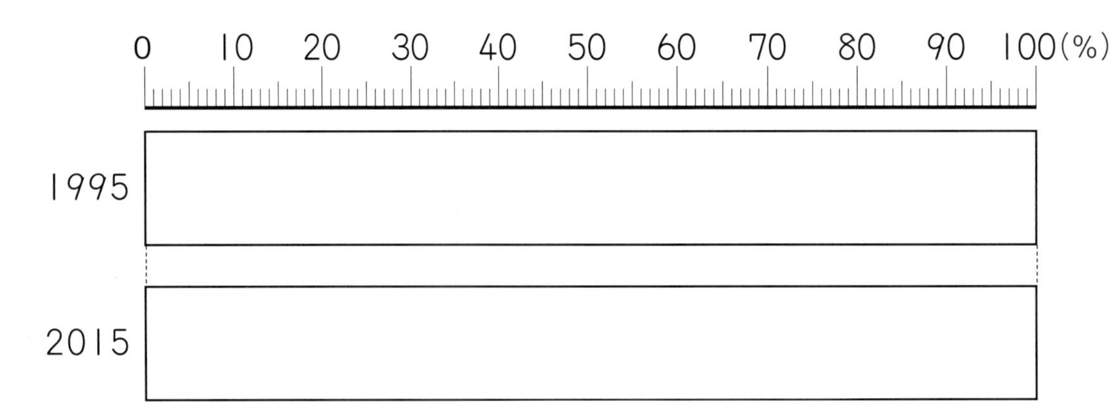

③ To make sweet soy bean flour, 35 g of soy bean flour and 14 g of sugar were mixed.

ⓐ If the value of the sugar is set to 2, what is the value of the soy bean flour?

ⓑ Sweet soy bean flour will be made. If there are 140 g of soy bean flour, how many grams of sugar are needed?

Quantities changing together

13 Let's represent quantities in math sentences and graphs.

① Let's write a math sentence, using x, for the area of the following triangle and trapezoid. Also, let's find the number that applies for x.

ⓐ

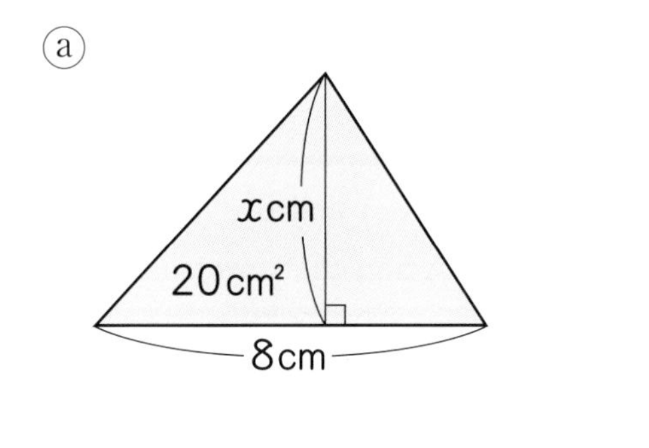

ⓑ

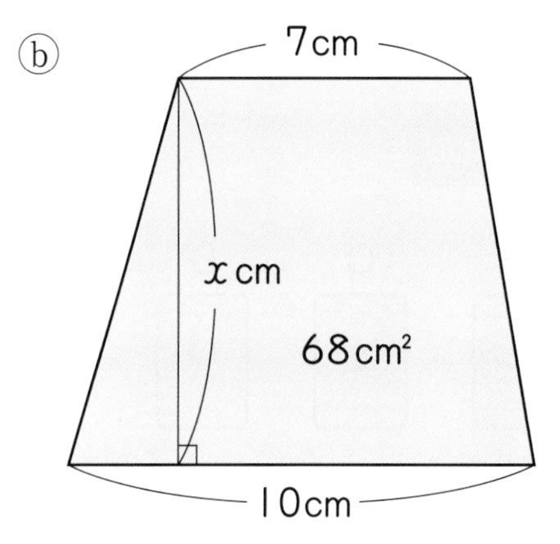

② Let's explore the relationship between x and y summarized in the table Ⓐ and table Ⓑ.

Ⓐ

Number of people by which a string is divided x (people)	2	3	4	6	8	
Length of the string per person y (m)	12	8	6	4	3	

Ⓑ

Length of the string x (m)	0	1	2	3	4	5	
Weight of the string y (g)	0	8	16	24	32	40	

ⓐ In which table is y proportional to x? Also, in which table is y inversely proportional to x?

ⓑ As for tables Ⓐ and Ⓑ, let's represent the relationship between x and y in a math sentence.

ⓒ Let's draw a graph for the proportional relationship.

Let's produce the algorithm.

Computational thinkin 

01607

Let's teach Robo "a method to align numbers or quantities in ascending order."

① There are four cards from 1 to 4. Let's try to think how to align four cards in ascending order after being placed in an scattered manner.

1 2 3 4

② Let's try to explain the following aligning method to a classmate.

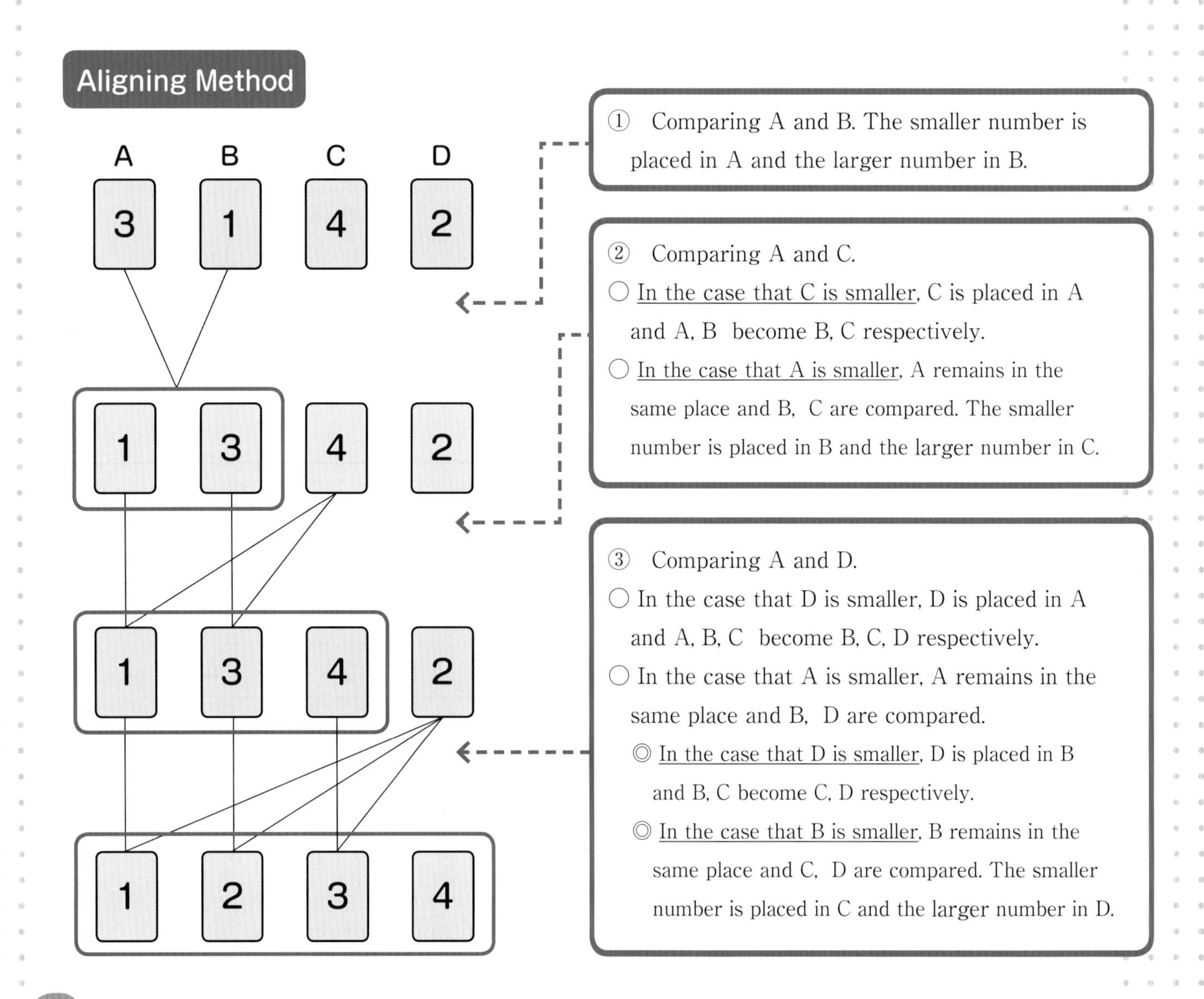

③ As shown in the method from the previous page, let's place the numerals from 1 to 8 inside the following ▢ to try to align a scattered combination in ascending order.

Let's produce a graduation ceremony.

We are about to graduate.

How to enter, giving thanks to the teacher,...

What kind of graduation will it be?

Do you want to propose something?

Let's create a "Thank you card."

Yui: I want to make a card using regular polygons and symmetry.

Hiroto: I want to put what I learned in mathematics on the card.

Daiki: I made cards using regular hexagons.

Front

DAIKI
March, 202X
Graduation

○× School

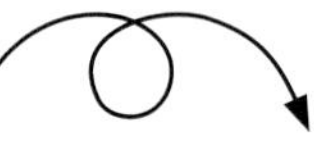

Back

I learned mathematics for 5 hours a week for 35 weeks.

Those are 7875 minutes.
Thanks for the fun mathematics.

○× School
Class ○ 6th Grade

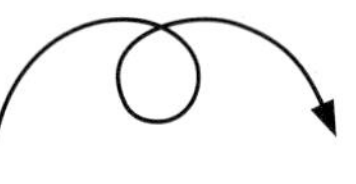

I used 12 notebooks in one year. Until 5th grade, there were about 6

notebooks, so I can write about twice as much. Thank you very much.

Nanami: I made a card with a regular pentagon. I wrote my name upsidedown.

Entry parade proposal.

This year I want to start from the entry parade. There are three classes in 6th grade. Since there are 30 students in each class...

I drew a reduced drawing of the gymnasium.

Hiroto

□ m

7m

A

10m

Class1

5m

Class2

5m

Class3

10m

30 seats for teachers

stage

30m

45m

5m

? What is the reduced scale of the gymnasium drawing?

? If 30 students are aligned as shown on the right, about how many meters will that be? Also, how many centimeters will be required if the rows are arranged as shown on the above reduced drawing?

If 30 students are aligned in a row, how many meters will that be?

? By class order, students enter from position A to the ● place in front of the stage. If they walk 1 m per second and pass the designated passage, how long does it take for them to finish? Let's measure the length of the passage. When one class arrives to the designated place, the next class will enter from point A. The row is assumed to move as it is.

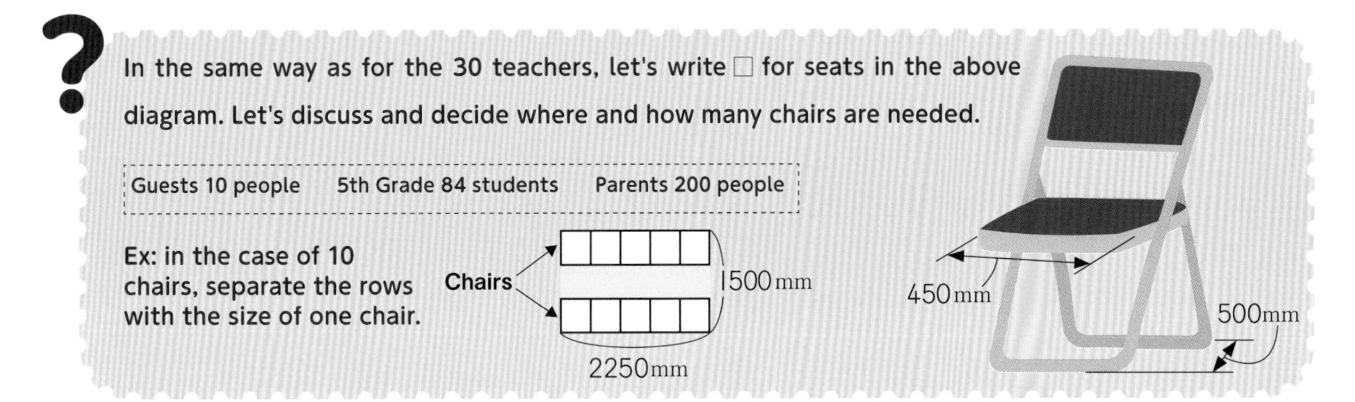

? In the same way as for the 30 teachers, let's write □ for seats in the above diagram. Let's discuss and decide where and how many chairs are needed.

Guests 10 people　5th Grade 84 students　Parents 200 people

Ex: in the case of 10 chairs, separate the rows with the size of one chair.

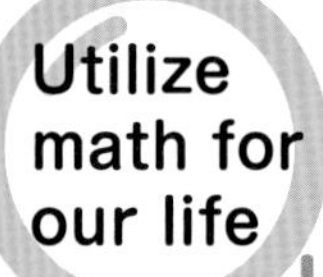

Let's produce a graduation ceremony.

1. Toward learning competency.

	Strongly agree	Agree	Don't agree
① It was fun working on it.			
② The learning contents were helpful.			
③ Concentrated our efforts to make something better.			

2. Thinking, deciding, and representing competency.

	Definitely did	so so	I didn't
① In the graduation ceremony proposal, I was able to discover where to use the mathematical knowledge.			
② I was able to confirm how to use the mathematical knowledge and decide whether the numerical values were correct.			
③ In the graduation ceremony proposal, I was able to represent mathematical knowledge with characters, pictures, figures, and tables.			

3. What I know and can do.

	Definitely did	so so	I didn't
① I was able to offer a better graduation ceremony.			
② Making a proposal for the graduation ceremony deepened my understanding of mathematical knowledge.			

4. Encouragement for myself.

	Strongly agree
① I think that I'm doing my best.	

Give yourself a compliment since you have worked so hard.

Let's try to work out what you were not able to accomplish and keep doing your best on what you were able to fulfill.

Problems

1 Ways of Ordering and Combinating

p.10~p.21

1 Four children, Tsukushi, Takuma, Yota, and Maimi sit on the four chairs shown on the right. Let's answer the following questions.

① Let's consider Tsukushi as Ⓐ, Takuma as Ⓑ, Yota as Ⓒ, and Maimi as Ⓓ, and draw the following diagram to show how to sit the children in the case that Tsukushi sits in the 1st seat.

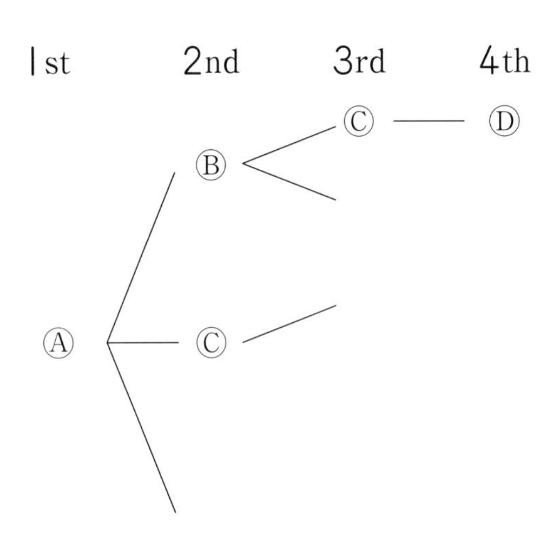

② As for how to sit children, how many ways are there when Tsukushi sits in the 1st seat?

③ As for how to sit children, how many ways are there when Takumi, Yota, and Maimi sit in the 1st seat respectively?

④ As for how to sit the four children, how many ways are there in total?

2 There is one card for each of the following numbers: 2 , 4 , 6 , 8 . From these 4 cards, use 3 cards to create 3-digit whole numbers. Let's answer the following questions.

① Let's write, in the table shown on the right, the whole numbers for the case in which 2 is located in the hundreds place.

Hundreds	Tens	Ones
2	4	6
2	4	□
2	6	□
□	□	□
□	□	□
□	□	□

② As for the 3-digit whole numbers, how many can you make in total?

3 There is one card for each of the following numbers: 0 , 3 , 6 , 9 . From these 4 cards, use 2 cards to create 2-digit whole numbers. Let's answer the following questions.

① Let's write, in the diagram shown on the right, the whole numbers for the case in which 3 is located in the tens place.

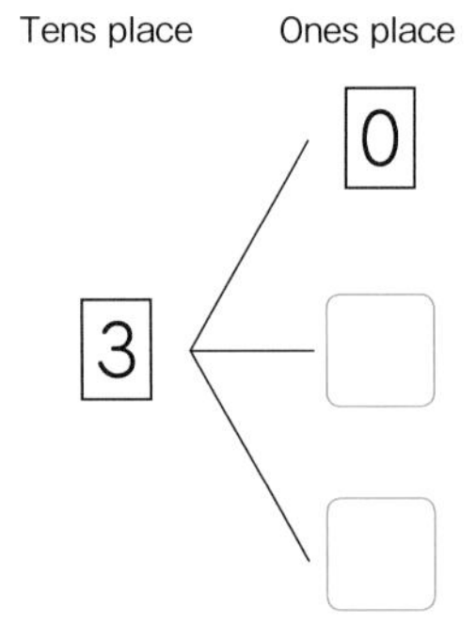

② Other than 3, what number card can be used in the tens place?

③ As for the 2-digit whole numbers, how many can you make in total?

4 Three coins, A, B, and C, are thrown. As for the front and back of the coins, let's answer the following questions.

① Let's consider front as ○ and back as ×. Let's draw the following diagram to show how the front and back of coins B and C come out when A is front.

A B C

○ — ○ — ○, □
○ — × — □, □

② As for how the front and back of coins come out, how many ways are there in total?

5 Five teams, A, B, C, D, and E, play soccer games. Let's answer the following questions when each team competes with the other teams only one time.

① Let's write the combination of games, expressed by ○, in the following table.

	A	B	C	D	E
A		○			
B					
C					
D					
E					

② How many games are there in total?

6 There are 4 sheets of colored paper: red, blue, yellow, and green. When you choose 3 sheets from these, how many ways of choosing are there in total?

2 Mathematical Letter and Sentence

p.22~p.33

1 Let's write a math expression to find the price when buying a icecreams that cost 70 yen per icecream.

2 Let's write a math expression to find the area of a rectangle that has a length of a cm and a width of 5 cm.

3 There are 3 bundles and 4 individual origami sheets. Let's answer the following questions.

① Let's represent the total number of origami sheets in a math expression by using x, considering that one bundle has x origami sheets.

② How many origami sheets are there in total when one bundle has 20 origami sheets?

4 There are 4 bottles and 3 dL of tea. Let's answer the following questions.

① Let's represent the total amount of tea in a math expression by using x, considering that one bottle has an amount of x dL.

② How many deciliters of tea are there in total when one bottle has an amount of 4 dL of tea?

5 There are 5 ribbons with an individual length of x cm and an additional ribbon with a length of 6 cm. Let's represent the total length of the ribbon in a math expression by using x.

6 Some deciliters of juice were drunk from a total amount of 14 dL. The remaining amount became 5 dL. Let's answer the following questions.

① Let's write a math sentence to find the original amount of juice, considering that x dL of juice were drunk.

② How many deciliters of juice were drunk?

7 There is a square with a surrounding length of 26 cm. Let's answer the following questions.

① Let's write a math sentence to find the surrounding length, considering that the length of one side of the square is x cm.

② How many centimeters is the length of one side of the square?

8 Let's find the number that applies for x.

① $x + 8 = 44$ ② $x + 19 = 27$

③ $27 + x = 50$ ④ $46 + x = 61$

⑤ $x - 7 = 15$ ⑥ $x - 18 = 22$

⑦ $x - 43 = 19$ ⑧ $x - 25 = 68$

⑨ $9 \times x = 54$ ⑩ $3 \times x = 21$

⑪ $x \times 8 = 32$ ⑫ $x \times 6 = 9$

9 There are 3 boxes and 5 cookies. Let's answer the following questions.

① Let's write a math expression that represents the total number, considering that 1 box has x cookies.

② There were 59 cookies in total. How many cookies did each box have? Let's explore the number that applies for x by placing 16, 17, 18, ... as x.

10 Let's find the number that applies for x by placing 6, 7, 8, 9, ... as x.

① $x \times 7 + 6 = 62$

② $5 \times x - 8 = 37$

11 The following math expressions ①~③ represent the area of the flowerbed shown on the right.

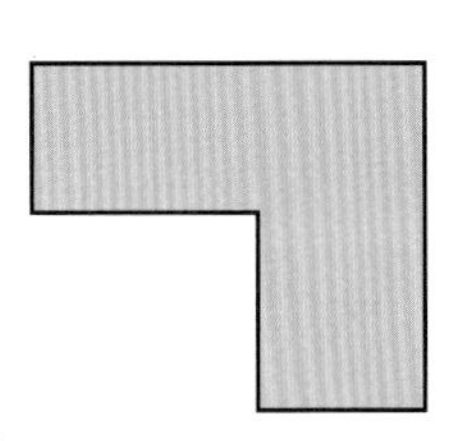

① $(14 - 8) \times 16 + 8 \times (16 - x)$

② $14 \times 16 - 8 \times x$

③ $6 \times x + 14 \times (16 - x)$

Which of the following diagrams Ⓐ~Ⓒ represent each of the above math expression?

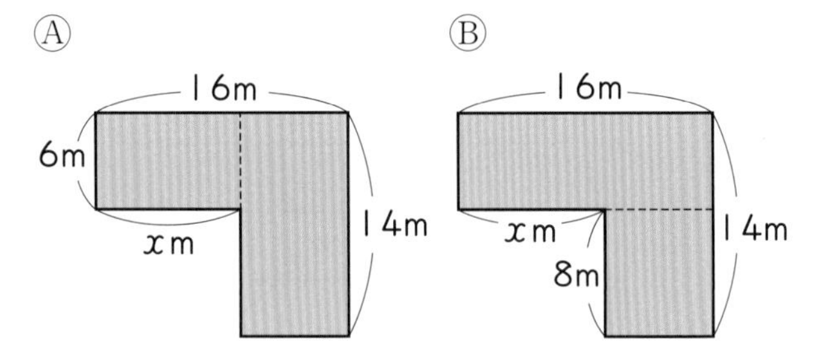

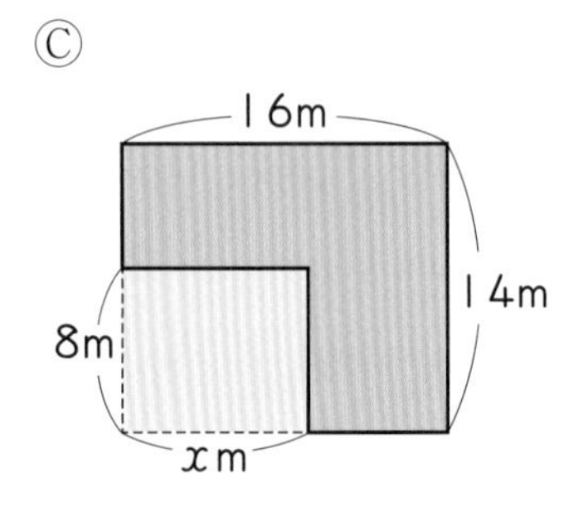

3 Multiplication and Division of Fractions and Whole Numbers

p.34~p.46

1 Let's solve the following calculations.

① $\frac{2}{7} \times 2$ ② $\frac{3}{10} \times 3$

③ $\frac{1}{5} \times 3$ ④ $\frac{3}{4} \times 5$

⑤ $\frac{4}{9} \times 4$ ⑥ $\frac{5}{8} \times 3$

2 Let's solve the following calculations.

① $\frac{5}{12} \times 2$ ② $\frac{4}{15} \times 5$

③ $\frac{1}{6} \times 4$ ④ $\frac{2}{9} \times 6$

⑤ $\frac{9}{8} \times 12$ ⑥ $\frac{6}{5} \times 10$

3 4 pieces of tape will be made, each with a length of $\frac{5}{8}$ m. How many meters of tape will be needed in total?

4 Let's solve the following calculations.

① $1\frac{2}{9} \times 4$ ② $1\frac{5}{11} \times 2$

③ $1\frac{1}{10} \times 4$ ④ $2\frac{4}{9} \times 3$

⑤ $2\frac{1}{4} \times 2$ ⑥ $2\frac{4}{15} \times 5$

5 How many square meters is the area of a rectangular field that has a length of $3\frac{1}{2}$ m and a width of 4 m?

6 Let's solve the following calculations.

① $\frac{1}{3} \div 3$ ② $\frac{5}{7} \div 4$

③ $\frac{3}{5} \div 2$ ④ $\frac{3}{8} \div 5$

⑤ $\frac{4}{9} \div 3$ ⑥ $\frac{3}{4} \div 7$

7 Let's solve the following calculations.

① $\frac{4}{7} \div 2$ ② $\frac{5}{16} \div 5$

③ $\frac{8}{9} \div 6$ ④ $\frac{8}{15} \div 4$

⑤ $\frac{4}{3} \div 2$ ⑥ $\frac{20}{7} \div 16$

8 $\frac{9}{8}$ L of juice will be equally poured into 3 bottles. How many liters of juice will each bottle contain?

9 Let's solve the following calculations.

① $2\frac{4}{9} \div 3$ ② $1\frac{7}{10} \div 2$

③ $2\frac{1}{2} \div 5$ ④ $3\frac{1}{5} \div 6$

⑤ $1\frac{5}{7} \div 4$ ⑥ $2\frac{8}{11} \div 4$

10 There is a pipe that weighs $3\frac{3}{4}$ kg and has a length of 5 m. How many kilograms is the weight of this pipe per meter?

4 Fraction × Fraction

p.47~p.60

1 Let's solve the following calculations.

① $\frac{5}{8}\times\frac{1}{9}$　② $\frac{7}{9}\times\frac{1}{2}$

③ $\frac{2}{3}\times\frac{2}{5}$　④ $\frac{1}{6}\times\frac{1}{4}$

⑤ $\frac{7}{6}\times\frac{5}{2}$　⑥ $\frac{9}{8}\times\frac{5}{4}$

⑦ $\frac{10}{9}\times\frac{2}{3}$　⑧ $\frac{3}{4}\times\frac{9}{7}$

2 Let's solve the following calculations.

① $\frac{5}{6}\times\frac{2}{9}$　② $\frac{7}{12}\times\frac{2}{3}$

③ $\frac{3}{10}\times\frac{5}{6}$　④ $\frac{8}{15}\times\frac{9}{10}$

⑤ $\frac{5}{3}\times\frac{9}{11}$　⑥ $\frac{3}{8}\times\frac{12}{5}$

⑦ $4\times\frac{3}{8}$　⑧ $5\times\frac{7}{10}$

⑨ $6\times\frac{2}{3}$　⑩ $9\times\frac{5}{12}$

3 $\frac{7}{10}$ m² of a wall can be painted per liter of paint. How many square meters can be painted with $\frac{5}{6}$ dL of paint?

4 Let's solve the following calculations.

① $1\frac{1}{2}\times1\frac{2}{5}$　② $5\frac{5}{6}\times2\frac{1}{7}$

③ $2\frac{2}{3}\times2\frac{1}{4}$　④ $4\frac{2}{9}\times6\frac{3}{4}$

⑤ $3\frac{1}{5}\times\frac{5}{8}$　⑥ $1\frac{5}{6}\times\frac{3}{10}$

⑦ $\frac{5}{12}\times1\frac{3}{5}$　⑧ $\frac{2}{3}\times2\frac{1}{4}$

5 There is an iron bar that weighs $1\frac{1}{8}$ kg per meter. How many kilograms is the weight for $2\frac{2}{3}$ m of this iron bar?

6 Let's write the equality or inequality sign that applies inside each □.

① $4\times\frac{9}{10}$ □ 4

② $\frac{4}{5}\times\frac{4}{3}$ □ $\frac{4}{5}$

③ $\frac{5}{9}\times1\frac{1}{8}$ □ $\frac{5}{9}\times\frac{9}{8}$

④ $3\times\frac{5}{4}$ □ $3\times\frac{3}{4}$

7 Let's find the area of the following figures.

① Rectangle

$\frac{3}{4}$ m, $\frac{3}{7}$ m

② Parallelogram

$\frac{5}{8}$ m, $\frac{2}{5}$ m

③ Trapezoid

$\frac{1}{2}$ m, $\frac{3}{4}$ m, $\frac{2}{3}$ m

8 Let's write the number that applies inside each □.

① $\frac{4}{5} \times \frac{3}{8} = \square \times \frac{4}{5}$

② $\left(\frac{5}{7} \times \frac{8}{9}\right) \times \frac{3}{4} = \frac{5}{7} \times \left(\square \times \frac{3}{4}\right)$

③ $\left(\frac{2}{3} + \frac{2}{5}\right) \times \square = \frac{2}{3} \times \frac{3}{10} + \frac{2}{5} \times \frac{3}{10}$

④ $\left(\frac{7}{8} - \frac{2}{3}\right) \times 24 = \frac{7}{8} \times 24 - \frac{2}{3} \times \square$

9 What is the product of a number and the reciprocal of that number?

10 Let's find the reciprocal of the following numbers.

① $\frac{3}{8}$ ② $\frac{7}{9}$

③ $1\frac{1}{2}$ ④ $2\frac{2}{5}$

⑤ $\frac{1}{6}$ ⑥ 0.8

⑦ 1.3 ⑧ 4

5 Fraction ÷ Fraction

p.61～71

1 Let's write the number that applies inside the □, concerning how to calculate $\frac{4}{7} \div \frac{3}{5}$.

In division, given that "if the same number is multiplied by the divisor and dividend then the quotient is the same as the original quotient," the following is valid:

$$\frac{4}{7} \div \frac{3}{5} = \left(\frac{4}{7} \times 5\right) \div \left(\frac{3}{5} \times \square\right)$$
$$= \frac{4 \times 5}{7} \div \square$$
$$= \frac{4 \times 5}{7 \times \square}$$

2 As for the division by fractions, the calculation is done by multiplying the dividend by what?

3 Let's solve the following calculations.

① $\frac{2}{9} \div \frac{3}{5}$ ② $\frac{5}{6} \div \frac{6}{7}$

③ $\frac{8}{3} \div \frac{1}{2}$ ④ $\frac{3}{4} \div \frac{8}{7}$

4 Let's solve the following calculations.

① $\frac{7}{8} \div \frac{1}{2}$ ② $\frac{3}{10} \div \frac{5}{8}$

③ $\frac{12}{5} \div \frac{8}{7}$ ④ $\frac{7}{4} \div \frac{14}{3}$

⑤ $\frac{5}{3} \div \frac{5}{6}$ ⑥ $\frac{5}{9} \div \frac{10}{3}$

⑦ $5 \div \frac{4}{7}$ ⑧ $3 \div \frac{2}{9}$

⑨ $6 \div \frac{3}{8}$ ⑩ $15 \div \frac{3}{5}$

5 There is a metal bar that is $\frac{4}{5}$ m long and weighs $\frac{8}{9}$ kg. How many kilograms is the weight per meter of this metal bar?

6 Let's solve the following calculations.

① $\frac{4}{7} \div 1\frac{4}{5}$ ② $\frac{5}{8} \div 2\frac{6}{7}$

③ $8 \div 1\frac{7}{10}$ ④ $11 \div 2\frac{3}{4}$

⑤ $2\frac{1}{4} \div \frac{3}{7}$ ⑥ $3\frac{3}{5} \div \frac{9}{10}$

⑦ $1\frac{7}{8} \div 1\frac{2}{3}$ ⑧ $5\frac{5}{6} \div 3\frac{1}{2}$

7 Let's write the inequality sign that applies inside each □.

① $9 \div \frac{2}{3}$ □ 9

② $\frac{5}{6} \div \frac{6}{5}$ □ $\frac{5}{6}$

③ $\frac{7}{10} \div \frac{7}{8}$ □ $\frac{7}{10}$

8 There are $3\frac{3}{4}$ kg of rice. If $\frac{3}{8}$ kg of rice are eaten each day, how many days can rice be eaten?

9 $2\frac{1}{4}$ dL of paint were used to paint each square meter of a wall. How many square meters can be painted with 18 dL of paint?

10 There is a rectangular flowerbed with a length of $1\frac{5}{7}$ m and an area of $3\frac{3}{5}$ m². How many meters is the width of this flowerbed?

11 Nozomi cut and used a piece of tape with a length of $1\frac{1}{3}$ m. This piece represented $\frac{3}{5}$ of the initial tape. How many meters was the length of the initial tape?

12 Keita drank only $\frac{2}{9}$ of the total amount of juice. The amount of juice that Keita drank was 180 mL. How many milliliters was the initial amount of juice?

6 Calculations with Decimal Numbers and Fractions

p.72~p.77

1 Let's solve the following calculations.

① $3.64+1.76$　② $7.4-2.53$

③ 4.8×1.63　④ $6.46\div3.8$

2 Let's solve the following calculations.

① $\frac{3}{8}+\frac{5}{12}$　② $\frac{5}{6}-\frac{3}{10}$

③ $\frac{4}{7}\times2\frac{5}{8}$　④ $1\frac{7}{9}\div1\frac{1}{3}$

3 From the following Ⓐ~Ⓓ calculations, which cannot be calculated exactly if it is changed to decimal numbers?

Ⓐ $0.7+\frac{2}{5}$　Ⓑ $1.3+\frac{2}{3}$

Ⓒ $\frac{5}{6}-0.2$　Ⓓ $0.92-\frac{3}{4}$

4 Let's solve the following calculations.

① $0.8+\frac{5}{6}$　② $0.15+\frac{3}{4}$

③ $\frac{7}{9}+0.5$　④ $\frac{3}{5}+0.24$

⑤ $\frac{3}{8}-0.2$　⑥ $\frac{3}{4}-0.7$

⑦ $\frac{2}{3}-0.16$　⑧ $0.45-\frac{1}{6}$

5 Let's calculate using fractions.

① $\frac{5}{8}\times0.6\div\frac{1}{2}$

② $0.25\div\frac{7}{10}\times0.9$

③ $0.9\div0.63\times1.75$

④ $10\div8\times6\div15$

6 How many square meters is the area of the following triangle?

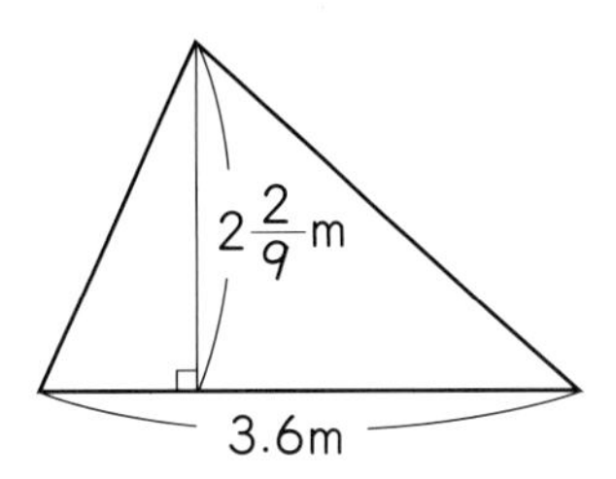

7 There is a car that uses 15 L of gasoline to run 250 km. How many liters of gasoline will this car use to run 150 km?

8 A stock item purchased at 600 yen is priced with a profit of 25 % from the stock value. What is the established price of this item?

9 I purchased a 900 yen item that had a discount of 15%. At how many yen did I buy it?

⑦ Symmetry

p.86~p.103

1 In the following diagram, which is a line symmetric figure?

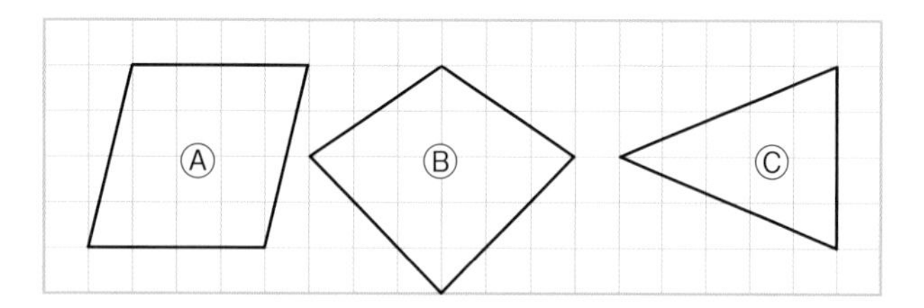

2 Let's answer about the line symmetric figure shown on the right.

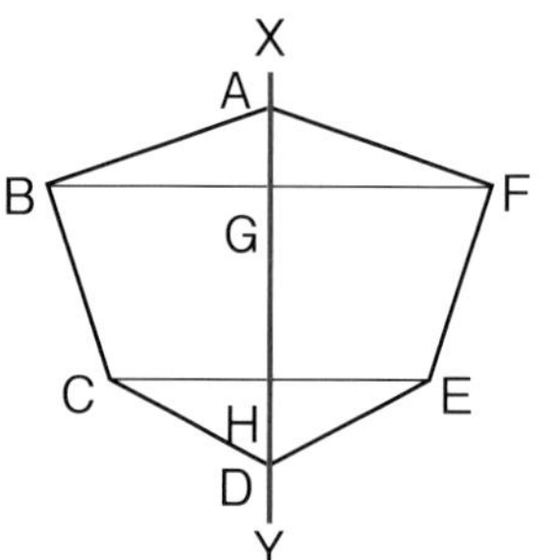

① Which is the corresponding side to side BC?

② How is the intersection between the straight line BF and the line of symmetry?

③ Which other straight line has the same length as straight line EH?

3 In the following diagram, let's complete the line symmetric figure considering straight line XY as the line of symmetry.

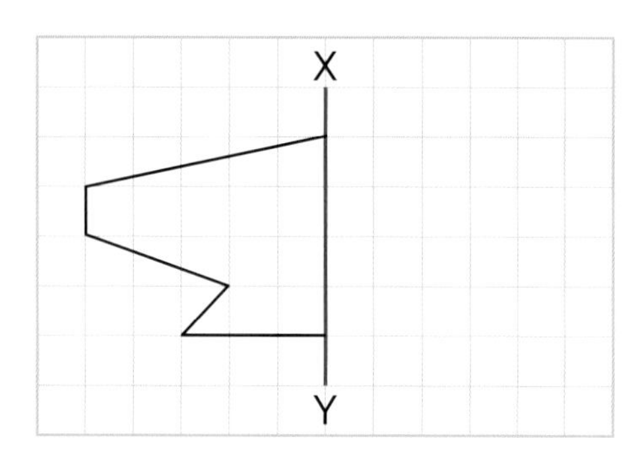

4 In the following diagram, let's complete the line symmetric figure considering straight line XY as the line of symmetry.

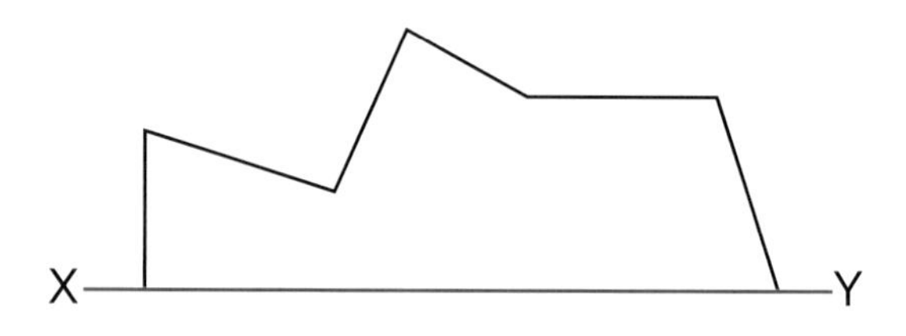

5 In the following diagram, which is a point symmetric figure?

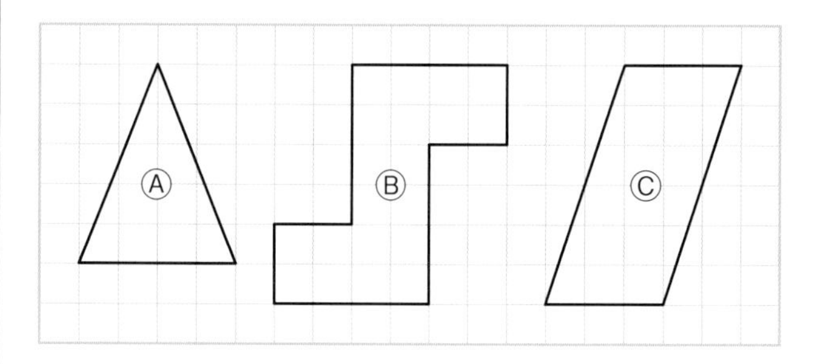

6 Let's answer about the point symmetric figure shown on the right.

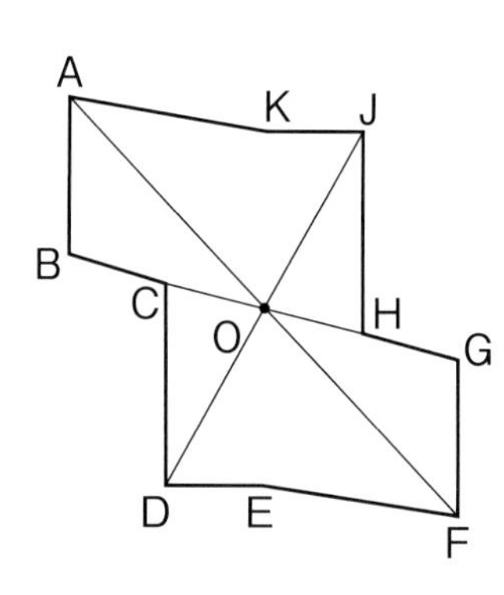

① Which point is the point of symmetry?

② Which is the corresponding point to point D?

③ Which is the corresponding side to side BC?

④ Which other straight line has the same length as straight line AO?

7 In the following diagram, let's complete the point symmetric figure considering point O as the point of symmetry.

① 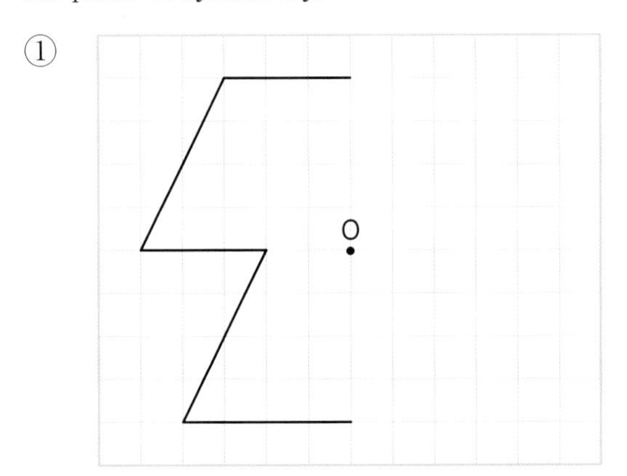

② 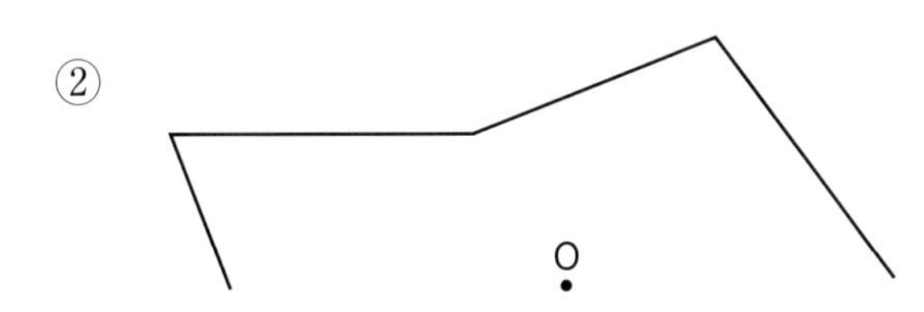

8 Let's answer the following questions about the figures below.

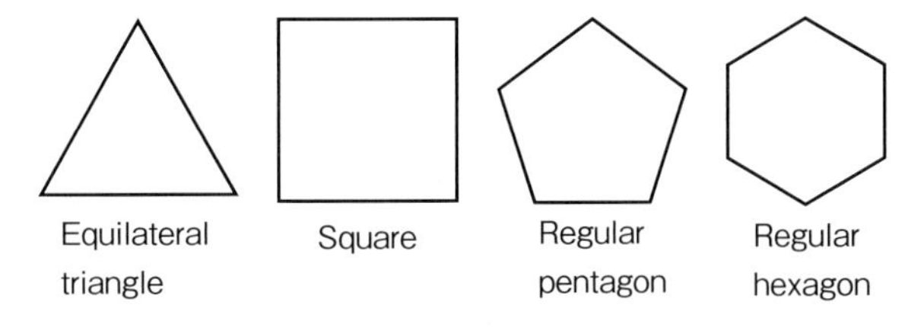

① Which is a line symmetric figure?

② Which is a point symmetric figure?

③ As for the line symmetric figures, how many lines of symmetry does each have?

④ Where is the point of symmetry in a square?

8 Area of a Circle

p.104~p.117

1 Let's find the circumference and area of the following circles.

① 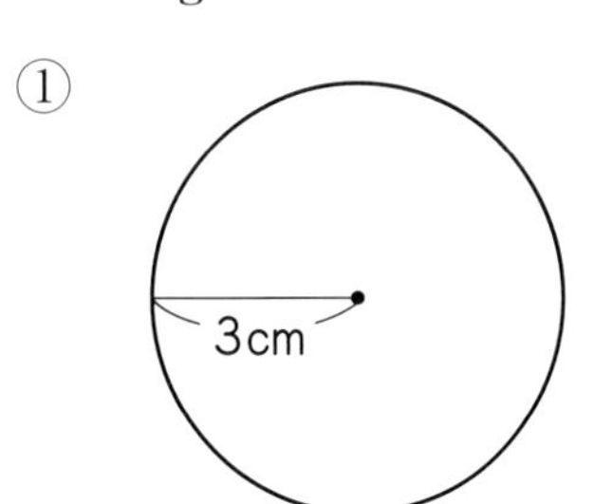

② 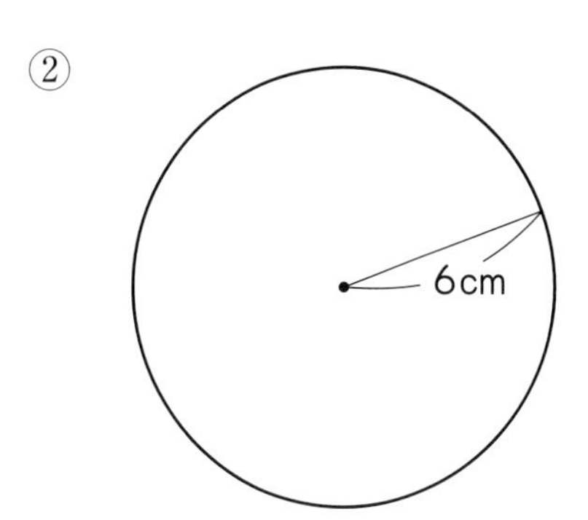

③ 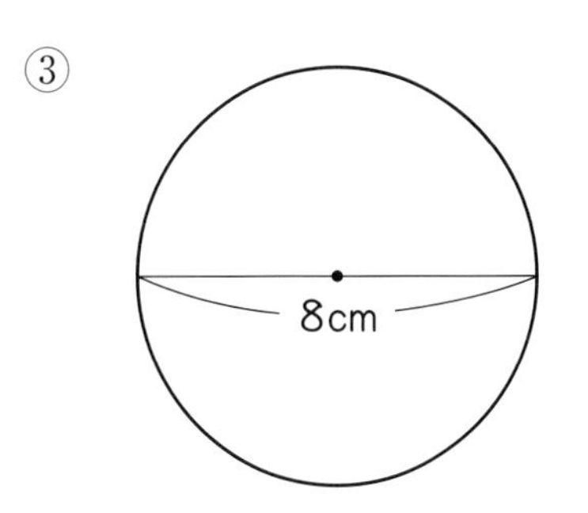

④ 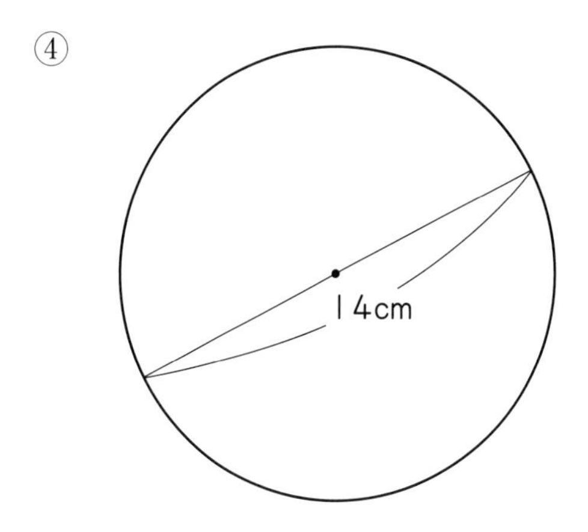

2 Let's answer the following questions about the length of the circumferences shown below.

ⓐ 81.64 cm

ⓑ 56.52 cm

ⓒ 94.2 cm

① Let's find the radius of each circle.

② Let's find the area of each circle.

3 For the following diagrams, let's find the area of the colored parts.

①

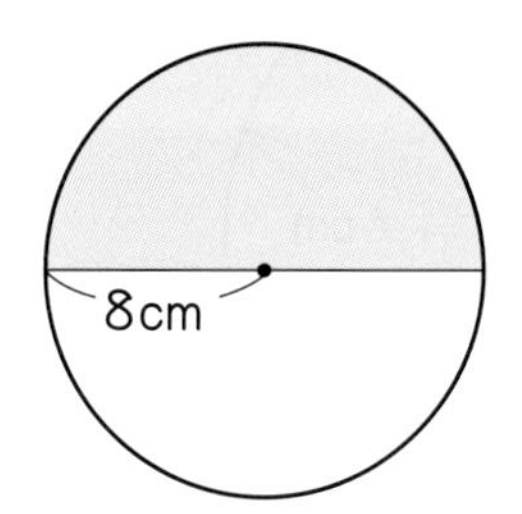

②

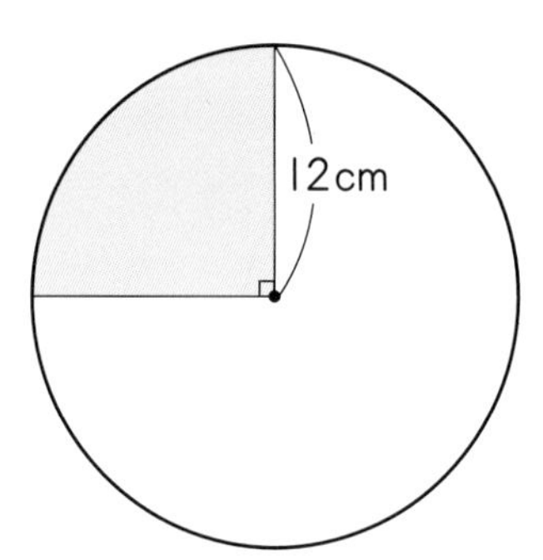

③

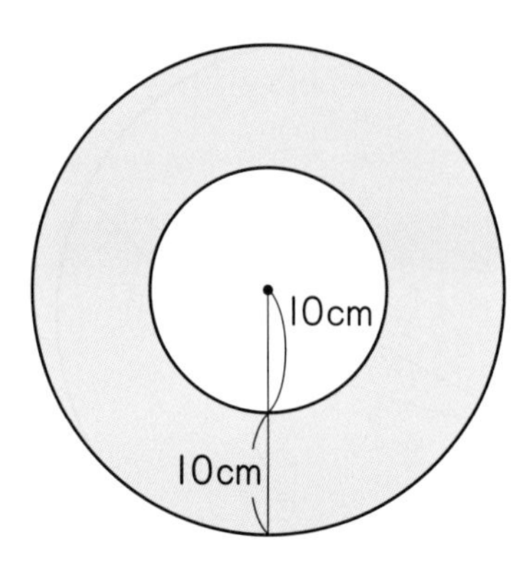

4 For the following diagrams, let's find the area of the colored parts.

①

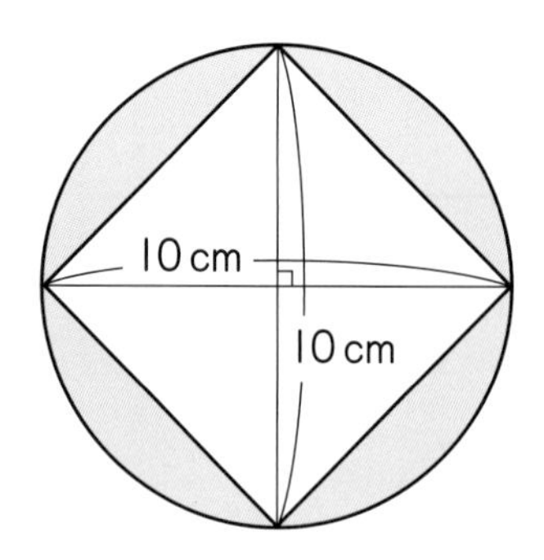

②

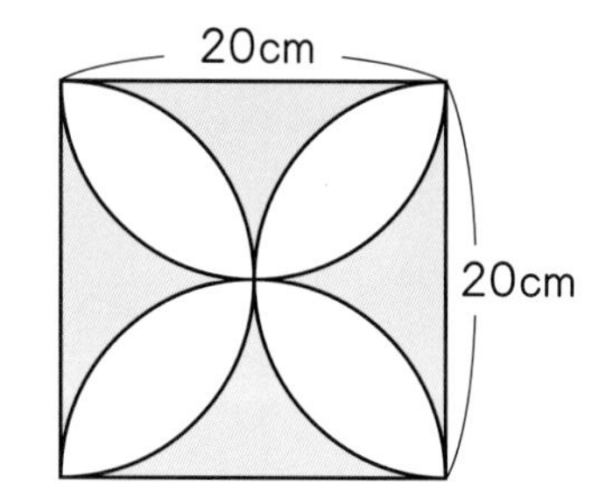

9 Volume of Solids

p.120~p.129

1 There is a quadrangular prism like the one shown on the right. Let's answer the following questions, when the rectangular face ⓐ is the base.

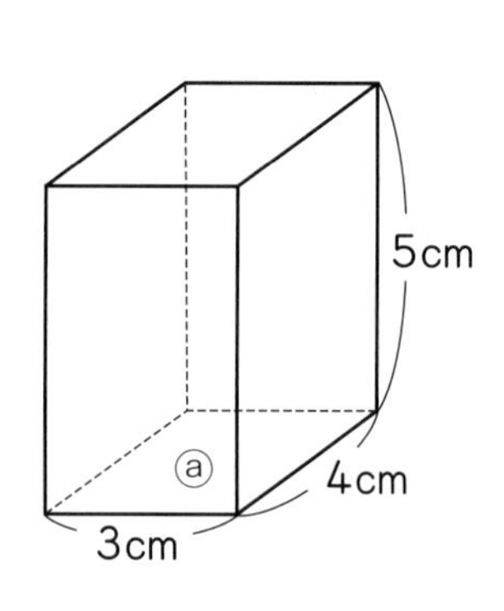

① How many square centimeters is the area of the base?

② How many centimeters is the height?

③ How many cubic centimeters is the volume?

2 Let's find the volume of the following prisms.

①

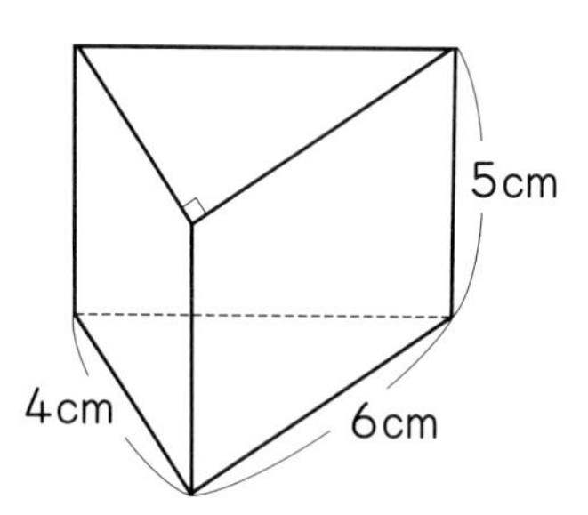

②

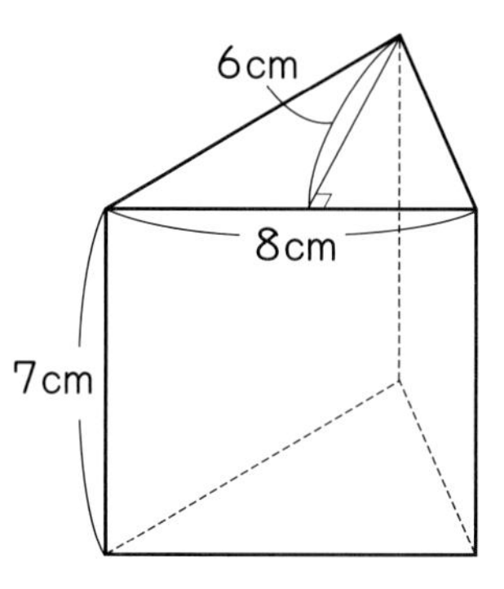

③

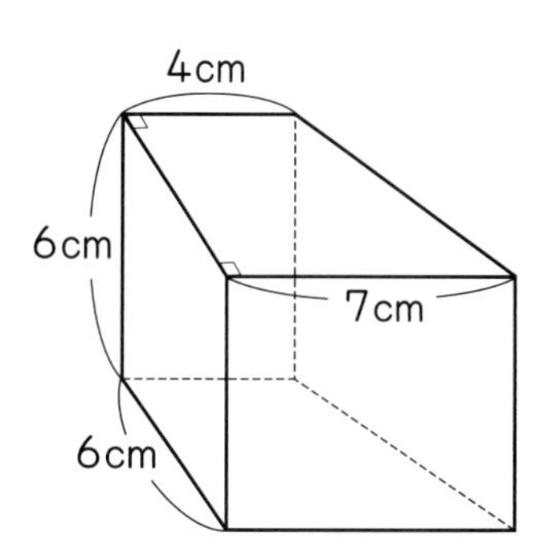

3 Let's find the volume of the following cylinders.

①

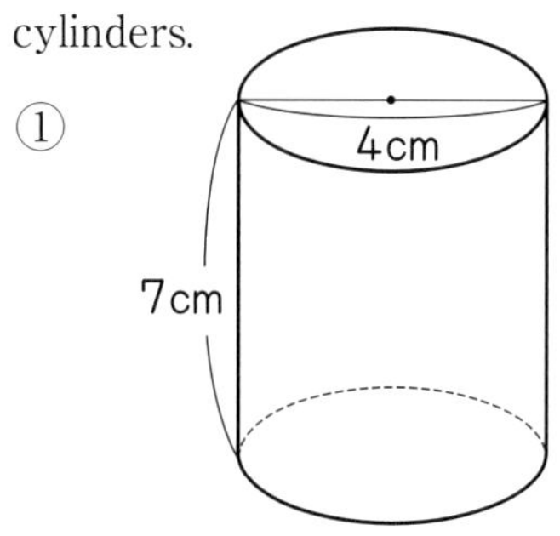

②

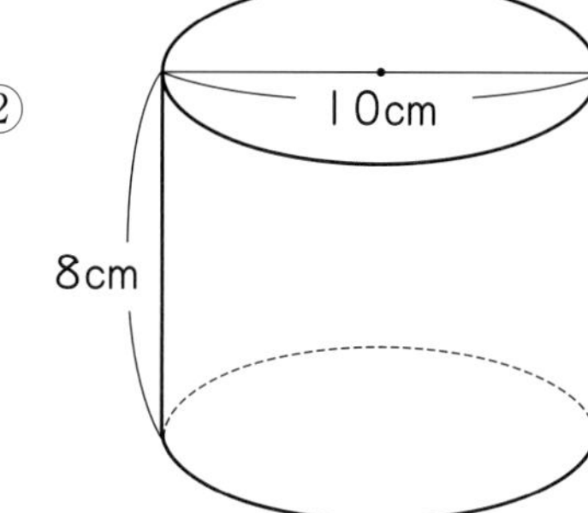

③

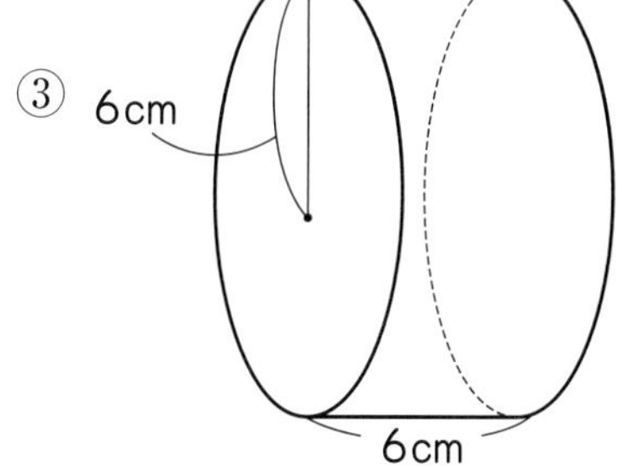

4 The following diagram is the net of a cylinder. Let's find the volume of the cylinder that can be assembled.

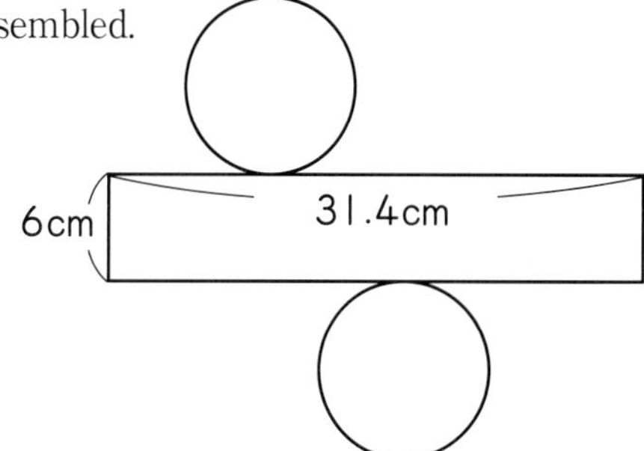

5 Let's find the volume of the following solids.

①

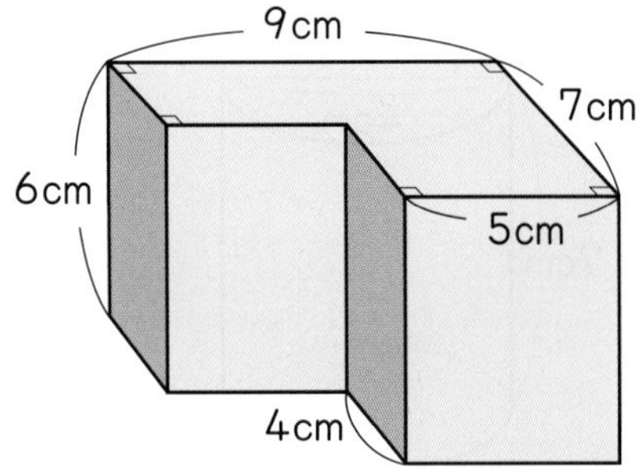

②

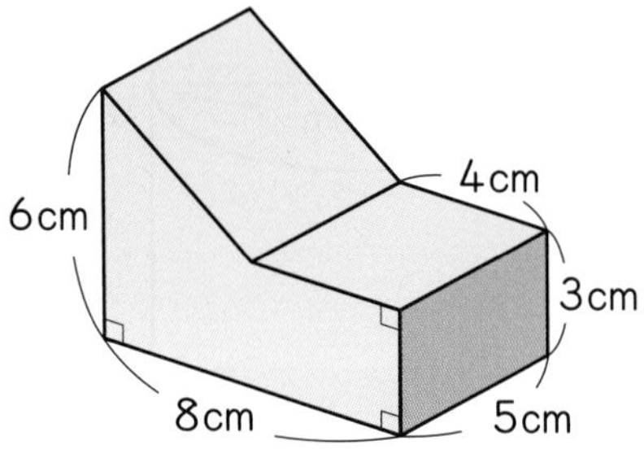

③

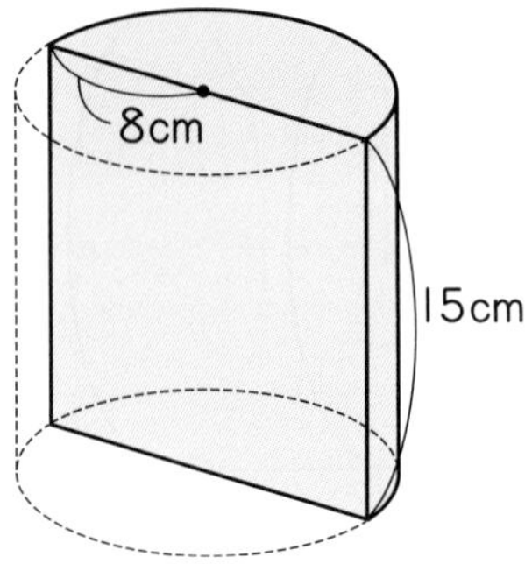

④ This solid has 2 cylindrical holes.

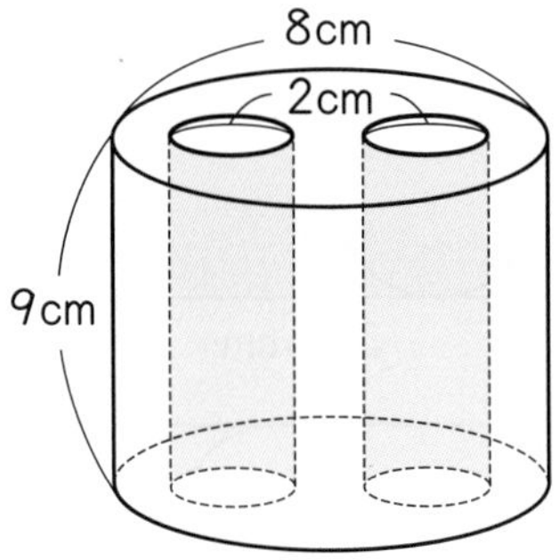

⑩ Ratio and its Application

p.132~p.143

1 Let's explore the rectangle shown on the right.

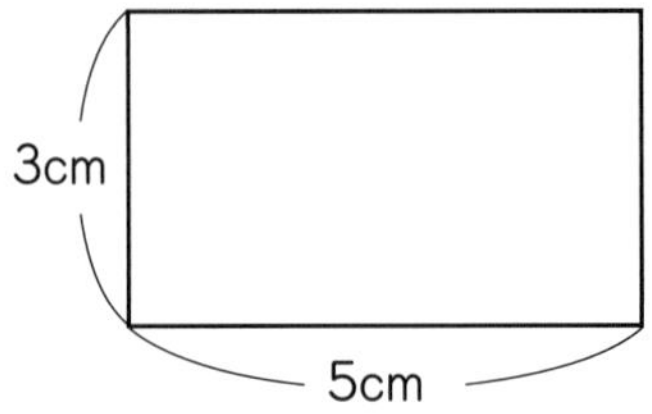

① Let's write the ratio between the length and width.

② Let's find the value of the ratio in ①.

2 Let's find the value of the following ratios.

① 4 : 9 ② 3 : 2

③ 5 : 10 ④ 9 : 12

3 Let's write the number that applies inside each □.

① 4 : 5 = (4 × 3) : (5 × □)
= 12 : □

② 2 : 7 = (2 × □) : (7 × □)
= 10 : 35

③ 30 : 80 = (30 ÷ 10) : (80 ÷ □)
= 3 : □

④ 36 : 30 = (36 ÷ □) : (30 ÷ □)
= 6 : □

4 From the following ratios, which is an equal ratio to 3 : 2?

Ⓐ 9 : 6 Ⓑ 4 : 6

Ⓒ 18 : 15 Ⓓ 15 : 10

Ⓔ 12 : 8 Ⓕ 24 : 12

5 Let's write three equal ratios to the following ratios.

① $4:1$　② $2:7$

③ $5:8$　④ $6:8$

6 Let's find the number that applies for x.

① $8:5=x:25$

② $7:6=21:x$

③ $4:9=x:36$

④ $3:10=18:x$

⑤ $5:x=1:6$

⑥ $x:20=5:4$

⑦ $28:x=7:3$

⑧ $x:45=8:15$

7 Let's simplify the following ratios. Let's write the number that applies inside each $\square$.

① $20:35=(20\div 5):(35\div\square)$
$=4:\square$

② $0.2:0.9=(0.2\times\square):(0.9\times 10)$
$=\square:9$

③ $1.5:2.4=(1.5\times 10):(2.4\times\square)$
$=15:\square$
$=5:\square$

④ $\frac{2}{3}:\frac{1}{2}=\left(\frac{2}{3}\times 6\right):\left(\frac{1}{2}\times\square\right)$
$=4:\square$

⑤ $\frac{3}{8}:\frac{5}{12}=\left(\frac{3}{8}\times 24\right):\left(\frac{5}{12}\times\square\right)$
$=9:\square$

8 Let's simplify the following ratios.

① $16:20$　② $36:32$

③ $150:200$　④ $144:12$

⑤ $2.4:3.2$　⑥ $2:0.8$

⑦ $\frac{1}{6}:\frac{3}{10}$　⑧ $\frac{2}{7}:\frac{4}{9}$

9 An elder and younger sister have sheets of colored paper. The ratio between the elder sister's number and the younger sister's number is $5:4$, and the elder sister has 40 sheets of colored paper. How many sheets of colored paper does the younger sister have?

10 The shadow of a tree that was measured in the school yard was 8 m. Also, a 0.9 m wooden stick was placed at the school yard and the shadow's length was 1.2 m. How many meters is the height of the tree?

11 An elder and younger brother need to divide 2000 yen, received from their mother, in a $3:2$ ratio. How many yen will the elder brother receive?

12 Yuna's group has 35 children and the ratio between boys and girls is $3:4$. How many boys and girls are there in this group?

11 Enlarged and Reduced Drawings

p.144~p.159

1 In the following diagram, triangle DEF is an enlarged drawing of triangle ABC. Let's answer the following questions.

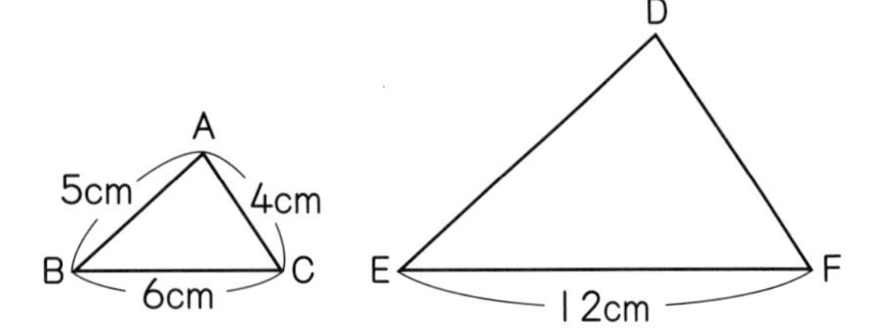

① Let's find the ratio between the length of side BC and side EF.

② Let's find the ratio between the length of side AB and side DE.

③ How many centimeters is the length of sides DE and DF?

④ Is the size of angle B equal to the size of angle E?

⑤ Which angle has the same size as angle C?

⑥ How many times of triangle ABC is enlarged triangle DEF?

⑦ Based on triangle DEF, by how much is triangle ABC reduced?

2 In the following diagram, triangle DEF is a reduced drawing of triangle ABC. Let's answer the following questions.

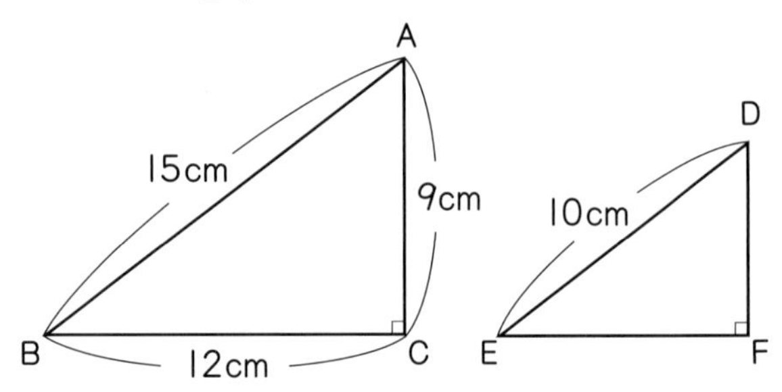

① Let's find the ratio between the length of side AB and side DE.

② How many centimeters is the length of side EF?

③ How many centimeters is the length of side DF?

3 Let's draw quadrilateral EFGH that is a 2 times enlarged drawing of quadrilateral ABCD shown below.

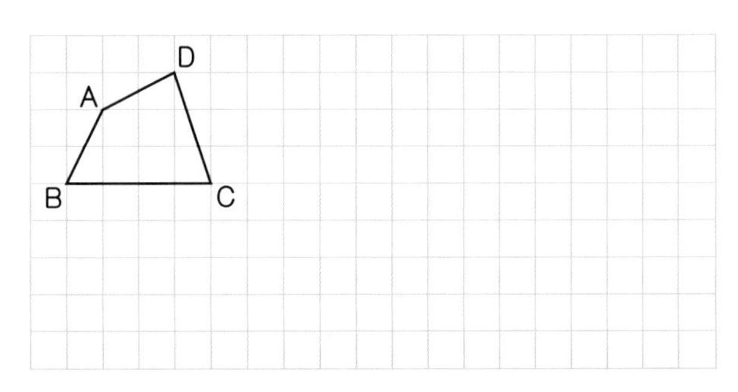

4 Let's draw triangle DEF that is a $\frac{1}{2}$ reduced drawing of triangle ABC shown below.

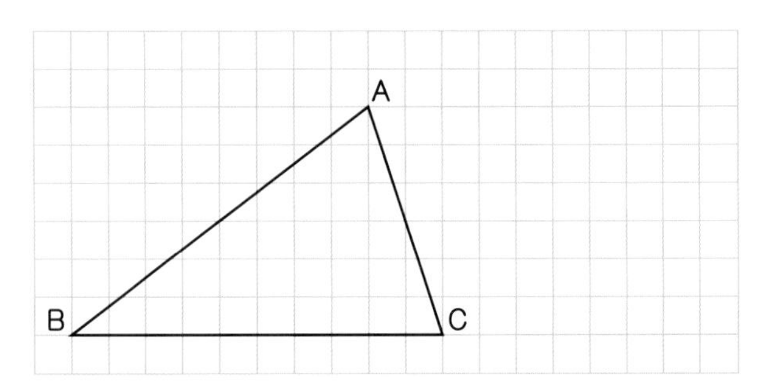

5 In the following diagram, point B is the center of enlarging. Let's draw an enlarged drawing that is 2 times triangle ABC.

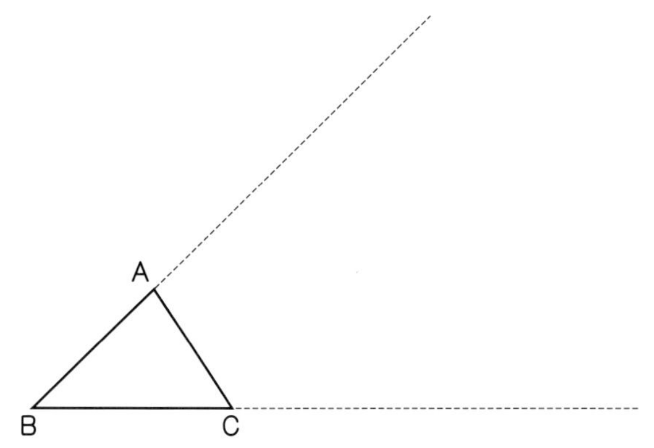

6 Let's use point E as the center point to draw a $\frac{1}{2}$ reduced drawing of quadrilateral ABCD.

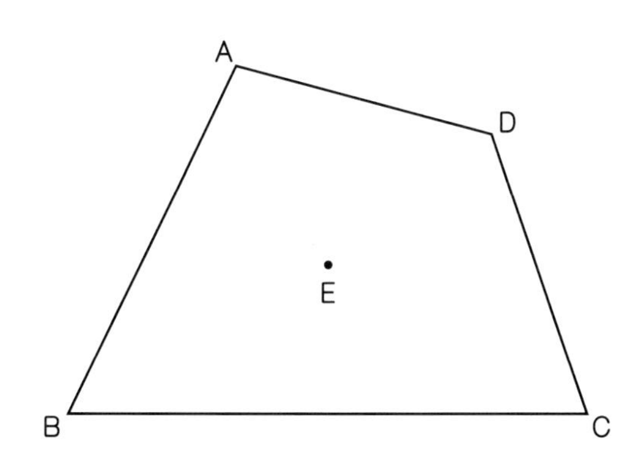

7 In the figure shown below, how many meters is the actual height of the tree? Let's find the answer by drawing a reduced triangle in $\frac{1}{100}$ reduced scale.

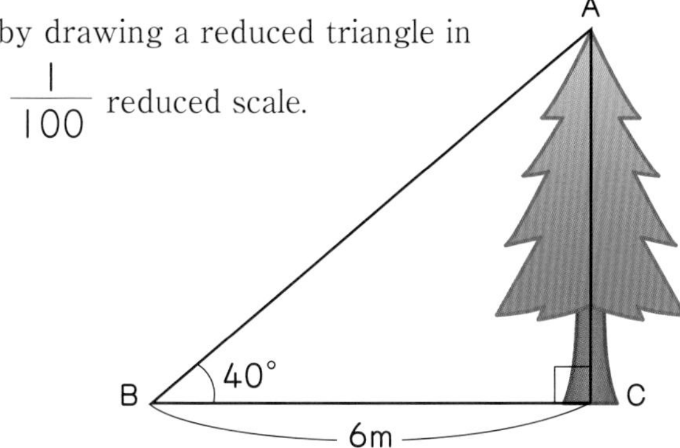

8 There is a rectangular park with a length of 200 m and a width of 300 m. If a reduced drawing of this park is done in $\frac{1}{5000}$ reduced scale, how many centimeters is the length and width?

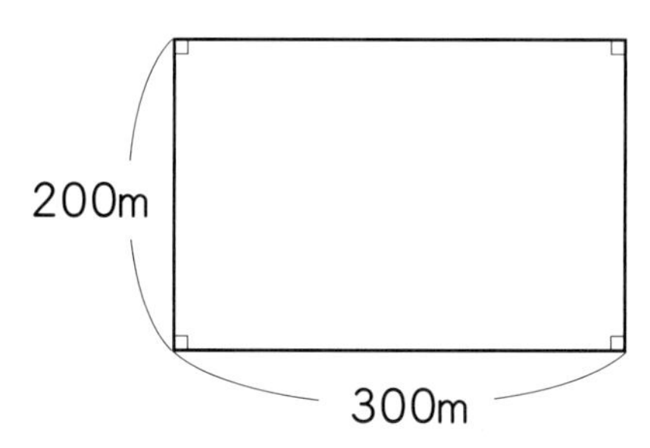

9 Let's answer about a $\frac{1}{2000}$ reduced scale.

① Using the above scale, how many meters is the actual length of the following?

ⓐ 4cm ⓑ 6.5cm

② The following lengths are represented by how many centimeters using the above scale?

ⓐ 60 m ⓑ 210 m

10 Let's answer about a $\frac{1}{10000}$ reduced map scale.

① Using the above map scale, how many meters is the actual length of the following?

ⓐ 7cm ⓑ 9.5cm

② The following lengths are represented by how many centimeters using the above reduced map scale?

ⓐ 0.5km ⓑ 1.2km

12 Proportion and Inverse Proportion

p.160~p.181

1 When you walk at 60 m per minute, y m can be traveled in x minutes. Considering this, let's answer the following questions.

① Let's fill in the following table with the number that applies.

Time and Distance Traveled at 60m per Minute

Time x (min)	1	2	3	4	5	6
Distance y (m)	60	120	180	240	300	

② When the value of x changes by 2 times, 3 times, ... and so on, how does the value of y change?

③ Is y proportional to x?

④ When the value of x changes by $\frac{1}{2}$ times, $\frac{1}{3}$ times, ... and so on, how does the value of y change?

2 The table below shows the relationship between the water depth y cm and time x minutes when water is poured into a cuboid. Let's answer the following questions.

Water Depth and Pouring Time

Time x (min)	1	2	3	4	5	6
Water depth y (cm)	3	6	9	12	15	18

① How much does the value of y increase when the value of x increases by 1?

② Let's represent the relationship between x and y in a math sentence.

③ How many centimeters is the water depth when the water is poured for 9 minutes?

④ Let's represent the relationship between x and y with a graph.

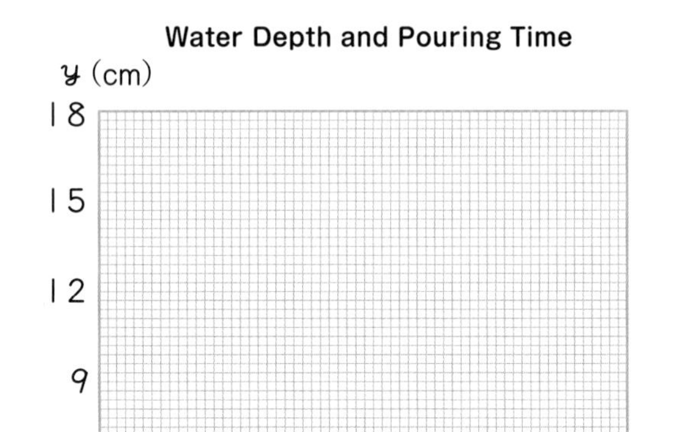

3 For a distance of 60 km, consider the speed as x km per hour and the traveling time as y hours. Let's answer the following questions.

① Let's fill in the following table with the number that applies.

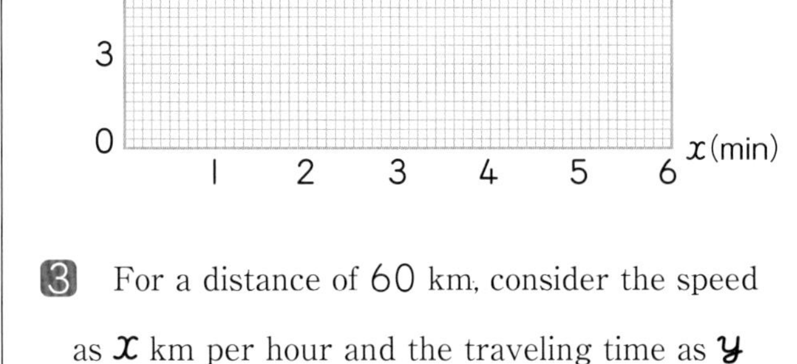

Speed and Traveling Time

Distance per hour x (km)	1	2	3	4	6	12
Time y (hours)	60	30	20	15	10	

② When the value of x changes by 2 times, 3 times, ... and so on, how does the value of y change?

③ Is y inversely proportional to x?

④ When the value of x changes by $\frac{1}{2}$ times, $\frac{1}{3}$ times, ... and so on, how does the value of y change?

4 From the following ⓐ, ⓑ, and ⓒ, choose the option in which y is inversely proportional to x.

ⓐ A rectangle with a fixed perimeter of 20 cm, length of x cm, and width of y cm.

ⓑ A rectangle with length of 4 cm, width of x cm, and an area of y cm².

ⓒ A rectangle with an area of 40 cm², length of x cm, and width of y cm.

13 Data Arrangement

p.184~p.198

1 Let's look at the records for throwing a softball, and answer the following questions.

Records for Throwing a Softball

Number	Distance (m)	Number	Distance (m)	Number	Distance (m)
①	28	⑧	29	⑮	31
②	33	⑨	43	⑯	23
③	20	⑩	34	⑰	37
④	26	⑪	46	⑱	19
⑤	16	⑫	27	⑲	28
⑥	35	⑬	26	⑳	44
⑦	24	⑭	21		

① Let's summarize the records in the following frequency distribution table.

Records for Throwing a Softball

Class (m)	Number of children
greater than or equal to 15 ~ less than 20	
20 ~ 25	
25 ~ 30	
30 ~ 35	
35 ~ 40	
40 ~ 45	
45 ~ 50	
Total	

② Let's represent the records with a histogram.

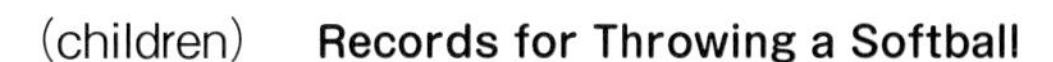

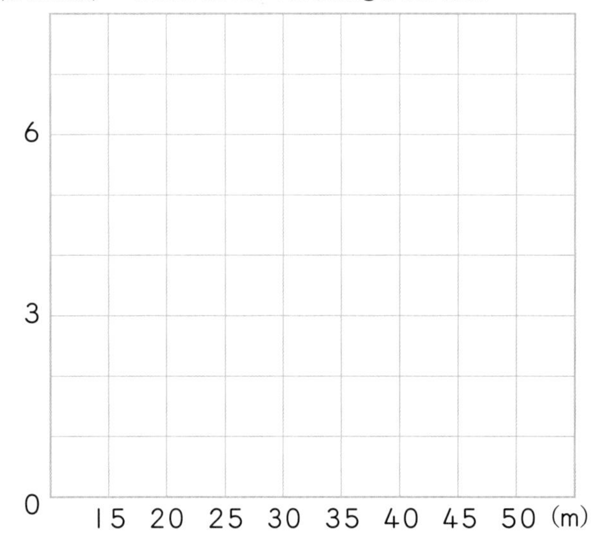

③ Which class has the most number of children?

④ Based on the child with the longest distance, which class does the 5th place child belong to?

2 The following 15 scores represent the mathematics test score of children based on a 10 point score.

Let's answer the following questions.

(points)

5	7	4	9	3	6	5	10
4	6	8	2	5	9	7	

① How many points is the mean?

② What is the median value?

③ What is the mode value?

1 Ways of Ordering and Combinating

1 ① 1st 2nd 3rd 4th

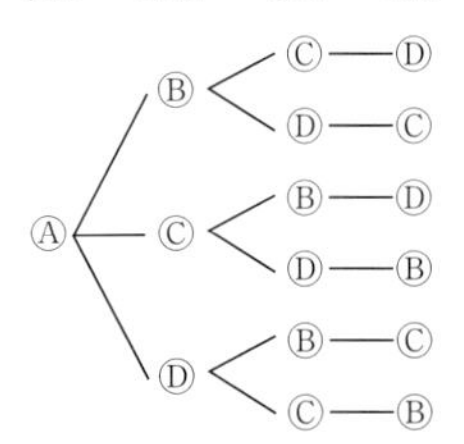

② 6 ways ③ 6 ways each ④ 24 ways

2 ①

Hundreds	Tens	Ones
2	4	6
2	4	8
2	6	4
2	6	8
2	8	4
2	8	6

② 24 ways

3 ① Tens place Ones place

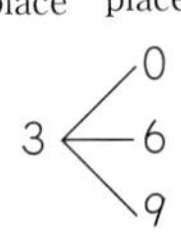

② 6, 9 ③ 9 ways

4 ① A B C

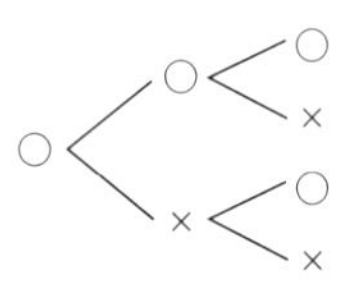

② 8 ways

5 ①

	A	B	C	D	E
A		○	○	○	○
B			○	○	○
C				○	○
D					○
E					

② 10 games

6 4 ways

2 Mathematical Letter and Sentence

1 $70\times a$

2 $a\times 5$

3 ① $x\times 3+4$ ② 64 origami sheets

4 ① $x\times 4+3$ ② 19dL

5 $x\times 5+6$

6 ① $x+5=14$ ② 9dL

7 ① $x\times 4=26$ ② 6.5cm

8 ① 36 ② 8 ③ 23 ④ 15 ⑤ 22 ⑥ 40 ⑦ 62 ⑧ 93 ⑨ 6 ⑩ 7 ⑪ 4 ⑫ 1.5

9 ① $x\times 3+5$ ② 18 cookies

10 ① 8 ② 9

11 ① Ⓑ ② Ⓒ ③ Ⓐ

3 Multiplication and Division of Fractions and Whole Numbers

1 ① $\frac{4}{7}$ ② $\frac{9}{10}$ ③ $\frac{3}{5}$ ④ $3\frac{3}{4}\left(\frac{15}{4}\right)$ ⑤ $1\frac{7}{9}\left(\frac{16}{9}\right)$ ⑥ $1\frac{7}{8}\left(\frac{15}{8}\right)$

2 ① $\frac{5}{6}$ ② $1\frac{1}{3}\left(\frac{4}{3}\right)$ ③ $\frac{2}{3}$ ④ $1\frac{1}{3}\left(\frac{4}{3}\right)$ ⑤ $13\frac{1}{2}\left(\frac{27}{2}\right)$ ⑥ 12

3 $\frac{5}{8}\times 4=2\frac{1}{2}$ $2\frac{1}{2}$ m$\left(\frac{5}{2}\text{ m}\right)$

4 ① $4\frac{8}{9}\left(\frac{44}{9}\right)$ ② $2\frac{10}{11}\left(\frac{32}{11}\right)$ ③ $4\frac{2}{5}\left(\frac{22}{5}\right)$ ④ $7\frac{1}{3}\left(\frac{22}{3}\right)$ ⑤ $4\frac{1}{2}\left(\frac{9}{2}\right)$ ⑥ $11\frac{1}{3}\left(\frac{34}{3}\right)$

5 $3\frac{1}{2}\times 4=14$ 14m²

6 ① $\frac{1}{9}$ ② $\frac{5}{28}$ ③ $\frac{3}{10}$ ④ $\frac{3}{40}$ ⑤ $\frac{4}{27}$ ⑥ $\frac{3}{28}$

7 ① $\frac{2}{7}$ ② $\frac{1}{16}$ ③ $\frac{4}{27}$ ④ $\frac{2}{15}$ ⑤ $\frac{2}{3}$ ⑥ $\frac{5}{28}$

8 $\frac{9}{8}\div 3=\frac{3}{8}$ $\frac{3}{8}$ L

9 ① $\frac{22}{27}$ ② $\frac{17}{20}$ ③ $\frac{1}{2}$ ④ $\frac{8}{15}$ ⑤ $\frac{3}{7}$ ⑥ $\frac{15}{22}$

10 $3\frac{3}{4}\div 5=\frac{3}{4}$ $\frac{3}{4}$ kg

4 Fraction × Fraction

1 ① $\frac{5}{72}$ ② $\frac{7}{18}$ ③ $\frac{4}{15}$ ④ $\frac{1}{24}$ ⑤ $2\frac{11}{12}\left(\frac{35}{12}\right)$ ⑥ $1\frac{13}{32}\left(\frac{45}{32}\right)$ ⑦ $\frac{20}{27}$ ⑧ $\frac{27}{28}$

2 ① $\frac{5}{27}$ ② $\frac{7}{18}$ ③ $\frac{1}{4}$ ④ $\frac{12}{25}$ ⑤ $1\frac{4}{11}\left(\frac{15}{11}\right)$ ⑥ $\frac{9}{10}$ ⑦ $1\frac{1}{2}\left(\frac{3}{2}\right)$ ⑧ $3\frac{1}{2}\left(\frac{7}{2}\right)$ ⑨ 4 ⑩ $3\frac{3}{4}\left(\frac{15}{4}\right)$

3 $\frac{7}{10}\times\frac{5}{6}=\frac{7}{12}$ $\frac{7}{12}$ m²

4 ① $2\frac{1}{10}\left(\frac{21}{10}\right)$ ② $12\frac{1}{2}\left(\frac{25}{2}\right)$ ③ 6 ④ $28\frac{1}{2}\left(\frac{57}{2}\right)$ ⑤ 2 ⑥ $\frac{11}{20}$ ⑦ $\frac{2}{3}$ ⑧ $1\frac{1}{2}\left(\frac{3}{2}\right)$

5 $1\frac{1}{8}\times 2\frac{2}{3}=3$ 3 kg

6 ① $<$ ② $>$ ③ $=$ ④ $>$

7 ① $\frac{9}{28}$ m² ② $\frac{1}{4}$ m² ③ $\frac{7}{16}$ m²

8 ① $\frac{3}{8}$ ② $\frac{8}{9}$ ③ $\frac{3}{10}$ ④ 24

9 1

10 ① $2\frac{2}{3}\left(\frac{8}{3}\right)$ ② $1\frac{2}{7}\left(\frac{9}{7}\right)$ ③ $\frac{2}{3}$ ④ $\frac{5}{12}$ ⑤ 6 ⑥ $1\frac{1}{4}\left(\frac{5}{4}\right)$ ⑦ $\frac{10}{13}$ ⑧ $\frac{1}{4}$

5 Fraction ÷ Fraction

1 5, 3, 3

2 Reciprocal of the divisor

3 ① $\frac{10}{27}$ ② $\frac{35}{36}$ ③ $5\frac{1}{3}\left(\frac{16}{3}\right)$ ④ $\frac{21}{32}$

4 ① $1\frac{3}{4}\left(\frac{7}{4}\right)$ ② $\frac{12}{25}$ ③ $2\frac{1}{10}\left(\frac{21}{10}\right)$ ④ $\frac{3}{8}$ ⑤ 2

⑥ $\frac{1}{6}$ ⑦ $8\frac{3}{4}\left(\frac{35}{4}\right)$ ⑧ $13\frac{1}{2}\left(\frac{27}{2}\right)$ ⑨ 16 ⑩ 25

5 $\frac{8}{9} \div \frac{4}{5} = 1\frac{1}{9}$ $\underline{1\frac{1}{9}\text{kg}\left(\frac{10}{9}\text{kg}\right)}$

6 ① $\frac{20}{63}$ ② $\frac{7}{32}$ ③ $4\frac{12}{17}\left(\frac{80}{17}\right)$ ④ 4

⑤ $5\frac{1}{4}\left(\frac{21}{4}\right)$ ⑥ 4 ⑦ $1\frac{1}{8}\left(\frac{9}{8}\right)$ ⑧ $1\frac{2}{3}\left(\frac{5}{3}\right)$

7 ① > ② < ③ >

8 $3\frac{3}{4} \div \frac{3}{8} = 10$ $\underline{10\text{ days}}$

9 $18 \div 2\frac{1}{4} = 8$ $\underline{8\text{ m}^2}$

10 $3\frac{3}{5} \div 1\frac{5}{7} = 2\frac{1}{10}$ $\underline{2\frac{1}{10}\text{m}\left(\frac{21}{10}\text{m}\right)}$

11 $1\frac{1}{3} \div \frac{3}{5} = 2\frac{2}{9}$ $\underline{2\frac{2}{9}\text{m}\left(\frac{20}{9}\text{m}\right)}$

12 $180 \div \frac{2}{9} = 810$ $\underline{810\text{mL}}$

6 Calculations with Decimal Numbers and Fractions

1 ① 5.4 ② 4.87 ③ 7.824 ④ 1.7

2 ① $\frac{19}{24}$ ② $\frac{8}{15}$ ③ $1\frac{1}{2}\left(\frac{3}{2}\right)$ ④ $1\frac{1}{3}\left(\frac{4}{3}\right)$

3 Ⓑ, Ⓒ

4 ① $1\frac{19}{30}\left(\frac{49}{30}\right)$ ② $\frac{9}{10}$ ③ $1\frac{5}{18}\left(\frac{23}{18}\right)$ ④ $\frac{21}{25}$

⑤ $\frac{7}{40}$ ⑥ $\frac{1}{20}$ ⑦ $\frac{38}{75}$ ⑧ $\frac{17}{60}$

5 ① $\frac{3}{4}$ ② $\frac{9}{28}$ ③ $2\frac{1}{2}\left(\frac{5}{2}\right)$ ④ $\frac{1}{2}$

6 4m²

7 $15 \div 250 = \frac{3}{50}$ $\frac{3}{50} \times 150 = 9$ $\underline{9\text{ L}}$

$\left(250 \div 15 = \frac{50}{3} \quad 150 \div \frac{50}{3} = 9\right)$

8 600 × (1 + 0.25) = 750 <u>750 yen</u>

(600 + 600 × 0.25 = 750)

9 900 × (1 − 0.15) = 765 <u>765 yen</u>

(900 − 900 × 0.15 = 765)

7 Symmetry

1 Ⓑ, Ⓒ

2 ① side FE ② perpendicular intersection

③ straight line CH

3

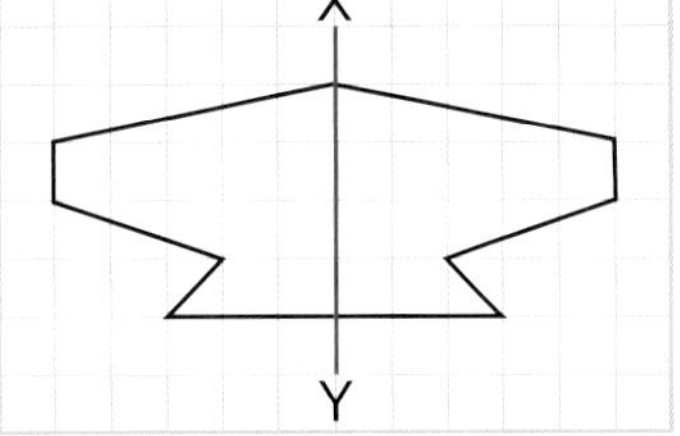

4

5 Ⓑ, Ⓒ

6 ① point O ② point J ③ side GH ④ straight line FO

7 ①

②

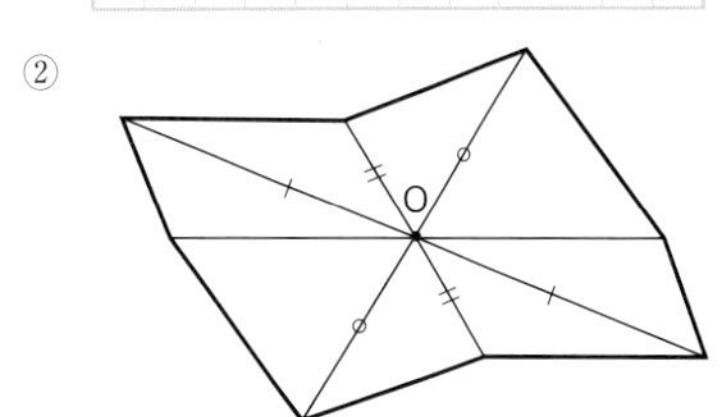

8 ① equilateral triangle, square, regular pentagon, regular hexagon

② square, regular hexagon

③ equilateral triangle…3 square…4 regular pentagon…5

regular hexagon…6

④ In the point where the two diagonals intersect

8 Area of a circle

1 ① circumference…18.84cm area…28.26cm²

② circumference…37.68cm area…113.04cm²

③ circumference…25.12cm area…50.24cm²

④ circumference…43.96cm area…153.86cm²

2 ① ⓐ 13cm ⓑ 9cm ⓒ 15cm

② ⓐ 530.66cm² ⓑ 254.34cm² ⓒ 706.5cm²

3 ① 100.48cm² ② 113.04cm² ③ 942cm²

4 ① 28.5cm² ② 172cm²

9 Volume of solids

1 ① $12cm^2$ ② 5cm ③ $60cm^3$

2 ① $60cm^3$ ② $168cm^3$ ③ $198cm^3$

3 ① $87.92cm^3$ ② $628cm^3$ ③ $678.24cm^3$

4 $471cm^3$

5 ① $282cm^3$ ② $150cm^3$ ③ $1507.2cm^3$
④ $395.64cm^3$

10 Ratio and its Application

1 ① 3 : 5 ② $\frac{3}{5}$

2 ① $\frac{4}{9}$ ② $\frac{3}{2}$ ③ $\frac{1}{2}$ ④ $\frac{3}{4}$

3 ① 3, 15 ② 5, 5 ③ 10, 8 ④ 6, 6, 5

4 Ⓐ, Ⓓ, Ⓔ

5 ① (examples) 8 : 2, 12 : 3, 16 : 4
② (examples) 4 : 14, 6 : 21, 8 : 28
③ (examples) 10 : 16, 15 : 24, 20 : 32
④ (examples) 3 : 4, 9 : 12, 12 : 16

6 ① 40 ② 18 ③ 16 ④ 60 ⑤ 30 ⑥ 25
⑦ 12 ⑧ 24

7 ① 5, 7 ② 10, 2 ③ 10, 24, 8 ④ 6, 3
⑤ 24, 10

8 ① 4 : 5 ② 9 : 8 ③ 3 : 4 ④ 12 : 1 ⑤ 3 : 4
⑥ 5 : 2 ⑦ 5 : 9 ⑧ 9 : 14

9 32 sheets of colored paper

10 6 m

11 1200 yen

12 boys...15 children girls...20 children

11 Enlarged and Reduced Drawings

1 ① 1 : 2 ② 1 : 2 ③ side DE···10cm side DF···8cm
④ equal ⑤ angle F ⑥ 2 times ⑦ $\frac{1}{2}$

2 ① 3 : 2 ② 8cm ③ 6cm

3

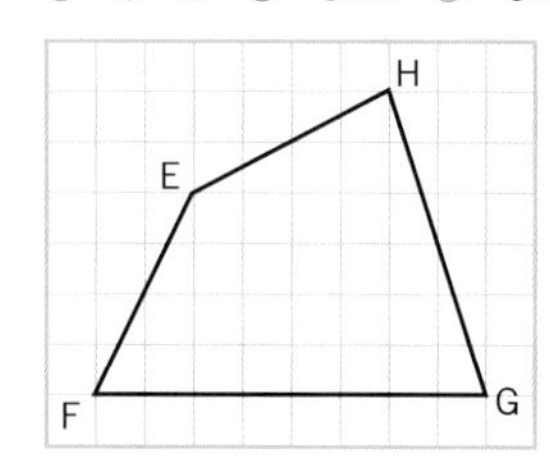

4

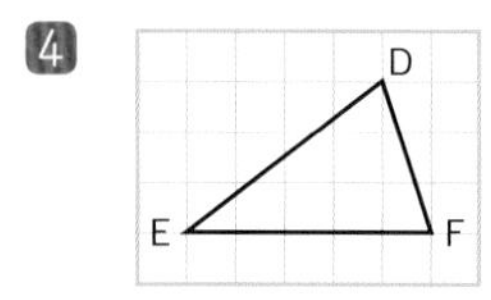

5

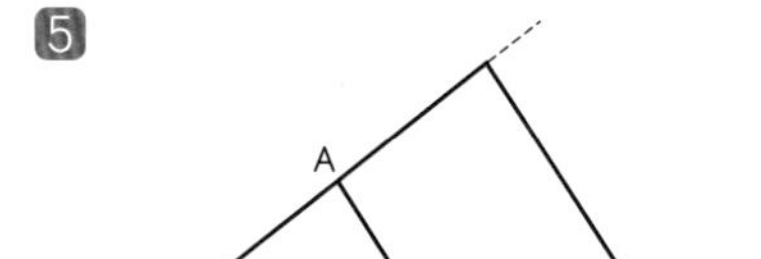

6

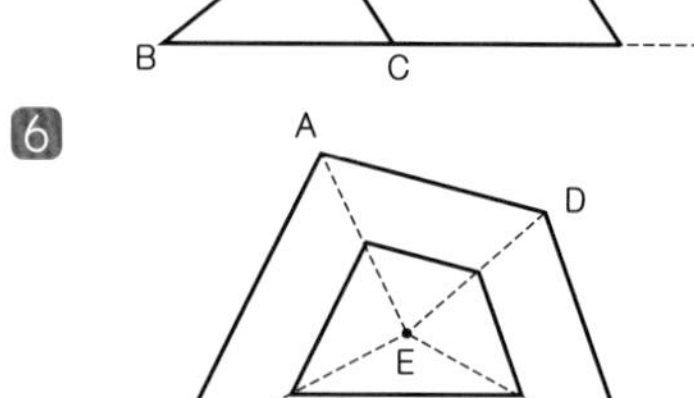

7 In the reduced drawing, the length of AC is about 5cm.
5 × 100 = 500 (cm) About 5m

8 length···4cm width···6cm

9 ① ⓐ 80m ⓑ 130m ② ⓐ 3cm ⓑ 10.5cm

10 ① ⓐ 700m ⓑ 950m ② ⓐ 5cm ⓑ 12cm

12 Proportion and Inverse Proportion

1 ① 360 ② It changes by 2 times, 3 times, and so on.
③ Yes.
④ It changes by $\frac{1}{2}$ times, $\frac{1}{3}$ times, and so on.

2 ① 3 ② $y = 3 \times x$ ③ 27cm
④

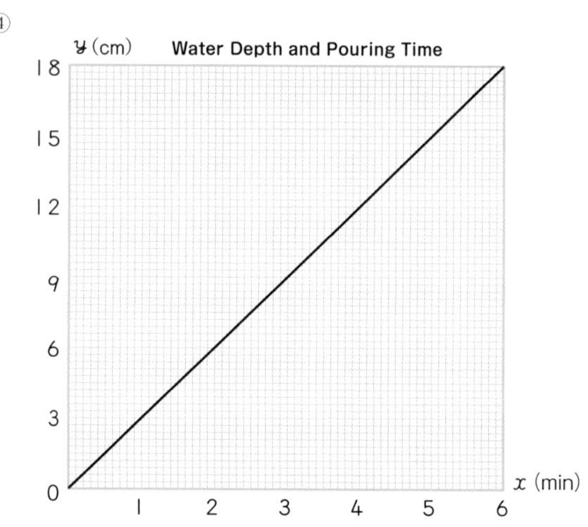

3 ① 5 ② It changes by $\frac{1}{2}$ times, $\frac{1}{3}$ times, and so on. ③ Yes.
④ It changes by 2 times, 3 times, and so on.

4 Ⓒ

⑬ Data Arrangement

1 ①

Records for Throwing a Softball

Class (m)	Number of children
greater than or equal to 15 ~ less than 20	2
20 ~ 25	4
25 ~ 30	6
30 ~ 35	3
35 ~ 40	2
40 ~ 45	2
45 ~ 50	1
Total	20

②

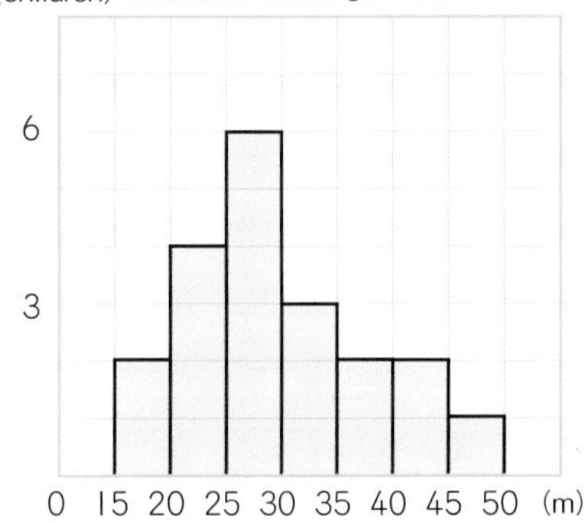

③ Greater than or equal to 25m and less than 30m

④ Greater than or equal to 35m and less than 40m

2 ① 6 points ② 6 points ③ 5 points

Words and symbols from this book.

Area of a circle

▼ will be used in pages 107 and 108.

16 equal parts

32 equal parts

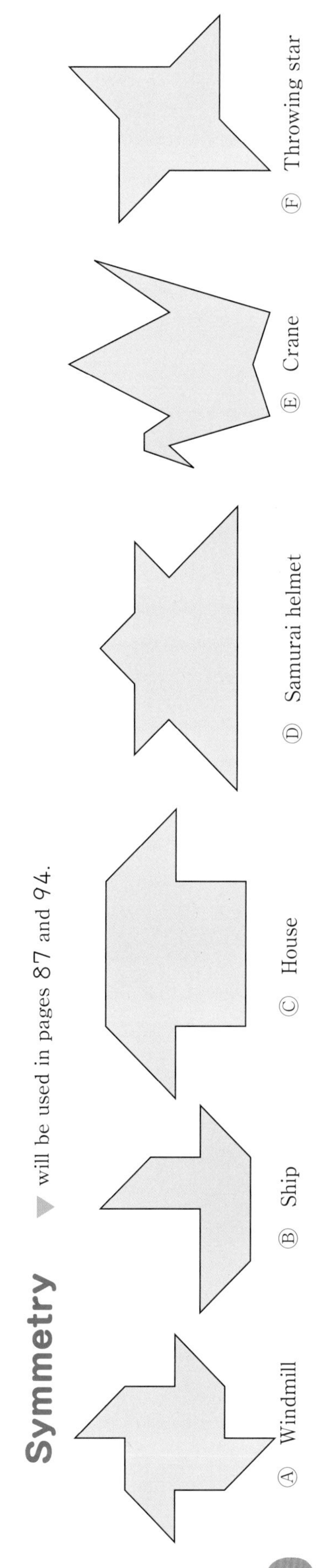

Memo

Editorial for English Edition:

Study with Your Friends, Mathematics for Elementary School

6th Grade, Gakko Tosho Co.,Ltd., Tokyo, Japan [2020]
